Two-dimensional Self and Product Cubic Systems, Vol. I

Albert C. J. Luo

Two-dimensional Self and Product Cubic Systems, Vol. I

Self-linear and Crossing-quadratic Product Vector Field

Springer

Albert C. J. Luo
Department of Mechanical and Mechatronics Engineering
Southern Illinois University Edwardsville
Edwardsville, IL, USA

ISBN 978-3-031-57095-7 ISBN 978-3-031-57096-4 (eBook)
https://doi.org/10.1007/978-3-031-57096-4

This Springer imprint is published by the registered company Springer Nature Switzerland AG
The registered company address is: Gewerbestrasse 11, 6330 Cham, Switzerland

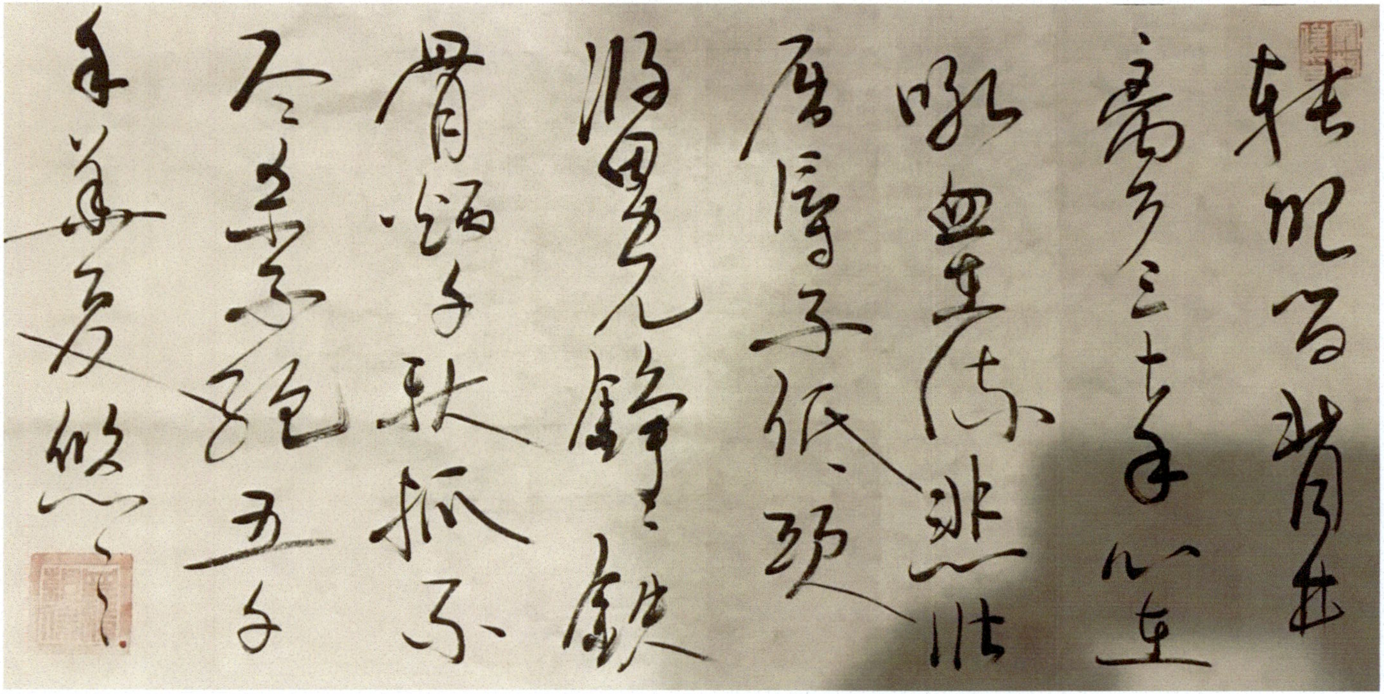

转眼间，背井离乡、三十年，心在吼，血在流，悲壮屈辱不低头。好男儿，铮铮铁骨、炳千秋，抓不尽，杀不决，五千年华夏、悠悠。 一朝俊【三十年流浪记】辛丑

Preface

In this book, consider self and product-cubic systems with a self-linear and crossing-quadratic product vector field. The saddle-nodes and inflection-source and sink flows exist in such cubic systems for the appearing bifurcations, and the third-order saddle, sink, and source exist for the appearing and switching bifurcations of saddle-nodes with saddle, source, and sink, and also for the appearing bifurcation of saddle, sink, and source. Equilibrium series with flow singularity are discussed, and the corresponding switching bifurcations are also presented. The equilibrium series with connected hyperbolic and hyperbolic-secant flows exist in such cubic systems, and the corresponding switching bifurcations are obtained through the inflection-source and sink infinite-equilibriums. In this book, there are five chapters to present the theory and illustrations.

In Chap. 1, the nonlinear dynamics and singularity of self and product cubic dynamical systems with a self-linear and crossing-quadratic product vector field are presented, and the corresponding appearing and switching bifurcations are dis-cussed through a theorem. Saddle-source (sink) equilibriums are the appearing bifurcations for saddle and source (sink) equilibriums, and inflection-source (sink) flows are the appearing bifurcations of the connected hyperbolic and hyperbolic-secant flows. The switching bifurcations are obtained by inflection-source and sink infinite-equilibriums, parabola-source and sink infinite-equilibriums, and up-down and down-up saddle infinite-equilibriums. The two paralleled double-switching bifurcations are obtained. Equilibrium series with connected hyperbolic and hyperbolic-secant flows are obtained for such cubic systems.

In Chap. 2, saddle-nodes (i.e., saddle-source and saddle-sink) and inflection-source and sink flows with saddle, source, and sink are discussed, and the corresponding switching bifurcations are discussed. The saddle-source (sink) is the appearing bifurcation of saddle and source (sink). Inflection-source and sink flows are for appearing bifurcations of the connected hyperbolic and hyperbolic-secant flows. In addition, the third-order saddles, source, and sink exist for the appearing bifurcations of saddle-nodes with source and sink and also for the appearing bifurcations of saddle, source, and sink. Switching bifurcations are through the inflection-saddle, parabola-source and sink, third-order trumpet source and sink infinite-equilibriums. The inflection-saddles are for the switching bifurcations of saddle with inflection-source and sink flows. The hyperbolic-source and sink are the switching bifurcations of the inflection-source flow with source and the inflection-sink flow with sink, respectively. The hyperbolic-secant source and sink are the switching bifurcations of the inflection source and sink flows with saddles. The trumpet sink and source infinite-equilibriums are the switching bifurcations of the inflection-source and sink flows with third-order saddle, sink, and source.

In Chap. 3, equilibrium series with inflection-source and sink flows are discussed, and the corresponding switching dynamics of the equilibrium series are discussed. Inflection-source and sink flows are for appearing bifurcations of the connected hyperbolic and hyperbolic-secant flows. The hyperbolic-source and sink are the switching bifurcations of the inflection-source flow with source and the inflection-sink flow with sink, respectively. The hyperbolic-secant source and sink are the switching bifurcations of the inflection source and sink flows with saddles.

In Chap. 4, equilibrium series with saddle-nodes and connected hyperbolic and hyperbolic-secant flows are discussed, and the corresponding switching dynamics of such equilibrium series are presented. The saddle-source (sink) is the appearing bifurcation of saddle and source (sink). The up-down and down-up saddle infinite-equilibriums are the switching bifurcations of the saddle-source (sink) and a hyperbolic flow. The inflection-saddles are for the switching bifurcations of saddle with inflection-source and sink flows. Up-down and down-up saddle infinite-equilibriums are the switching bifurcations of saddle-source and sinks with hyperbolic and hyperbolic-secant flows. The inflection-source and sink infinite-equilibriums are the switching bifurcations of a hyperbolic flow with saddle and a hyperbolic-secant flow with source or sink. Increasing and decreasing concave source (sink) infinite-equilibriums are for the switching bifurcations of the third-order saddles, sink, and source with a connected hyperbolic or hyperbolic-secant flow. In addition, double switching bifurcations are given by the inflection-source and sink infinite-equilibriums and the up-down and down-up saddle infinite-equilibriums.

In Chap. 5, simple equilibrium series with connected hyperbolic flows are presented, and the corresponding switching dynamics are discussed through the inflection-source and sink infinite-equilibriums. The inflection-source is the switching bifurcation of a hyperbolic-secant flow with a connected source and for a hyperbolic flow with a connected saddle. The inflection-sink is the switching bifurcation of a hyperbolic-secant flow with connected sink and for the hyperbolic flow with a connected saddle. The double switching bifurcations are generated by the two inflection-source and sink infinite-equilibriums.

Finally, the author hopes the materials presented herein can provide a better understanding of cubic nonlinear systems in science and engineering.

Edwardsville, IL, USA Albert C. J. Luo

Contents

Chapter 1
Self and Product Cubic Systems

In this chapter, the nonlinear dynamics and singularity of self and product-cubic dynamical systems with a self-linear and crossing-quadratic product vector field are presented, and the corresponding appearing and switching bifurcations are discussed through a theorem. The saddle-source (sink) equilibriums are the appearing bifurcations for saddle and source (sink) equilibriums, and inflection-source (sink) flows are the appearing bifurcations of the connected hyperbolic and hyperbolic-secant flows. The switching bifurcations are obtained by inflection-source and sink infinite-equilibriums, parabola-source and source infinite-equilibriums, and up-down and down-up saddle infinite-equilibriums. The two paralleled double-switching bifurcations are obtained. Equilibrium series with connected hyperbolic and hyperbolic-secant flows are obtained for such cubic systems.

1.1 Self-Linear and Crossing-Quadratic Product Vector Fields

In this section, cubic dynamical systems with a self-linear and crossing-quadratic product vector field and a self-cubic vector field will be discussed. The corresponding dynamical behaviors will be presented through the following theorem.

Theorem 1.1 *Consider a two-dimensional, cubic dynamical system with a product-cubic vector field and a self-cubic vector field as*

$$\begin{aligned}\dot{x}_{j_2} &= a_{j_2j_20}(x_{j_2} - b_{j_2j_21})(x_{j_2}^2 + B_{j_2j_2}x_{j_2} + C_{j_2j_2}),\\ \dot{x}_{j_1} &= a_{j_1j_10}(x_{j_1} - a_{j_1j_11})(x_{j_2}^2 + B_{j_1j_2}x_{j_2} + C_{j_1j_2})\end{aligned} \tag{1.1}$$

with

A. C. J. Luo, *Two-dimensional Self and Product Cubic Systems, Vol. I*,
https://doi.org/10.1007/978-3-031-57096-4_1

$$\Delta_{j_1j_2} = B_{j_1j_2}^2 - 4C_{j_1j_2}, \Delta_{j_2j_2} = B_{j_2j_2}^2 - 4C_{j_2j_2}. \tag{1.2}$$

(i) *For* $\Delta_{j_1j_2} < 0$ and $\Delta_{j_2j_2} < 0$, the standard form is

$$\begin{aligned} \dot{x}_{j_2} &= a_{j_2j_20}(x_{j_2} - a_{j_2j_21})\left[(x_{j_2} - a_{j_2j_2})^2 + b_{j_2j_2}\right], \\ \dot{x}_{j_1} &= a_{j_1j_10}(x_{j_1} - a_{j_1j_11})\left[(x_{j_2} - a_{j_1j_2})^2 + b_{j_1j_2}\right] \end{aligned} \tag{1.3}$$

where

$$\begin{aligned} &a_{j_2j_21} = b_{j_2j_21}, a_{j_1j_2} = -\frac{1}{2}B_{j_1j_2}, b_{j_1j_2} = \frac{1}{4}(-\Delta_{j_1j_2}); \\ &a_{j_2j_2} = -\frac{1}{2}B_{j_2j_2}, b_{j_2j_2} = \frac{1}{4}(-\Delta_{j_2j_2}). \end{aligned} \tag{1.4}$$

The first integral manifold is

$$\begin{aligned} \ln\frac{|x_{j_1} - a_{j_1j_11}|}{|x_{j_10} - a_{j_1j_11}|} = \frac{a_{j_1j_10}}{a_{j_2j_20}}\Big[&A\ln\frac{|x_{j_2} - a_{j_2j_21}|}{|x_{j_20} - a_{j_2j_21}|} \\ &+\frac{1}{2}(1-A)\ln\frac{|(x_{j_2} - a_{j_2j_2})^2 + b_{j_2j_2}|}{|(x_{j_20} - a_{j_2j_2})^2 + b_{j_2j_2}|} \\ &+\frac{B}{\sqrt{b_{j_2j_2}}}\Big(\arctan\frac{x_{j_2} - a_{j_2j_2}}{\sqrt{b_{j_2j_2}}} - \arctan\frac{x_{j_20} - a_{j_2j_2}}{\sqrt{b_{j_2j_2}}}\Big)\Big] \end{aligned} \tag{1.5}$$

where

$$A = \frac{(a_{j_2j_21} - a_{j_1j_2})^2 + b_{j_1j_2}}{(a_{j_2j_21} - a_{j_2j_2})^2 + b_{j_2j_2}}, B = a_{j_2j_2} + a_{j_2j_21} - 2a_{j_1j_2} - (a_{j_2j_21} - a_{j_2j_2})A. \tag{1.6}$$

The equilibrium of $(x_{j_1}^*, x_{j_2}^*) = (a_{j_1j_11}, a_{j_2j_21})$ *has the following properties.*

- *For* $a_{j_1j_10} > 0$ *and* $a_{j_2j_20} > 0$,

$$(a_{j_1j_11}, a_{j_2j_21}) = \underbrace{(\mathrm{SO}, \mathrm{SO})}_{\text{source}}. \tag{1.7}$$

The equilibrium of $(x_{j_1}^*, x_{j_2}^*) = (a_{j_1j_11}, a_{j_2j_21})$ *is an* (SO,SO)*-source.*

- *For* $a_{j_1j_10} < 0$ *and* $a_{j_2j_20} > 0$,

$$(a_{j_1j_11}, a_{j_2j_21}) = \underbrace{(\mathrm{SI}, \mathrm{SO})}_{\text{saddle}}. \tag{1.8}$$

The equilibrium of $(x_{j_1}^*, x_{j_2}^*) = (a_{j_1j_11}, a_{j_2j_21})$ *is an* (SI,SO)*-saddle.*

- *For* $a_{j_1j_10} > 0$ *and* $a_{j_2j_20} < 0$,

$$(a_{j_1j_11}, a_{j_2j_21}) = \underbrace{(\mathrm{SO}, \mathrm{SI})}_{\text{saddle}}. \tag{1.9}$$

The equilibrium of $(x^*_{j_1}, x^*_{j_2}) = (a_{j_1j_11}, a_{j_2j_21})$ *is an* (SO,SI)*-saddle.*

- *For* $a_{j_1j_10} < 0$ *and* $a_{j_2j_20} < 0$,

$$(a_{j_1j_11}, a_{j_2j_21}) = \underbrace{(\mathrm{SI}, \mathrm{SI})}_{\text{sink}}. \tag{1.10}$$

The equilibrium of $(x^*_{j_1}, x^*_{j_2}) = (a_{j_1j_11}, a_{j_2j_21})$ *is an* (SI,SI)*-sink.*

(ii) *For* $\Delta_{j_1j_2} < 0$ and $\Delta_{j_2j_2} = 0$, the standard form is

$$\begin{aligned}\dot{x}_{j_2} &= a_{j_2j_20}(x_{j_2} - a_{j_2j_2s_1})(x_{j_2} - a_{j_2j_2s_2})^2,\\ \dot{x}_{j_1} &= a_{j_1j_10}(x_{j_1} - a_{j_1j_11})\left[(x_{j_2} - a_{j_1j_2})^2 + b_{j_1j_2}\right]\end{aligned} \tag{1.11}$$

where

$$\begin{aligned}a_{j_1j_2} &= -\frac{1}{2}B_{j_1j_2}, b_{j_1j_2} = \frac{1}{4}(-\Delta_{j_1j_2}),\\ a_{j_2j_2s_2} &= -\frac{1}{2}B_{j_2j_2}, a_{j_2j_2s_1} = b_{j_2j_21}.\end{aligned} \tag{1.12}$$

(ii₁) The first integral manifold for $a_{j_2j_2s_1} \neq a_{j_2j_2s_2}$ *is*

$$\begin{aligned}\ln\frac{|x_{j_1} - a_{j_1j_11}|}{|x_{j_10} - a_{j_1j_11}|} = \frac{a_{j_1j_10}}{a_{j_2j_20}}\Bigg\{&\frac{(a_{j_2j_2s_1} - a_{j_1j_2})^2 + b_{j_1j_2}}{(a_{j_2j_2s_1} - a_{j_2j_2s_2})^2}\ln\frac{|x_{j_2} - a_{j_2j_2s_1}|}{|x_{j_20} - a_{j_2j_2s_1}|}\\ &+\left[\frac{2(a_{j_2j_2s_2} - a_{j_1j_2})}{a_{j_2j_2s_2} - a_{j_2j_2s_1}} - \frac{(a_{j_2j_2s_1} - a_{j_1j_2})^2 + b_{j_1j_2}}{(a_{j_2j_2s_1} - a_{j_2j_2s_2})^2}\right]\ln\frac{|x_{j_2} - a_{j_2j_2s_2}|}{|x_{j_20} - a_{j_2j_2s_2}|}\\ &-\frac{(a_{j_2j_2s_2} - a_{j_1j_2})^2 + b_{j_1j_2}}{a_{j_2j_2s_2} - a_{j_2j_2s_2}}\left(\frac{1}{x_{j_2} - a_{j_2j_2s_2}} - \frac{1}{x_{j_20} - a_{j_2j_2s_2}}\right)\Bigg\}.\end{aligned} \tag{1.13}$$

(ii₁ₐ) The equilibrium of $(x^*_{j_1}, x^*_{j_2}) = (a_{j_1j_11}, a_{j_2j_2s_1})$ *has the following properties.*

- *For* $a_{j_1j_10} > 0$ *and* $a_{j_2j_20}(a_{j_2j_2s_1} - a_{j_2j_2s_2})^2 > 0$,

$$(a_{j_1j_11}, a_{j_2j_2s_1}) = \underbrace{(\mathrm{SO}, \mathrm{SO})}_{\text{source}}. \tag{1.14}$$

The equilibrium of $(x^*_{j_1}, x^*_{j_2}) = (a_{j_1j_11}, a_{j_2j_2s_1})$ *is an* (SO,SO)*-source.*

- *For $a_{j_1j_10}<0$ and $a_{j_2j_20}(a_{j_2j_2s_1}-a_{j_2j_2s_2})^2>0$,*

$$(a_{j_1j_11},a_{j_2j_2s_1})=\underbrace{(\text{SI},\text{SO})}_{\text{saddle}}. \tag{1.15}$$

The equilibrium of $(x^*_{j_1},x^*_{j_2})=(a_{j_1j_11},a_{j_2j_2s_1})$ is an (SI,SO)*-saddle.*

- *For $a_{j_1j_10}>0$ and $a_{j_2j_20}(a_{j_2j_2s_1}-a_{j_2j_2s_2})^2<0$,*

$$(a_{j_1j_11},a_{j_2j_2s_1})=\underbrace{(\text{SO},\text{SI})}_{\text{saddle}}. \tag{1.16}$$

The equilibrium of $(x^*_{j_1},x^*_{j_2})=(a_{j_1j_11},a_{j_2j_21})$ is an (SO,SI)*-saddle.*

- *For $a_{j_1j_10}<0$ and $a_{j_2j_20}(a_{j_2j_2s_1}-a_{j_2j_2s_2})^2<0$,*

$$(a_{j_1j_11},a_{j_2j_2s_1})=\underbrace{(\text{SI},\text{SI})}_{\text{sink}}. \tag{1.17}$$

The equilibrium of $(x^*_{j_1},x^*_{j_2})=(a_{j_1j_11},a_{j_2j_2s_1})$ is an (SI,SI)*-sink.*

(ii_{1b}) The equilibrium of $(x^*_{j_1},x^*_{j_2})=(a_{j_1j_11},a_{j_2j_2s_1})$ has the following properties.

- *For $a_{j_1j_10}>0$ and $a_{j_2j_20}(a_{j_2j_2s_2}-a_{j_2j_2s_1})>0$,*

$$(a_{j_1j_11},a_{j_2j_2s_2})=\underbrace{(\text{SO},\text{US})}_{\text{upper-saddle source}}. \tag{1.18}$$

The equilibrium of $(x^*_{j_1},x^*_{j_2})=(a_{j_1j_11},a_{j_2j_2s_2})$ is an (SO,US)*-upper-saddle source.*

- *For $a_{j_1j_10}<0$ and $a_{j_2j_20}(a_{j_2j_2s_2}-a_{j_2j_2s_1})>0$,*

$$(a_{j_1j_11},a_{j_2j_2s_2})=\underbrace{(\text{SI},\text{US})}_{\text{upper-saddle sink}}. \tag{1.19}$$

The equilibrium of $(x^*_{j_1},x^*_{j_2})=(a_{j_1j_11},a_{j_2j_2s_2})$ is an (SI,US)*-upper-saddle sink.*

- *For $a_{j_1j_10}>0$ and $a_{j_2j_20}(a_{j_2j_2s_2}-a_{j_2j_2s_1})<0$,*

$$(a_{j_1j_11},a_{j_2j_2s_2})=\underbrace{(\text{SO},\text{LS})}_{\text{lower-saddle source}}. \tag{1.20}$$

The equilibrium of $(x^*_{j_1},x^*_{j_2})=(a_{j_1j_11},a_{j_2j_2s_2})$ is an (SO,LS)*-lower-saddle source.*

- *For* $a_{j_1j_10} < 0$ *and* $a_{j_2j_20}(a_{j_2j_2s_2} - a_{j_2j_2s_1}) < 0$,

$$(a_{j_1j_11}, a_{j_2j_2s_2}) = \underbrace{(\mathrm{SI}, \mathrm{LS})}_{\text{lower-saddle source}}. \tag{1.21}$$

The equilibrium of $(x_{j_1}^*, x_{j_2}^*) = (a_{j_1j_11}, a_{j_2j_2s_2})$ *is an* (SI,LS)-*lower-saddle sink.*

– *The saddle-source and saddle-sink are the appearing bifurcations of saddle-source and saddle-sink.*

(ii₂) The first integral manifold for $a_{j_2j_2s_1} = a_{j_2j_2s_2} = a_{j_2j_21}$ *is*

$$\begin{aligned}
\ln \frac{|x_{j_1} - a_{j_1j_11}|}{|x_{j_10} - a_{j_1j_11}|} = \frac{a_{j_1j_10}}{a_{j_2j_20}} \Big\{ & \ln \frac{|x_{j_2} - a_{j_2j_21}|}{|x_{j_20} - a_{j_2j_21}|} \\
& - 2(a_{j_2j_21} - a_{j_1j_2}) \Big[\frac{1}{x_{j_2} - a_{j_2j_21}} - \frac{1}{x_{j_20} - a_{j_2j_21}}\Big] \\
& - \frac{1}{2}\big[(a_{j_2j_21} - a_{j_1j_2})^2 + b_{j_1j_2}\big] \Big[\frac{1}{(x_{j_2} - a_{j_2j_21})^2} - \frac{1}{(x_{j_20} - a_{j_2j_21})^2}\Big] \Big\}.
\end{aligned} \tag{1.22}$$

The equilibrium of $(x_{j_1}^*, x_{j_2}^*) = (a_{j_1j_11}, a_{j_2j_21})$ *has the following properties.*

- *For* $a_{j_1j_10} > 0$ *and* $a_{j_2j_20} > 0$,

$$(a_{j_1j_11}, a_{j_2j_21}) = \underbrace{(\mathrm{SO}, 3^{\mathrm{rd}}\mathrm{SO})}_{\text{third-order source}}. \tag{1.23}$$

The equilibrium of $(x_{j_1}^*, x_{j_2}^*) = (a_{j_1j_11}, a_{j_2j_21})$ *is an* (SO,3rdSO)-*third-order source, which is for an appearing bifurcation from source to source, saddle, and source.*

- *For* $a_{j_1j_10} < 0$ *and* $a_{j_2j_20} > 0$,

$$(a_{j_1j_11}, a_{j_2j_21}) = \underbrace{(\mathrm{SI}, 3^{\mathrm{rd}}\mathrm{SO})}_{\text{third-order saddle}}. \tag{1.24}$$

The equilibrium of $(x_{j_1}^*, x_{j_2}^*) = (a_{j_1j_11}, a_{j_2j_21})$ *is an* (SI,3rdSO)-*third-order saddle, which is for an appearing bifurcation from saddle to saddle, sink, and saddle.*

- *For* $a_{j_1j_10} > 0$ *and* $a_{j_2j_20} < 0$,

$$(a_{j_1j_11}, a_{j_2j_21}) = \underbrace{(\mathrm{SO}, 3^{\mathrm{rd}}\mathrm{SI})}_{\text{third-order saddle}}. \tag{1.25}$$

The equilibrium of $(x_{j_1}^*, x_{j_2}^*) = (a_{j_1j_11}, a_{j_2j_21})$ *is an* (SO,3rdSI)-*third-order saddle, which is for an appearing bifurcation from saddle to saddle, source, and saddle.*

- *For* $a_{j_1j_10}<0$ *and* $a_{j_2j_20}<0$,

$$(a_{j_1j_11},a_{j_2j_21})=\underbrace{(\mathrm{SI},3^{\mathrm{rd}}\mathrm{SI})}_{\text{third-order sink}}. \tag{1.26}$$

The equilibrium of $(x_{j_1}^*,x_{j_2}^*)=(a_{j_1j_11},a_{j_2j_21})$ *is an* (SI,3rdSI)-*third-order sink, which is for an appearing bifurcation from sink to sink, saddle, and sink.*

(iii) *For* $\Delta_{j_1j_2}<0$ *and* $\Delta_{j_2j_2}>0$, *the standard form is*

$$\begin{aligned}\dot{x}_{j_2}&=a_{j_2j_20}(x_{j_2}-a_{j_2j_21})(x_{j_2}-a_{j_2j_22})(x_{j_2}-a_{j_2j_23}),\\ \dot{x}_{j_1}&=a_{j_1j_10}(x_{j_1}-a_{j_1j_11})\big[(x_{j_2}-a_{j_1j_2})^2+b_{j_1j_2}\big]\end{aligned} \tag{1.27}$$

where

$$\begin{aligned}&a_{j_1j_2}=-\frac{1}{2}B_{j_1j_2},\,b_{j_1j_2}=\frac{1}{4}(-\Delta_{j_1j_2}),\\ &b_{j_2j_22},b_{j_2j_23}=-\frac{1}{2}(B_{j_2j_2}\pm\sqrt{\Delta_{j_2j_2}}),\\ &\{a_{j_2j_21},a_{j_2j_22},a_{j_2j_23}\}=\mathrm{sort}\{b_{j_2j_21},b_{j_2j_22},b_{j_2j_23}\},\\ &a_{j_2j_2s_1}<a_{j_2j_2s_2},s_1,s_2\in\{1,2,3\},s_1<s_2.\end{aligned} \tag{1.28}$$

The first integral manifold is

$$\begin{aligned}&\ln\frac{|x_{j_1}-a_{j_1j_11}|}{|x_{j_10}-a_{j_1j_11}|}\\ &=\frac{a_{j_1j_10}}{a_{j_2j_20}}\sum\nolimits_{s_1=1}^{3}\frac{(a_{j_2j_2s_1}-a_{j_1j_2})^2+b_{j_1j_2}}{\prod_{s_2=1,s_2\neq s_1}^{3}(a_{j_2j_2s_1}-a_{j_1j_2s_2})}\ln\frac{|x_{j_2}-a_{j_2j_2s_1}|}{|x_{j_20}-a_{j_2j_2s_1}|}.\end{aligned} \tag{1.29}$$

The equilibrium of $(x_{j_1}^*,x_{j_2}^*)=(a_{j_1j_11},a_{j_2j_2s_1})$ $(s_1,s_2\in\{1,2\},s_1\neq s_2)$ *has the following properties.*

- *For* $a_{j_1j_10}>0$ *and* $a_{j_2j_20}\prod_{s_2=1,s_2\neq s_1}^{3}(a_{j_2j_2s_1}-a_{j_1j_2s_2})>0$,

$$(a_{j_1j_11},a_{j_2j_2s_1})=\underbrace{(\mathrm{SO},\mathrm{SO})}_{\text{source}}. \tag{1.30}$$

The equilibrium of $(x_{j_1}^*,x_{j_2}^*)=(a_{j_1j_11},a_{j_2j_2s_1})$ *is an* (SO,SO)-*source.*

- *For* $a_{j_1j_10}<0$ *and* $a_{j_2j_20}\prod_{s_2=1,s_2\neq s_1}^{3}(a_{j_2j_2s_1}-a_{j_1j_2s_2})>0$,

$$(a_{j_1j_11},a_{j_2j_2s_1})=\underbrace{(\mathrm{SI},\mathrm{SO})}_{\text{saddle}}. \tag{1.31}$$

The equilibrium of $(x_{j_1}^*,x_{j_2}^*)=(a_{j_1j_11},a_{j_2j_2s_1})$ *is an* (SI,SO)-*saddle.*

- *For* $a_{j_1j_10} > 0$ *and* $a_{j_2j_20}\prod_{s_2=1,s_2\neq s_1}^{3}(a_{j_2j_2s_1} - a_{j_1j_2s_2}) < 0$,

$$(a_{j_1j_11}, a_{j_2j_2s_1}) = \underbrace{(\text{SO}, \text{SI})}_{\text{saddle}}. \tag{1.32}$$

The equilibrium of $(x_{j_1}^*, x_{j_2}^*) = (a_{j_1j_11}, a_{j_2j_2s_1})$ *is an* (SO,SI)*-saddle.*

- *For* $a_{j_1j_10} < 0$ *and* $a_{j_2j_20}\prod_{s_2=1,s_2\neq s_1}^{3}(a_{j_2j_2s_1} - a_{j_1j_2s_2}) < 0$,

$$(a_{j_1j_11}, a_{j_2j_2s_1}) = \underbrace{(\text{SI}, \text{SI})}_{\text{sink}}. \tag{1.33}$$

The equilibrium of $(x_{j_1}^*, x_{j_2}^*) = (a_{j_1j_11}, a_{j_2j_2s_1})$ *is an* (SI,SI)*-sink.*

(iv) *For* $\Delta_{j_1j_2} = 0$ *and* $\Delta_{j_2j_2} < 0$, *the standard form is*

$$\begin{aligned} \dot{x}_{j_2} &= a_{j_2j_20}(x_{j_2} - b_{j_2j_21})\left[(x_{j_2} - a_{j_2j_2})^2 + b_{j_2j_2}\right], \\ \dot{x}_{j_1} &= a_{j_1j_10}(x_{j_1} - a_{j_1j_11})(x_{j_2} - a_{j_1j_21})^2 \end{aligned} \tag{1.34}$$

where

$$\begin{aligned} a_{j_1j_21} &= -\frac{1}{2}B_{j_1j_2}, a_{j_2j_21} = b_{j_2j_21}, \\ a_{j_2j_2} &= -\frac{1}{2}B_{j_2j_2}, b_{j_2j_2} = \frac{1}{4}(-\Delta_{j_2j_2}). \end{aligned} \tag{1.35}$$

(iv$_1$) The first integral manifold for $a_{j_1j_21} \neq a_{j_2j_21}$ *is*

$$\begin{aligned} \ln\frac{|x_{j_1} - a_{j_1j_11}|}{|x_{j_10} - a_{j_1j_11}|} = \frac{a_{j_1j_10}}{a_{j_2j_20}}\Bigg[& A\ln\frac{|x_{j_2} - a_{j_2j_21}|}{|x_{j_20} - a_{j_2j_21}|} \\ & + \frac{1}{2}(1 - A)\ln\frac{|(x_{j_2} - a_{j_2j_2})^2 + b_{j_2j_2}|}{|(x_{j_20} - a_{j_2j_2})^2 + b_{j_2j_2}|} \\ & + \frac{B}{\sqrt{b_{j_2j_2}}}\left(\arctan\frac{x_{j_2} - a_{j_2j_2}}{\sqrt{b_{j_2j_2}}} - \arctan\frac{x_{j_20} - a_{j_2j_2}}{\sqrt{b_{j_2j_2}}}\right)\Bigg] \end{aligned} \tag{1.36}$$

where

$$A = \frac{(a_{j_2j_21} - a_{j_1j_21})^2}{(a_{j_2j_21} - a_{j_2j_2})^2 + b_{j_2j_2}}, B = a_{j_2j_2} + a_{j_2j_21} - 2a_{j_1j_21} - (a_{j_2j_21} - a_{j_2j_2})A. \tag{1.37}$$

(iv_{1a}) The equilibrium of $(x_{j_1}^*, x_{j_2}^*) = (a_{j_1j_1}1, a_{j_2j_2}1)$ *has the following properties.*

- *For* $a_{j_1j_1}0(a_{j_2j_2}1 - a_{j_1j_2}1)^2 > 0$ *and* $a_{j_2j_2}0 > 0$,

$$(a_{j_1j_1}1, a_{j_2j_2}1) = \underbrace{(\text{SO}, \text{SO})}_{\text{source}}. \tag{1.38}$$

The equilibrium of $(x_{j_1}^*, x_{j_2}^*) = (a_{j_1j_1}1, a_{j_2j_2}1)$ *is an* (SO,SO)*-source.*

- *For* $a_{j_1j_1}0(a_{j_2j_2}1 - a_{j_1j_2}1)^2 < 0$ *and* $a_{j_2j_2}0 > 0$,

$$(a_{j_1j_1}1, a_{j_2j_2}1) = \underbrace{(\text{SI}, \text{SO})}_{\text{saddle}}. \tag{1.39}$$

The equilibrium of $(x_{j_1}^*, x_{j_2}^*) = (a_{j_1j_1}1, a_{j_2j_2}1)$ *is an* (SI,SO)*-saddle.*

- *For* $a_{j_1j_1}0(a_{j_2j_2}1 - a_{j_1j_2}1)^2 > 0$ *and* $a_{j_2j_2}0 < 0$,

$$(a_{j_1j_1}1, a_{j_2j_2}1) = \underbrace{(\text{SO}, \text{SI})}_{\text{saddle}}. \tag{1.40}$$

The equilibrium of $(x_{j_1}^*, x_{j_2}^*) = (a_{j_1j_1}1, a_{j_2j_2}1)$ *is an* (SO,SI)*-saddle.*

- *For* $a_{j_1j_1}0(a_{j_2j_2}1 - a_{j_1j_2}1)^2 < 0$ *and* $a_{j_2j_2}0 < 0$,

$$(a_{j_1j_1}1, a_{j_2j_2}1) = \underbrace{(\text{SI}, \text{SI})}_{\text{sink}}. \tag{1.41}$$

The equilibrium of $(x_{j_1}^*, x_{j_2}^*) = (a_{j_1j_1}1, a_{j_2j_2}1)$ *is an* (SI,SI)*-sink.*

(iv_{1b}) The equilibrium of $(x_{j_1}^*, x_{j_2}^*) = (a_{j_1j_1}1, a_{j_1j_2}1)$ *has the following properties.*

- *For* $a_{j_1j_1}0 > 0$ *and* $a_{j_2j_2}0(a_{j_1j_2}1 - a_{j_2j_2}1) > 0$,

$$(a_{j_1j_1}1, a_{j_1j_2}1) = \underbrace{({}_{\text{DI:II}}\text{SO}, \text{pF})}_{\text{inflection-source flow }(+)}. \tag{1.42}$$

The equilibrium of $(x_{j_1}^*, x_{j_2}^*) = (a_{j_1j_1}1, a_{j_1j_2}1)$ *is a* (${}_{\text{DI:II}}$SO, pF)*-positive inflection-source flow.*

- *For* $a_{j_1j_1}0 < 0$ *and* $a_{j_2j_2}0(a_{j_1j_2}1 - a_{j_2j_2}1) > 0$,

$$(a_{j_1j_1}1, a_{j_1j_2}1) = \underbrace{({}_{\text{II:DI}}\text{SI}, \text{pF})}_{\text{inflection-sink flow }(+)}. \tag{1.43}$$

The equilibrium of $(x_{j_1}^*, x_{j_2}^*) = (a_{j_1j_1}1, a_{j_1j_2}1)$ *is an* (${}_{\text{II:DI}}$SI, pF)*-positive inflection-sink flow.*

- *For* $a_{j_1j_10} > 0$ *and* $a_{j_2j_20}(a_{j_1j_21} - a_{j_2j_21}) < 0$,

$$(a_{j_1j_11}, a_{j_1j_21}) = \underbrace{({}_{\text{II:DI}}\text{SO, nF})}_{\text{inflection-source flow }(-)}. \tag{1.44}$$

The equilibrium of $(x^*_{j_1}, x^*_{j_2}) = (a_{j_1j_11}, a_{j_1j_21})$ *is an* $({}_{\text{II:DI}}\text{SO, nF})$*-negative inflection-source flow.*

- *For* $a_{j_1j_10} < 0$ *and* $a_{j_2j_20}(a_{j_1j_21} - a_{j_2j_21}) < 0$,

$$(a_{j_1j_11}, a_{j_1j_21}) = \underbrace{({}_{\text{DI:II}}\text{SI, nF})}_{\text{inflection-sink flow }(-)}. \tag{1.45}$$

The equilibrium of $(x^*_{j_1}, x^*_{j_2}) = (a_{j_1j_11}, a_{j_1j_21})$ *is a* $({}_{\text{DI:II}}\text{SI, nF})$*-negative inflection-sink flow.*

(iv$_2$) The first integral manifold for $a_{j_1j_21} = a_{j_2j_21}$ *is*

$$\begin{aligned}\ln\frac{|x_{j_1} - a_{j_1j_11}|}{|x_{j_10} - a_{j_1j_11}|} = \frac{a_{j_1j_10}}{a_{j_2j_20}}\Big[\frac{1}{2}\ln\frac{|(x_{j_2} - a_{j_2j_2})^2 + b_{j_2j_2}|}{|(x_{j_20} - a_{j_2j_2})^2 + b_{j_2j_2}|} \\ + \frac{a_{j_2j_2} - a_{j_1j_21}}{\sqrt{b_{j_2j_2}}}(\arctan\frac{x_{j_2} - a_{j_2j_2}}{\sqrt{b_{j_2j_2}}} - \arctan\frac{x_{j_20} - a_{j_2j_2}}{\sqrt{b_{j_2j_2}}})\Big].\end{aligned} \tag{1.46}$$

(iv$_{2a}$) The infinite-equilibrium of $x^*_{j_2} = a_{j_1j_21} = a_{j_2j_21}$ *with* $\overline{x}_{j_1} \neq a_{j_1j_11}$ *has the following properties.*

- *For* $a_{j_1j_10}(\overline{x}_{j_1} - a_{j_1j_11}) > 0$ *and* $a_{j_2j_20} > 0$,

$$(\overline{x}_{j_1}, a_{j_2j_21}) = \underbrace{(\text{UP, SO})}_{\text{up-parabola source}}. \tag{1.47}$$

The infinite-equilibrium of $x^*_{j_2} = a_{j_1j_21} = a_{j_2j_21}$ *is a* (UP,SO)*-up-parabola source.*

- *For* $a_{j_1j_10}(\overline{x}_{j_1} - a_{j_1j_11}) < 0$ *and* $a_{j_2j_20} > 0$,

$$(\overline{x}_{j_1}, a_{j_2j_21}) = \underbrace{(\text{DP, SO})}_{\text{down-parabola source}}. \tag{1.48}$$

The infinite-equilibrium of $x^*_{j_2} = a_{j_1j_21} = a_{j_2j_21}$ *is a* (DP,SO)*-down-parabola source.*

- *For* $a_{j_1j_10}(\overline{x}_{j_1} - a_{j_1j_11}) > 0$ *and* $a_{j_2j_20} < 0$,

$$(\overline{x}_{j_1}, a_{j_2j_21}) = \underbrace{(\text{DP}, \text{SI})}_{\text{down-parabola sink}}. \tag{1.49}$$

The infinite-equilibrium of $x^*_{j_2} = a_{j_1j_21} = a_{j_2j_21}$ *is a* (DP,SI)-*down-parabola sink.*

- *For* $a_{j_1j_10}(\overline{x}_{j_1} - a_{j_1j_11}) < 0$ *and* $a_{j_2j_20} < 0$,

$$(\overline{x}_{j_1}, a_{j_2j_21}) = \underbrace{(\text{UP}, \text{SI})}_{\text{up-parabola sink}}. \tag{1.50}$$

The infinite-equilibrium of $x^*_{j_2} = a_{j_1j_21} = a_{j_2j_21}$ *is a* (UP,SI)-*up-parabola sink.*

(iv$_{2b}$) The equilibrium of $(x^*_{j_1}, x^*_{j_2}) = (a_{j_1j_11}, a_{j_2j_21})$ *with* $a_{j_1j_21} = a_{j_2j_21}$ *has the following properties.*

- *For* $a_{j_1j_10} > 0$ *and* $a_{j_2j_20} > 0$,

$$(a_{j_1j_11}, a_{j_2j_21}) = \underbrace{(\text{DP} : \text{UP}, \text{SO})}_{\text{hyperbolic source}}. \tag{1.51}$$

The equilibrium of $(x^*_{j_1}, x^*_{j_2}) = (a_{j_1j_11}, a_{j_2j_21})$ *is a* (DP:UP,SO)-*hyperbolic source.*

- *For* $a_{j_1j_10} < 0$ *and* $a_{j_2j_20} > 0$,

$$(a_{j_1j_11}, a_{j_2j_21}) = \underbrace{(\text{UP} : \text{DP}, \text{SO})}_{\text{hyperbolic-secant source}}. \tag{1.52}$$

The equilibrium of $(x^*_{j_1}, x^*_{j_2}) = (a_{j_1j_11}, a_{j_2j_21})$ *is a* (UP:DP,SO)-*hyperbolic-secant source.*

- *For* $a_{j_1j_10} > 0$ *and* $a_{j_2j_20} < 0$,

$$(a_{j_1j_11}, a_{j_2j_21}) = \underbrace{(\text{UP} : \text{DP}, \text{SI})}_{\text{hyperbolic-secant sink}}. \tag{1.53}$$

The equilibrium of $(x^*_{j_1}, x^*_{j_2}) = (a_{j_1j_11}, a_{j_2j_21})$ *is a* (UP:DP,SI)-*hyperbolic-secant sink.*

- *For* $a_{j_1j_10} < 0$ *and* $a_{j_2j_20} < 0$,

$$(a_{j_1j_11}, a_{j_2j_21}) = \underbrace{(\text{DP} : \text{UP}, \text{SI})}_{\text{hyperbolic sink}}. \tag{1.54}$$

The equilibrium of $(x^*_{j_1}, x^*_{j_2}) = (a_{j_1j_11}, a_{j_2j_21})$ *is a* (DP:UP,SI)-hyperbolic sink.

(v) *For* $\Delta_{j_1j_2}=0$ and $\Delta_{j_2j_2}=0$, the standard form is

$$\begin{aligned}\dot{x}_{j_2} &= a_{j_2j_20}(x_{j_2}-a_{j_2j_2s_1})(x_{j_2}-a_{j_2j_2s_2})^2,\\ \dot{x}_{j_1} &= a_{j_1j_10}(x_{j_1}-a_{j_1j_11})(x_{j_2}-a_{j_1j_21})^2\end{aligned} \tag{1.55}$$

where

$$\begin{aligned}&a_{j_1j_21}=-\frac{1}{2}B_{j_1j_2}, a_{j_2j_2s_2}=-\frac{1}{2}B_{j_2j_2}, a_{j_2j_2s_1}=b_{j_2j_21}\\ &s_1,s_2\in\{1,2\}; s_1\neq s_2.\end{aligned} \tag{1.56}$$

(v_1) The first integral manifold for $a_{j_2j_2s}\neq a_{j_1j_21}(s=s_1,\ s_2)$ *is*

$$\begin{aligned}\ln\frac{|x_{j_1}-a_{j_1j_11}|}{|x_{j_10}-a_{j_1j_11}|} &= \frac{a_{j_1j_10}}{a_{j_2j_20}}\left\{\frac{(a_{j_2j_2s_1}-a_{j_1j_21})^2}{(a_{j_2j_2s_1}-a_{j_2j_2s_2})^2}\ln\frac{|x_{j_2}-a_{j_2j_2s_1}|}{|x_{j_20}-a_{j_2j_2s_1}|}\right.\\ &\quad+\left[\frac{2(a_{j_2j_2s_2}-a_{j_1j_21})}{(a_{j_2j_2s_2}-a_{j_2j_2s_1})}-\frac{(a_{j_2j_2s_2}-a_{j_1j_21})^2}{(a_{j_2j_2s_2}-a_{j_2j_2s_1})^2}\right]\ln\frac{|x_{j_2}-a_{j_2j_2s_1}|}{|x_{j_20}-a_{j_2j_2s_1}|}\\ &\quad\left.-\frac{(a_{j_2j_2s_2}-a_{j_1j_21})^2}{(a_{j_2j_2s_2}-a_{j_2j_2s_1})}\left(\frac{1}{x_{j_2}-a_{j_2j_2s_2}}-\frac{1}{x_{j_20}-a_{j_2j_2s_2}}\right)\right\}.\end{aligned} \tag{1.57}$$

(v_{1a}) The equilibrium of $(x^*_{j_1},x^*_{j_2})=(a_{j_1j_11},a_{j_2j_2s_1})$ *has the following properties.*

- *For* $a_{j_1j_10}(a_{j_2j_2s_1}-a_{j_1j_21})^2>0$ *and* $a_{j_2j_20}(a_{j_2j_2s_1}-a_{j_2j_2s_2})^2>0$,

$$(a_{j_1j_11},a_{j_2j_2s_1})=\underbrace{(\text{SO},\text{SO})}_{\text{source}}. \tag{1.58}$$

  The equilibrium of $(x^*_{j_1},x^*_{j_2})=(a_{j_1j_11},a_{j_2j_2s_1})$ *is an* (SO,SO)*-source.*

- *For* $a_{j_1j_10}(a_{j_2j_2s_1}-a_{j_1j_21})^2<0$ *and* $a_{j_2j_20}(a_{j_2j_2s_1}-a_{j_2j_2s_2})^2>0$,

$$(a_{j_1j_11},a_{j_2j_2s_1})=\underbrace{(\text{SI},\text{SO})}_{\text{saddle}}. \tag{1.59}$$

  The equilibrium of $(x^*_{j_1},x^*_{j_2})=(a_{j_1j_11},a_{j_2j_2s_1})$ *is an* (SI,SO)*-saddle.*

- *For* $a_{j_1j_10}(a_{j_2j_2s_1}-a_{j_1j_21})^2>0$ *and* $a_{j_2j_20}(a_{j_2j_2s_1}-a_{j_2j_2s_2})^2<0$,

$$(a_{j_1j_11},a_{j_2j_2s_1})=\underbrace{(\text{SO},\text{SI})}_{\text{saddle}}. \tag{1.60}$$

  The equilibrium of $(x^*_{j_1},x^*_{j_2})=(a_{j_1j_11},a_{j_2j_21})$ *is an* (SO,SI)*-saddle.*

- *For* $a_{j_1j_10}(a_{j_2j_2s_1}-a_{j_1j_21})^2<0$ *and* $a_{j_2j_20}(a_{j_2j_2s_1}-a_{j_2j_2s_2})^2<0$,

$$(a_{j_1j_11},a_{j_2j_2s_1})=\underbrace{(\mathrm{SI},\mathrm{SI})}_{\text{sink}}. \tag{1.61}$$

The equilibrium of $(x^*_{j_1},x^*_{j_2})=(a_{j_1j_11},a_{j_2j_2s_1})$ *is an* (SI,SI)-*sink.*

(v_{1b}) The equilibrium of $(x^*_{j_1},x^*_{j_2})=(a_{j_1j_11},a_{j_2j_2s_2})$ *with* $a_{j_2j_21}\neq a_{j_1j_21}$ *has the following properties.*

- *For* $a_{j_1j_10}(a_{j_2j_2s_2}-a_{j_1j_21})^2>0$ *and* $a_{j_2j_20}(a_{j_2j_2s_2}-a_{j_1j_2s_1})>0$,

$$(a_{j_1j_11},a_{j_2j_2s_2})=\underbrace{(\mathrm{SO},\mathrm{US})}_{\text{upper-saddle source}}. \tag{1.62}$$

The equilibrium of $(x^*_{j_1},x^*_{j_2})=(a_{j_1j_11},a_{j_2j_2s_2})$ *is an* (SO,US)-*upper-saddle source.*

- *For* $a_{j_1j_10}(a_{j_2j_2s_2}-a_{j_1j_21})^2<0$ *and* $a_{j_2j_20}(a_{j_2j_2s_2}-a_{j_1j_2s_1})>0$,

$$(a_{j_1j_11},a_{j_2j_2s_2})=\underbrace{(\mathrm{SI},\mathrm{US})}_{\text{upper-saddle-sink}}. \tag{1.63}$$

The equilibrium of $(x^*_{j_1},x^*_{j_2})=(a_{j_1j_11},a_{j_2j_2s_2})$ *is an* (SI,US)-*upper-saddle sink.*

- *For* $a_{j_1j_10}(a_{j_2j_2s_2}-a_{j_1j_21})^2>0$ *and* $a_{j_2j_20}(a_{j_2j_2s_2}-a_{j_1j_2s_1})<0$,

$$(a_{j_1j_11},a_{j_2j_2s_2})=\underbrace{(\mathrm{SO},\mathrm{LS})}_{\text{lower-saddle source}}. \tag{1.64}$$

The equilibrium of $(x^*_{j_1},x^*_{j_2})=(a_{j_1j_11},a_{j_2j_2s_2})$ *is an* (SO,LS)-*lower-saddle source.*

- *For* $a_{j_1j_10}(a_{j_2j_2s_2}-a_{j_1j_21})^2<0$ *and* $a_{j_2j_20}(a_{j_2j_2s_2}-a_{j_1j_2s_1})<0$,

$$(a_{j_1j_11},a_{j_2j_2s_2})=\underbrace{(\mathrm{SI},\mathrm{LS})}_{\text{lower-saddle sink}}. \tag{1.65}$$

The equilibrium of $(x^*_{j_1},x^*_{j_2})=(a_{j_1j_11},a_{j_2j_2s_2})$ *is an* (SI,LS)-*lower-saddle sink.*

– *The saddle-source and saddle-sink are the appearing bifurcations of saddle-source and saddle-sink.*

(v_{1c}) The equilibrium of $(x^*_{j_1},x^*_{j_2})=(a_{j_1j_11},a_{j_1j_21})$ *with* $a_{j_2j_2s}\neq a_{j_1j_21}(s=s_1,\,s_2)$ *has the following properties.*

- *For* $a_{j_1j_10}>0$ *and* $a_{j_2j_20}(a_{j_1j_21}-a_{j_2j_2s_1})(a_{j_1j_21}-a_{j_2j_2s_2})^2>0$,

$$(a_{j_1j_11}, a_{j_1j_21}) = \underbrace{({}_{\text{DI:II}}\text{SO, pF})}_{\text{inflection-source flow }(+)}. \tag{1.66}$$

The equilibrium of $(x_{j_1}^*, x_{j_2}^*) = (a_{j_1j_11}, a_{j_1j_21})$ *is a* $({}_{\text{DI:II}}\text{SO, pF})$*-positive inflection-source flow.*

- *For* $a_{j_1j_10} < 0$ *and* $a_{j_2j_20}(a_{j_1j_21} - a_{j_2j_2s_1})(a_{j_1j_21} - a_{j_2j_2s_2})^2 > 0$,

$$(a_{j_1j_11}, a_{j_1j_21}) = \underbrace{({}_{\text{II:DI}}\text{SI, pF})}_{\text{inflection-sink flow }(+)}. \tag{1.67}$$

The equilibrium of $(x_{j_1}^*, x_{j_2}^*) = (a_{j_1j_11}, a_{j_1j_21})$ *is an* $({}_{\text{II:DI}}\text{SI, pF})$*-positive inflection-sink flow.*

- *For* $a_{j_1j_10} > 0$ *and* $a_{j_2j_20}(a_{j_1j_21} - a_{j_2j_2s_1})(a_{j_1j_21} - a_{j_2j_2s_2})^2 < 0$,

$$(a_{j_1j_11}, a_{j_1j_21}) = \underbrace{({}_{\text{II:DI}}\text{SO, nF})}_{\text{inflection-source flow }(-)}. \tag{1.68}$$

The equilibrium of $(x_{j_1}^*, x_{j_2}^*) = (a_{j_1j_11}, a_{j_1j_21})$ *is an* $({}_{\text{II:DI}}\text{SO, nF})$*-negative inflection-source flow.*

- *For* $a_{j_1j_10} < 0$ *and* $a_{j_2j_20}(a_{j_1j_21} - a_{j_2j_2s_1})(a_{j_1j_21} - a_{j_2j_2s_2})^2 < 0$,

$$(a_{j_1j_11}, a_{j_1j_21}) = \underbrace{({}_{\text{DI:II}}\text{SI, nF})}_{\text{inflection-sink flow }(-)}. \tag{1.69}$$

The equilibrium of $(x_{j_1}^*, x_{j_2}^*) = (a_{j_1j_11}, a_{j_1j_21})$ *is a* $({}_{\text{DI:II}}\text{SI, nF})$*-negative inflection-sink flow.*

(v₂) For $a_{j_2j_2s_1} = a_{j_2j_2s_2} = a_{j_2j_21}$, the standard form is

$$\begin{aligned} \dot{x}_{j_2} &= a_{j_2j_20}(x_{j_2} - a_{j_2j_21})^3, \\ \dot{x}_{j_1} &= a_{j_1j_10}(x_{j_1} - a_{j_1j_11})(x_{j_2} - a_{j_1j_21})^2 \end{aligned} \tag{1.70}$$

The first integral manifold for is

$$\begin{aligned} \ln\frac{|x_{j_1} - a_{j_1j_11}|}{|x_{j_10} - a_{j_1j_11}|} = \frac{a_{j_1j_10}}{a_{j_2j_20}}\Big\{ &\ln\frac{|x_{j_2} - a_{j_2j_21}|}{|x_{j_20} - a_{j_2j_21}|} \\ &- 2(a_{j_2j_21} - a_{j_1j_21})(\frac{1}{x_{j_2} - a_{j_2j_21}} - \frac{1}{x_{j_20} - a_{j_2j_21}}) \\ &- \frac{1}{2}(a_{j_2j_21} - a_{j_1j_21})^2[\frac{1}{(x_{j_2} - a_{j_2j_21})^2} - \frac{1}{(x_{j_20} - a_{j_2j_21})^2}]\Big\}. \end{aligned} \tag{1.71}$$

The equilibrium of $(x_{j_1}^*, x_{j_2}^*) = (a_{j_1j_11}, a_{j_2j_21})$ *has the following properties.*

- *For* $a_{j_1j_10}(a_{j_2j_21}-a_{j_1j_21})^2>0$ *and* $a_{j_2j_20}>0$,

$$(a_{j_1j_11},a_{j_2j_21})=\underbrace{(\text{SO},3^{\text{rd}}\text{SO})}_{\text{third-order source}}. \tag{1.72}$$

The equilibrium of $(x^*_{j_1},x^*_{j_2})=(a_{j_1j_11},a_{j_2j_21})$ *is an* (SO,3rdSO)-*third-order source, which is for an appearing bifurcation from source to source, saddle, and source.*

- *For* $a_{j_1j_10}(a_{j_2j_21}-a_{j_1j_21})^2<0$ *and* $a_{j_2j_20}>0$,

$$(a_{j_1j_11},a_{j_2j_21})=\underbrace{(\text{SI},3^{\text{rd}}\text{SO})}_{\text{third-order saddle}}. \tag{1.73}$$

The equilibrium of $(x^*_{j_1},x^*_{j_2})=(a_{j_1j_11},a_{j_2j_21})$ *is an* (SI,3rdSO)-*third-order saddle, which is for an appearing bifurcation from saddle to saddle, sink, and saddle.*

- *For* $a_{j_1j_10}(a_{j_2j_21}-a_{j_1j_21})^2>0$ *and* $a_{j_2j_20}<0$,

$$(a_{j_1j_11},a_{j_2j_21})=\underbrace{(\text{SO},3^{\text{rd}}\text{SI})}_{\text{third-order saddle}}. \tag{1.74}$$

The equilibrium of $(x^*_{j_1},x^*_{j_2})=(a_{j_1j_11},a_{j_2j_21})$ *is an* (SO,3rdSI)-*third-order saddle, which is for an appearing bifurcation from saddle to saddle, source, and saddle.*

- *For* $a_{j_1j_10}(a_{j_2j_21}-a_{j_1j_21})^2<0$ *and* $a_{j_2j_20}<0$,

$$(a_{j_1j_11},a_{j_2j_21})=\underbrace{(\text{SI},3^{\text{rd}}\text{SI})}_{\text{sink}}. \tag{1.75}$$

The equilibrium of $(x^*_{j_1},x^*_{j_2})=(a_{j_1j_11},a_{j_2j_21})$ *is an* (SI,3rdSI)-*third-order sink, which is for an appearing bifurcation form sink to sink, saddle, and sink.*

(v_3) The first integral manifold for $a_{j_2j_2s_1}=a_{j_1j_21}$ *is*

$$\begin{aligned}\ln\frac{|x_{j_1}-a_{j_1j_11}|}{|x_{j_10}-a_{j_1j_11}|}=\frac{a_{j_1j_10}}{a_{j_2j_20}}\Bigg[&\ln\frac{|x_{j_2}-a_{j_2j_2s_2}|}{|x_{j_20}-a_{j_2j_2s_2}|}\\&-(a_{j_2j_2s_2}-a_{j_1j_21})(\frac{1}{x_{j_2}-a_{j_2j_2s_2}}-\frac{1}{x_{j_20}-a_{j_2j_2s_2}})\Bigg].\end{aligned} \tag{1.76}$$

(v_{3a}) The infinite-equilibrium of $x_{j_2}^* = a_{j_1 j_2 1} = a_{j_2 j_2 s_1}$ with $\bar{x}_{j_1} \neq a_{j_1 j_1 1}$ has the following properties.

- *For $a_{j_1 j_1 0}(\bar{x}_{j_1} - a_{j_1 j_1 1}) > 0$ and $a_{j_2 j_2 0}(a_{j_2 j_2 s_1} - a_{j_2 j_2 s_2})^2 > 0$,*

$$(\bar{x}_{j_1}, a_{j_2 j_2 s_1}) = \underbrace{(\text{UP}, \text{SO})}_{\text{up-parabola source}}. \tag{1.77}$$

The infinite-equilibrium of $x_{j_2}^* = a_{j_1 j_2 1} = a_{j_2 j_2 s_1}$ is a (UP,SO)-*up-parabola source.*

- *For $a_{j_1 j_1 0}(\bar{x}_{j_1} - a_{j_1 j_1 1}) < 0$ and $a_{j_2 j_2 0}(a_{j_2 j_2 s_1} - a_{j_2 j_2 s_2})^2 > 0$,*

$$(\bar{x}_{j_1}, a_{j_2 j_2 s_1}) = \underbrace{(\text{DP}, \text{SO})}_{\text{down-parabols saddle}}. \tag{1.78}$$

The infinite-equilibrium of $x_{j_2}^* = a_{j_1 j_2 1} = a_{j_2 j_2 s_1}$ is a (DP,SO)-*down-parabola source.*

- *For $a_{j_1 j_1 0}(\bar{x}_{j_1} - a_{j_1 j_1 1}) > 0$ and $a_{j_2 j_2 0}(a_{j_2 j_2 s_1} - a_{j_2 j_2 s_2})^2 < 0$,*

$$(\bar{x}_{j_1}, a_{j_2 j_2 s_1}) = \underbrace{(\text{DP}, \text{SI})}_{\text{down-parabola sink}}. \tag{1.79}$$

The infinite-equilibrium of $x_{j_2}^* = a_{j_1 j_2 1} = a_{j_2 j_2 s_1}$ is a (DP,SI)-*down-parabola sink.*

- *For $a_{j_1 j_1 0}(\bar{x}_{j_1} - a_{j_1 j_1 1}) < 0$ and $a_{j_2 j_2 0}(a_{j_2 j_2 s_1} - a_{j_2 j_2 s_2})^2 < 0$,*

$$(\bar{x}_{j_1}, a_{j_2 j_2 s_1}) = \underbrace{(\text{UP}, \text{SI})}_{\text{up-parabola sink}}. \tag{1.80}$$

The infinite-equilibrium of $x_{j_2}^* = a_{j_1 j_2 1} = a_{j_2 j_2 s_1}$ is a (UP,SI)-*up-parabola sink.*

(v_{3b}) The equilibrium of $(x_{j_1}^*, x_{j_2}^*) = (a_{j_1 j_1 1}, a_{j_2 j_2 s_1})$ with $a_{j_1 j_2 1} = a_{j_2 j_2 s_1}$ has the following properties.

- *For $a_{j_1 j_1 0} > 0$ and $a_{j_2 j_2 0}(a_{j_2 j_2 s_1} - a_{j_2 j_2 s_2})^2 > 0$,*

$$(a_{j_1 j_1 1}, a_{j_2 j_2 s_1}) = \underbrace{(\text{DP}:\text{UP}, \text{SO})}_{\text{hyperbolic source}}. \tag{1.81}$$

The equilibrium of $(x_{j_1}^*, x_{j_2}^*) = (a_{j_1 j_1 1}, a_{j_2 j_2 s_1})$ is a (DP:UP,SO)-*hyperbolic source.*

- *For $a_{j_1 j_1 0} < 0$ and $a_{j_2 j_2 0}(a_{j_2 j_2 s_1} - a_{j_2 j_2 s_2})^2 > 0$,*

$$(a_{j_1 j_1 1}, a_{j_2 j_2 s_1}) = \underbrace{(\text{UP}:\text{DP}, \text{SO})}_{\text{hyperbolic-secant source}}. \tag{1.82}$$

The equilibrium of $(x_{j_1}^*, x_{j_2}^*) = (a_{j_1 j_1 1}, a_{j_2 j_2 s_1})$ *is a* (UP:DP,SO)*-hyperbolic-secant source.*

- *For* $a_{j_1 j_1 0} > 0$ *and* $a_{j_2 j_2 0}(a_{j_2 j_2 s_1} - a_{j_2 j_2 s_2})^2 < 0$,

$$(a_{j_1 j_1 1}, a_{j_2 j_2 s_1}) = \underbrace{(\text{UP : DP, SI})}_{\text{hyperbolic-secant sink}}. \tag{1.83}$$

The equilibrium of $(x_{j_1}^*, x_{j_2}^*) = (a_{j_1 j_1 1}, a_{j_2 j_2 s_1})$ *is a* (UP:DP,SI)*-hyperbolic-secant sink.*

- *For* $a_{j_1 j_1 0} < 0$ *and* $a_{j_2 j_2 0}(a_{j_2 j_2 s_1} - a_{j_2 j_2 s_2})^2 < 0$,

$$(a_{j_1 j_1 1}, a_{j_2 j_2 s_1}) = \underbrace{(\text{DP : UP, SI})}_{\text{hyperbolic sink}}. \tag{1.84}$$

The equilibrium of $(x_{j_1}^*, x_{j_2}^*) = (a_{j_1 j_1 1}, a_{j_2 j_2 s_1})$ *is a* (DP:UP,SI)*-hyperbolic sink.*

(v_4) The first integral manifold for $a_{j_2 j_2 s_2} = a_{j_1 j_2 1}$ *is*

$$\ln \frac{|x_{j_1} - a_{j_1 j_1 1}|}{|x_{j_1 0} - a_{j_1 j_1 1}|} = \frac{a_{j_1 j_1 0}}{a_{j_2 j_2 0}} \ln \frac{|x_{j_2} - a_{j_2 j_2 s_1}|}{|x_{j_2 0} - a_{j_2 j_2 s_1}|}. \tag{1.85}$$

(v_{4a}) The infinite-equilibrium of $x_{j_2}^* = a_{j_1 j_2 1} = a_{j_2 j_2 s_2}$ *with* $\overline{x}_{j_1} \neq a_{j_1 j_1 1}$ *has the following properties.*

- *For* $a_{j_1 j_1 0}(\overline{x}_{j_1} - a_{j_1 j_1 1}) > 0$ *and* $a_{j_2 j_2 0}(a_{j_2 j_2 s_2} - a_{j_2 j_2 s_1}) > 0$,

$$(\overline{x}_{j_1}, a_{j_1 j_2 1}) = \underbrace{(\text{II, US})}_{\text{increasing-inflection upper-saddle}}. \tag{1.86}$$

The infinite-equilibrium of $x_{j_2}^* = a_{j_1 j_2 1} = a_{j_2 j_2 s_2}$ *is an* (II,US)*-increasing-inflection upper-saddle.*

- *For* $a_{j_1 j_1 0}(\overline{x}_{j_1} - a_{j_1 j_1 1}) < 0$ *and* $a_{j_2 j_2 0}(a_{j_2 j_2 s_2} - a_{j_2 j_2 s_1}) > 0$,

$$(\overline{x}_{j_1}, a_{j_1 j_2 1}) = \underbrace{(\text{DI, US})}_{\text{decreasing-inflection upper-saddle}}. \tag{1.87}$$

The infinite-equilibrium of $x_{j_2}^* = a_{j_1 j_2 1} = a_{j_2 j_2 s_2}$ *is a* (DI,US)*-decreasing-inflection upper-saddle.*

- *For* $a_{j_1 j_1 0}(\overline{x}_{j_1} - a_{j_1 j_1 1}) > 0$ *and* $a_{j_2 j_2 0}(a_{j_2 j_2 s_2} - a_{j_2 j_2 s_1}) < 0$,

$$(\overline{x}_{j_1}, a_{j_1 j_2 1}) = \underbrace{(\text{DI, LS})}_{\text{decreasing-inflection lower-saddle}}. \tag{1.88}$$

The infinite-equilibrium of $x_{j_2}^* = a_{j_1 j_2 1} = a_{j_2 j_2 s_2}$ *is a* (DI,LS)-*decreasing-inflection lower-saddle.*

- *For* $a_{j_1 j_1 0}(\bar{x}_{j_1} - a_{j_1 j_1 1}) < 0$ *and* $a_{j_2 j_2 0}(a_{j_2 j_2 s_2} - a_{j_2 j_2 s_1}) < 0$,

$$(\bar{x}_{j_1}, a_{j_1 j_2 1}) = \underbrace{(\mathrm{II}, \mathrm{LS})}_{\text{increasing-inflection lower-saddle}}. \tag{1.89}$$

The infinite-equilibrium of $x_{j_2}^* = a_{j_1 j_2 1} = a_{j_2 j_2 s_2}$ *is an* (II,LS)-*increasing-inflection lower-saddle.*

(v_{4b}) *The equilibrium of* $(x_{j_1}^*, x_{j_2}^*) = (a_{j_1 j_1 1}, a_{j_2 j_2 s_2})$ *with* $a_{j_1 j_2 1} = a_{j_2 j_2 s_2}$ *has the following properties.*

- *For* $a_{j_1 j_1 0} > 0$ *and* $a_{j_2 j_2 0}(a_{j_2 j_2 s_2} - a_{j_2 j_2 s_1}) > 0$,

$$(a_{j_1 j_1 1}, a_{j_2 j_2 s_2}) = \underbrace{(\mathrm{DI} : \mathrm{II}, \mathrm{US})}_{\text{inflection upper-saddle}}. \tag{1.90}$$

The equilibrium of $(x_{j_1}^*, x_{j_2}^*) = (a_{j_1 j_1 1}, a_{j_2 j_2 s_2})$ *is a* (DI:II,US)-*inflection upper-saddle.*

- *For* $a_{j_1 j_1 0} < 0$ *and* $a_{j_2 j_2 0}(a_{j_2 j_2 s_2} - a_{j_2 j_2 s_1}) > 0$,

$$(a_{j_1 j_1 1}, a_{j_2 j_2 s_2}) = \underbrace{(\mathrm{II} : \mathrm{DI}, \mathrm{US})}_{\text{inflection upper-saddle}}. \tag{1.91}$$

The equilibrium of $(x_{j_1}^*, x_{j_2}^*) = (a_{j_1 j_1 1}, a_{j_2 j_2 s_2})$ *is an* (II:DI,US)-*inflection upper-saddle.*

- *For* $a_{j_1 j_1 0} > 0$ *and* $a_{j_2 j_2 0}(a_{j_2 j_2 s_2} - a_{j_2 j_2 s_1}) < 0$,

$$(a_{j_1 j_1 1}, a_{j_2 j_2 s_2}) = \underbrace{(\mathrm{II} : \mathrm{DI}, \mathrm{LS})}_{\text{inflection lower-saddle}}. \tag{1.92}$$

The equilibrium of $(x_{j_1}^*, x_{j_2}^*) = (a_{j_1 j_1 1}, a_{j_2 j_2 s_2})$ *is an* (II:DI,LS)-*inflection lower-saddle.*

- *For* $a_{j_1 j_1 0} < 0$ *and* $a_{j_2 j_2 0}(a_{j_2 j_2 s_2} - a_{j_2 j_2 s_1}) < 0$,

$$(a_{j_1 j_1 1}, a_{j_2 j_2 s_2}) = \underbrace{(\mathrm{DI} : \mathrm{II}, \mathrm{LS})}_{\text{inflection lower-saddle}}. \tag{1.93}$$

The equilibrium of $(x_{j_1}^*, x_{j_2}^*) = (a_{j_1 j_1 1}, a_{j_2 j_2 s_2})$ *is a* (DI:II,LS)-*inflection lower-saddle.*

(v_5) For $a_{j_2j_2s_1} = a_{j_2j_2s_2} = a_{j_2j_21} = a_{j_1j_21}$*, the first integral manifold for is*

$$\ln\frac{|x_{j_1} - a_{j_1j_11}|}{|x_{j_10} - a_{j_1j_11}|} = \frac{a_{j_1j_10}}{a_{j_2j_20}}\ln\frac{|x_{j_2} - a_{j_2j_21}|}{|x_{j_20} - a_{j_2j_21}|}. \tag{1.94}$$

(v_{5a}) The infinite-equilibrium of $x_{j_2}^* = a_{j_1j_21} = a_{j_2j_21}$ *with* $\overline{x}_{j_1} \neq a_{j_1j_11}$ *has the following properties.*

- *For* $a_{j_1j_10}(\overline{x}_{j_1} - a_{j_1j_11}) > 0$ *and* $a_{j_2j_20} > 0$*,*

$$(\overline{x}_{j_1}, a_{j_2j_21}) = \underbrace{(\mathrm{UT}, 3^{\mathrm{rd}}\mathrm{SO})}_{\text{up-trumpet source}}. \tag{1.95}$$

  The infinite-equilibrium of $x_{j_2}^* = a_{j_1j_21} = a_{j_2j_21}$ *is a* (UT,3^rd^SO)*-third-order up-trumpet source.*
- *For* $a_{j_1j_10}(\overline{x}_{j_1} - a_{j_1j_11}) < 0$ *and* $a_{j_2j_20} > 0$*,*

$$(\overline{x}_{j_1}, a_{j_2j_21}) = \underbrace{(\mathrm{DT}, 3^{\mathrm{rd}}\mathrm{SO})}_{\text{down-trumpet source}}. \tag{1.96}$$

  The infinite-equilibrium of $x_{j_2}^* = a_{j_1j_21} = a_{j_2j_2s_2}$ *is a* $(\mathrm{DT}, 3^{\mathrm{rd}}\mathrm{SO})$*-third-order down-trumpet source.*
- *For* $a_{j_1j_10}(\overline{x}_{j_1} - a_{j_1j_11}) > 0$ *and* $a_{j_2j_20} < 0$*,*

$$(\overline{x}_{j_1}, a_{j_2j_21}) = \underbrace{(\mathrm{DT}, 3^{\mathrm{rd}}\mathrm{SI})}_{\text{up-trumpet sink}}. \tag{1.97}$$

  The infinite-equilibrium of $x_{j_2}^* = a_{j_1j_21} = a_{j_2j_2s_2}$ *is a* $(\mathrm{DT}, 3^{\mathrm{rd}}\mathrm{SI})$*-third-order down-trumpet sink.*
- *For* $a_{j_1j_10}(\overline{x}_{j_1} - a_{j_1j_11}) < 0$ *and* $a_{j_2j_20} < 0$*,*

$$(\overline{x}_{j_1}, a_{j_2j_21}) = \underbrace{(\mathrm{UT}, 3^{\mathrm{rd}}\mathrm{SI})}_{\text{up-trumpet sink}}. \tag{1.98}$$

  The infinite-equilibrium of $x_{j_2}^* = a_{j_1j_21} = a_{j_2j_2s_2}$ *is a* $(\mathrm{UT}, 3^{\mathrm{rd}}\mathrm{SO})$*-third-order up-trumpet sink.*

(v_{5b}) The equilibrium of $(x_{j_1}^*, x_{j_2}^*) = (a_{j_1j_11}, a_{j_2j_21})$ *with* $a_{j_1j_21} = a_{j_2j_21}$ *has the following properties.*

- *For* $a_{j_1j_10} > 0$ *and* $a_{j_2j_20} > 0$,

$$(a_{j_1j_1 1}, a_{j_2j_2 1}) = \underbrace{(\text{DT} : \text{UT}, 3^{\text{rd}}\text{SO})}_{\text{down-up trumpet source}}. \tag{1.99}$$

The equilibrium of $(x^*_{j_1}, x^*_{j_2}) = (a_{j_1j_1 1}, a_{j_2j_2 1})$ *is a* (DT:UT,3rdSO)-*third-order down-up trumpet source.*

- *For* $a_{j_1j_10} < 0$ *and* $a_{j_2j_20} > 0$,

$$(a_{j_1j_1 1}, a_{j_2j_2 1}) = \underbrace{(\text{UT} : \text{DT}, 3^{\text{rd}}\text{SO})}_{\text{up-down trumpet source}}. \tag{1.100}$$

The equilibrium of $(x^*_{j_1}, x^*_{j_2}) = (a_{j_1j_1 1}, a_{j_2j_2 1})$ *is a* (UT:DT,3rdSO)-*third-order up-down trumpet source.*

- *For* $a_{j_1j_10} > 0$ *and* $a_{j_2j_20} < 0$,

$$(a_{j_1j_1 1}, a_{j_2j_2 1}) = \underbrace{(\text{UT} : \text{DT}, 3^{\text{rd}}\text{SI})}_{\text{up-down trumpet sink}}. \tag{1.101}$$

The equilibrium of $(x^*_{j_1}, x^*_{j_2}) = (a_{j_1j_1 1}, a_{j_2j_2 1})$ *is a* (UT:DT,3rdSI)-*third-order up-down trumpet sink.*

- *For* $a_{j_1j_10} < 0$ *and* $a_{j_2j_20} < 0$,

$$(a_{j_1j_1 1}, a_{j_2j_2 1}) = \underbrace{(\text{DT} : \text{UT}, 3^{\text{rd}}\text{SI})}_{\text{down-up trumpet sink}}. \tag{1.102}$$

The equilibrium of $(x^*_{j_1}, x^*_{j_2}) = (a_{j_1j_1 1}, a_{j_2j_2 1})$ *is a* (DP:UP,3rdSI)-*third-order down-up trumpet sink.*

(*vi*) *For* $\Delta_{j_1j_2} = 0$ and $\Delta_{j_2j_2} > 0$, the standard form is

$$\begin{aligned} \dot{x}_{j_2} &= a_{j_2j_20}(x_{j_2} - a_{j_2j_2 1})(x_{j_2} - a_{j_2j_2 2})(x_{j_2} - a_{j_2j_2 3}), \\ \dot{x}_{j_1} &= a_{j_1j_10}(x_{j_1} - a_{j_1j_1 1})(x_{j_2} - a_{j_1j_2 1})^2 \end{aligned} \tag{1.103}$$

where

$$\begin{aligned} &a_{j_1j_2 1} = -\frac{1}{2}B_{j_1j_2}, \\ &b_{j_2j_2 2}, b_{j_2j_2 3} = -\frac{1}{2}(B_{j_2j_2} \pm \sqrt{\Delta_{j_2j_2}}), \\ &\{a_{j_2j_2 1}, a_{j_2j_2 2}, a_{j_2j_2 3}\} = \text{sort}\{b_{j_2j_2 1}, b_{j_2j_2 2}, b_{j_2j_2 3}\}, \\ &a_{j_2j_2 s_1} < a_{j_2j_2 s_2}; s_1, s_2 \subset \{1, 2, 3\}, s_1 < s_2. \end{aligned} \tag{1.104}$$

(vi_1) The first integral manifold ($a_{j_2j_2s_1} \neq a_{j_1j_2 1}$) is

$$\begin{aligned}
&\ln \frac{|x_{j_1} - a_{j_1j_1 1}|}{|x_{j_1 0} - a_{j_1j_1 1}|} \\
&= \frac{a_{j_1j_1 0}}{a_{j_2j_2 0}} \sum\nolimits_{s_1=1}^{3} \frac{(a_{j_2j_2s_1} - a_{j_1j_2 1})^2}{\prod\nolimits_{s_2=1, s_2 \neq s_1}^{3} (a_{j_2j_2s_1} - a_{j_2j_2s_2})} \ln \frac{|x_{j_2} - a_{j_2j_2s_1}|}{|x_{j_2 0} - a_{j_2j_2s_1}|}.
\end{aligned} \tag{1.105}$$

(vi_{1a}) The equilibrium of $(x_{j_1}^*, x_{j_2}^*) = (a_{j_1j_1 1}, a_{j_2j_2s_1})$ ($s_1, s_2 \in \{1,2,3\}$, $s_1 \neq s_2$) has the following properties.

- *For $a_{j_1j_1 0}(a_{j_2j_2s_1} - a_{j_1j_2 1})^2 > 0$ and $a_{j_2j_2 0}\prod\nolimits_{s_2=1, s_2 \neq s_1}^{3}(a_{j_2j_2s_1} - a_{j_2j_2s_2}) > 0$,*

$$(a_{j_1j_1 1}, a_{j_2j_2s_1}) = \underbrace{(\text{SO, SO})}_{\text{source}}. \tag{1.106}$$

  The equilibrium of $(x_{j_1}^*, x_{j_2}^*) = (a_{j_1j_1 1}, a_{j_2j_2s_1})$ is an (SO,SO)-source.

- *For $a_{j_1j_1 0}(a_{j_2j_2s_1} - a_{j_1j_2 1})^2 < 0$ and $a_{j_2j_2 0}\prod\nolimits_{s_2=1, s_2 \neq s_1}^{3}(a_{j_2j_2s_1} - a_{j_2j_2s_2}) > 0$,*

$$(a_{j_1j_1 1}, a_{j_2j_2s_1}) = \underbrace{(\text{SI, SO})}_{\text{saddle}}. \tag{1.107}$$

  The equilibrium of $(x_{j_1}^*, x_{j_2}^*) = (a_{j_1j_1 1}, a_{j_2j_2s_1})$ is an (SI,SO)-saddle.

- *For $a_{j_1j_1 0}(a_{j_2j_2s_1} - a_{j_1j_2 1})^2 > 0$ and $a_{j_2j_2 0}\prod\nolimits_{s_2=1, s_2 \neq s_1}^{3}(a_{j_2j_2s_1} - a_{j_2j_2s_2}) < 0$,*

$$(a_{j_1j_1 1}, a_{j_2j_2s_1}) = \underbrace{(\text{SO, SI})}_{\text{saddle}}. \tag{1.108}$$

  The equilibrium of $(x_{j_1}^*, x_{j_2}^*) = (a_{j_1j_1 1}, a_{j_2j_2s_1})$ is an (SO,SI)-saddle.

- *For $a_{j_1j_1 0}(a_{j_2j_2s_1} - a_{j_1j_2 1})^2 < 0$ and $a_{j_2j_2 0}\prod\nolimits_{s_2=1, s_2 \neq s_1}^{3}(a_{j_2j_2s_1} - a_{j_2j_2s_2}) < 0$,*

$$(a_{j_1j_1 1}, a_{j_2j_2s_1}) = \underbrace{(\text{SI, SI})}_{\text{sink}}. \tag{1.109}$$

  The equilibrium of $(x_{j_1}^*, x_{j_2}^*) = (a_{j_1j_1 1}, a_{j_2j_2s_1})$ is an (SI,SI)-sink.

(vi_{1b}) The equilibrium of $(x_{j_1}^*, x_{j_2}^*) = (a_{j_1j_1 1}, a_{j_1j_2 1})$ with $a_{j_1j_2 1} \neq a_{j_2j_2 1}, a_{j_2j_2 2}$ has the following properties.

- *For $a_{j_1j_1 0} > 0$ and $a_{j_2j_2 0}\prod\nolimits_{s_1=1}^{3}(a_{j_1j_2 1} - a_{j_2j_2s_1}) > 0$,*

$$(a_{j_1j_11}, a_{j_1j_21}) = \underbrace{({}_{\text{DI:II}}\text{SO}, \text{pF})}_{\text{inflection-source flow }(+)}. \tag{1.110}$$

The equilibrium of $(x^*_{j_1}, x^*_{j_2}) = (a_{j_1j_11}, a_{j_1j_21})$ *is a* $({}_{\text{DI:II}}\text{SO}, \text{pF})$*-positive inflection-source flow.*

- *For* $a_{j_1j_10} < 0$ *and* $a_{j_2j_20}\prod_{s_1=1}^{3}(a_{j_1j_21} - a_{j_2j_2s_1}) > 0$,

$$(a_{j_1j_11}, a_{j_1j_21}) = \underbrace{({}_{\text{II:DI}}\text{SI}, \text{pF})}_{\text{inflection-sink flow }(+)}. \tag{1.111}$$

The equilibrium of $(x^*_{j_1}, x^*_{j_2}) = (a_{j_1j_11}, a_{j_1j_21})$ *is an* $({}_{\text{II:DI}}\text{SI}, \text{pF})$*-positive inflection-sink flow.*

- *For* $a_{j_1j_10} > 0$ *and* $a_{j_2j_20}\prod_{s_1=1}^{3}(a_{j_1j_21} - a_{j_2j_2s_1}) < 0$,

$$(a_{j_1j_11}, a_{j_1j_21}) = \underbrace{({}_{\text{II:DI}}\text{SO}, \text{nF})}_{\text{inflection-source flow }(-)}. \tag{1.112}$$

The equilibrium of $(x^*_{j_1}, x^*_{j_2}) = (a_{j_1j_11}, a_{j_1j_21})$ *is an* $({}_{\text{II:DI}}\text{SO}, \text{nF})$*-negative inflection-source flow.*

- *For* $a_{j_1j_10} < 0$ *and* $a_{j_2j_20}\prod_{s_1=1}^{3}(a_{j_1j_21} - a_{j_2j_2s_1}) < 0$,

$$(a_{j_1j_11}, a_{j_1j_21}) = \underbrace{({}_{\text{DI:II}}\text{SI}, \text{nF})}_{\text{inflection-sink flow }(-)}. \tag{1.113}$$

The equilibrium of $(x^*_{j_1}, x^*_{j_2}) = (a_{j_1j_11}, a_{j_1j_21})$ *is a* $({}_{\text{DI:II}}\text{SI}, \text{nF})$*-negative inflection-sink flow.*

(vi$_2$*) The first integral manifold* $(a_{j_2j_2s_1} = a_{j_1j_21})$ *is*

$$\ln\frac{|x_{j_1} - a_{j_1j_11}|}{|x_{j_10} - a_{j_1j_11}|} = \frac{a_{j_1j_10}}{a_{j_2j_20}}\sum_{s_2=1, s_2\neq s_1, s_3}^{3}\frac{(a_{j_2j_2s_2} - a_{j_1j_21})}{(a_{j_2j_2s_2} - a_{j_1j_2s_3})}\ln\frac{|x_{j_2} - a_{j_2j_2s_2}|}{|x_{j_20} - a_{j_2j_2s_2}|}. \tag{1.114}$$

(vi$_{2a}$*) The infinite-equilibrium of* $x^*_{j_2} = a_{j_1j_21} = a_{j_2j_2s_1}$ *with* $\overline{x}_{j_1} \neq a_{j_1j_11}$ *has the following properties.*

- *For* $a_{j_1j_10}(\overline{x}_{j_1} - a_{j_1j_11}) > 0$ *and* $a_{j_2j_20}\prod_{s_2=1, s_2\neq s_1}^{3}(a_{j_2j_2s_1} - a_{j_2j_2s_2}) > 0$,

$$(\overline{x}_{j_1}, a_{j_2j_21}) = \underbrace{(\text{UP}, \text{SO})}_{\text{up-parabola source}}. \tag{1.115}$$

The infinite-equilibrium of $x^*_{j_2} = a_{j_1j_21} = a_{j_2j_2s_1}$ *is a* (UP,SO)*-up-parabola source.*

- *For* $a_{j_1j_10}(\overline{x}_{j_1}-a_{j_1j_11})<0$ *and* $a_{j_2j_20}\prod_{s_2=1,s_2\neq s_1}^{3}(a_{j_2j_2s_1}-a_{j_2j_2s_2})>0$,

$$(\overline{x}_{j_1},a_{j_2j_2s_1})=\underbrace{(\text{DP},\text{SO})}_{\text{down-parabola source}}. \tag{1.116}$$

The infinite-equilibrium of $x_{j_2}^*=a_{j_1j_21}=a_{j_2j_2s_1}$ *is a* (DP,SO)-*down-parabola source.*

- *For* $a_{j_1j_10}(\overline{x}_{j_1}-a_{j_1j_11})>0$ *and* $a_{j_2j_20}\prod_{s_2=1,s_2\neq s_1}^{3}(a_{j_2j_2s_1}-a_{j_2j_2s_2})<0$,

$$(\overline{x}_{j_1},a_{j_2j_2s_1})=\underbrace{(\text{DP},\text{SI})}_{\text{down-parabola sink}}. \tag{1.117}$$

The infinite-equilibrium of $x_{j_2}^*=a_{j_1j_21}=a_{j_2j_2s_1}$ *is a* (DP,SI)-*down-parabola sink.*

- *For* $a_{j_1j_10}(\overline{x}_{j_1}-a_{j_1j_11})<0$ *and* $a_{j_2j_20}\prod_{s_2=1,s_2\neq s_1}^{3}(a_{j_2j_2s_1}-a_{j_2j_2s_2})<0$,

$$(\overline{x}_{j_1},a_{j_2j_2s_1})=\underbrace{(\text{UP},\text{SI})}_{\text{up-parabola sink}}. \tag{1.118}$$

The infinite-equilibrium of $x_{j_2}^*=a_{j_1j_21}=a_{j_2j_2s_1}$ *is a* (UP,SI)-*up-parabola sink.*

(vi$_{2b}$*) The equilibrium of* $(x_{j_1}^*,x_{j_2}^*)=(a_{j_1j_11},a_{j_2j_2s_1})$ *with* $a_{j_2j_2s_1}=a_{j_1j_21}$ *has the following properties.*

- *For* $a_{j_1j_11}>0$ *and* $a_{j_2j_20}\prod_{s_2=1,s_2\neq s_1}^{3}(a_{j_2j_2s_1}-a_{j_2j_2s_2})>0$,

$$(a_{j_1j_11},a_{j_2j_2s_1})=\underbrace{(\text{DP}:\text{UP},\text{SO})}_{\text{hyperbolic source}}. \tag{1.119}$$

The equilibrium of $(x_{j_1}^*,x_{j_2}^*)=(a_{j_1j_11},a_{j_2j_2s_1})$ *is a* (DP:UP,SO)-hyperbolic source.

- *For* $a_{j_1j_11}<0$ *and* $a_{j_2j_20}\prod_{s_2=1,s_2\neq s_1}^{3}(a_{j_2j_2s_1}-a_{j_2j_2s_2})>0$,

$$(a_{j_1j_11},a_{j_1j_21})=\underbrace{(\text{UP}:\text{DP},\text{SO})}_{\text{hyperbolic-secant source}}. \tag{1.120}$$

The equilibrium of $(x_{j_1}^*,x_{j_2}^*)=(a_{j_1j_11},a_{j_2j_2s_1})$ *is a* (UP:DP,SO)-hyperbolic-secant source.

- *For* $a_{j_1j_10}>0$ *and* $a_{j_2j_20}\prod_{s_2=1,s_2\neq s_1}^{3}(a_{j_2j_2s_1}-a_{j_2j_2s_2})<0$,

$$(a_{j_1j_11},a_{j_1j_21})=\underbrace{(\text{UP}:\text{DP},\text{SI})}_{\text{hyperbolic-secant sink}}. \tag{1.121}$$

The equilibrium of $(x^*_{j_1}, x^*_{j_2}) = (a_{j_1j_1}1, a_{j_2j_2s_1})$ *is a* (UP:DP,SI)-hyperbolic-secant sink.

- *For* $a_{j_1j_1}0 < 0$ *and* $a_{j_2j_2}0\prod_{s_2=1,s_2\neq s_1}^{3}(a_{j_2j_2s_1} - a_{j_2j_2s_2}) < 0$,

$$(a_{j_1j_1}1, a_{j_1j_2}1) = \underbrace{(\text{DP}:\text{UP},\text{SI})}_{\text{hyperbolic sink}}. \tag{1.122}$$

The equilibrium of $(x^*_{j_1}, x^*_{j_2}) = (a_{j_1j_1}1, a_{j_2j_2s_1})$ *is a* (DP:UP,SI)-hyperbolic sink.

(vii) *For* $\Delta_{j_1j_2} > 0$ and $\Delta_{j_2j_2} < 0$, the standard form is

$$\begin{aligned}\dot{x}_{j_2} &= a_{j_2j_2}0(x_{j_2} - a_{j_2j_2}1)\left[(x_{j_2} - a_{j_2j_2})^2 + b_{j_2j_2}\right],\\ \dot{x}_{j_1} &= a_{j_1j_1}0(x_{j_1} - a_{j_1j_1}1)(x_{j_2} - a_{j_1j_2}1)(x_{j_2} - a_{j_1j_2}2)\end{aligned} \tag{1.123}$$

where

$$\begin{aligned}&b_{j_1j_2}1, b_{j_1j_2}2 = -\frac{1}{2}(B_{j_1j_2} \pm \sqrt{\Delta_{j_1j_2}}),\\ &\{a_{j_1j_2}1, a_{j_1j_2}2\} = \text{sort}\{b_{j_1j_2}1, b_{j_1j_2}2\}, a_{j_1j_2}1 < a_{j_1j_2}2;\\ &a_{j_2j_2} = -\frac{1}{2}B_{j_2j_2}, b_{j_2j_2} = \frac{1}{4}(-\Delta_{j_2j_2}), a_{j_2j_2}1 = b_{j_2j_2}1.\end{aligned} \tag{1.124}$$

(vii$_1$) *The first integral manifold for* $a_{j_2j_2}1 \neq a_{j_1j_2l_1} (l_1 = 1, 2)$ *is*

$$\begin{aligned}\ln\frac{|x_{j_1} - a_{j_1j_1}1|}{|x_{j_1}0 - a_{j_1j_1}1|} = \frac{a_{j_1j_1}0}{a_{j_2j_2}0}&\left[A\ln\frac{|x_{j_2} - a_{j_2j_2}1|}{|x_{j_2}0 - a_{j_2j_2}1|}\right.\\ &+\frac{1}{2}(1-A)\ln\frac{|(x_{j_2} - a_{j_2j_2})^2 + b_{j_2j_2}|}{|(x_{j_2}0 - a_{j_2j_2})^2 + b_{j_2j_2}|}\\ &\left.+\frac{B}{\sqrt{b_{j_2j_2}}}\left(\arctan\frac{x_{j_2} - a_{j_2j_2}}{\sqrt{b_{j_2j_2}}} - \arctan\frac{x_{j_2}0 - a_{j_2j_2}}{\sqrt{b_{j_2j_2}}}\right)\right]\end{aligned} \tag{1.125}$$

where

$$\begin{aligned}A &= \frac{(a_{j_2j_2}1 - a_{j_1j_2}1)(a_{j_2j_2}1 - a_{j_1j_2}2)}{(a_{j_2j_2}1 - a_{j_2j_2})^2 + b_{j_2j_2}},\\ B &= a_{j_2j_2} + a_{j_2j_2}1 - a_{j_1j_2}1 - a_{j_1j_2}2 - (a_{j_2j_2}1 - a_{j_2j_2})A.\end{aligned} \tag{1.126}$$

(vii$_{1a}$) The equilibrium of $(x^*_{j_1}, x^*_{j_2}) = (a_{j_1j_11}, a_{j_2j_21})$ *has the following properties.*

- *For* $a_{j_1j_10}\prod_{l_1=1}^{2}(a_{j_2j_21} - a_{j_1j_2l_1}) > 0$ *and* $a_{j_2j_20} > 0$,

$$(a_{j_1j_11}, a_{j_2j_21}) = \underbrace{(\mathrm{SO}, \mathrm{SO})}_{\text{source}}. \tag{1.127}$$

  The equilibrium of $(x^*_{j_1}, x^*_{j_2}) = (a_{j_1j_11}, a_{j_2j_21})$ *is an* (SO,SO)*-source.*

- *For* $a_{j_1j_10}\prod_{l_1=1}^{2}(a_{j_2j_21} - a_{j_1j_2l_1}) < 0$ *and* $a_{j_2j_20} > 0$,

$$(a_{j_1j_11}, a_{j_2j_21}) = \underbrace{(\mathrm{SI}, \mathrm{SO})}_{\text{saddle}}. \tag{1.128}$$

  The equilibrium of $(x^*_{j_1}, x^*_{j_2}) = (a_{j_1j_11}, a_{j_2j_21})$ *is an* (SI,SO)*-saddle.*

- *For* $a_{j_1j_10}\prod_{l_1=1}^{2}(a_{j_2j_21} - a_{j_1j_2l_1}) > 0$ *and* $a_{j_2j_20} < 0$,

$$(a_{j_1j_11}, a_{j_2j_21}) = \underbrace{(\mathrm{SO}, \mathrm{SI})}_{\text{saddle}}. \tag{1.129}$$

  The equilibrium of $(x^*_{j_1}, x^*_{j_2}) = (a_{j_1j_11}, a_{j_2j_21})$ *is an* (SO,SI)*-saddle.*

- *For* $a_{j_1j_10}\prod_{l_1=1}^{2}(a_{j_2j_21} - a_{j_1j_2l_1}) < 0$ *and* $a_{j_2j_20} < 0$,

$$(a_{j_1j_11}, a_{j_2j_21}) = \underbrace{(\mathrm{SI}, \mathrm{SI})}_{\text{sink}}. \tag{1.130}$$

  The equilibrium of $(x^*_{j_1}, x^*_{j_2}) = (a_{j_1j_11}, a_{j_2j_21})$ *is an* (SI,SI)*-sink.*

(vii$_{1b}$) The equilibrium of $(x^*_{j_1}, x^*_{j_2}) = (a_{j_1j_11}, a_{j_1j_2l_1})$ $(l_1, l_2 \in \{1, 2\}, l_1 \neq l_2)$ *has the following properties.*

- *For* $a_{j_1j_10}(a_{j_1j_2l_1} - a_{j_1j_2l_2}) > 0$ *and* $a_{j_2j_20}(a_{j_1j_2l_1} - a_{j_2j_21}) > 0$,

$$(a_{j_1j_11}, a_{j_1j_2l_1}) = \underbrace{(\mathrm{DP}:\mathrm{UP}, \mathrm{pF})}_{\text{hyperbolic flow }(+)}. \tag{1.131}$$

  The equilibrium of $(x^*_{j_1}, x^*_{j_2}) = (a_{j_1j_11}, a_{j_1j_2l_1})$ *is a* (DP:UP,pF)*-positive hyperbolic flow.*

- *For* $a_{j_1j_10}(a_{j_1j_2l_1} - a_{j_1j_2l_2}) < 0$ *and* $a_{j_2j_20}(a_{j_1j_2l_1} - a_{j_2j_21}) > 0$,

$$(a_{j_1j_11}, a_{j_1j_2l_1}) = \underbrace{(\mathrm{UP}:\mathrm{DP}, \mathrm{pF})}_{\text{hyperbolic-secant flow }(+)}. \tag{1.132}$$

The equilibrium of $(x_{j_1}^*, x_{j_2}^*) = (a_{j_1 j_1 1}, a_{j_1 j_2 l_1})$ *is a* (UP:DP,pF)-*positive hyperbolic-secant flow.*

- *For* $a_{j_1 j_1 0}(a_{j_1 j_2 l_1} - a_{j_1 j_2 l_2}) > 0$ *and* $a_{j_2 j_2 0}(a_{j_1 j_2 l_1} - a_{j_2 j_2 1}) < 0$,

$$(a_{j_1 j_1 1}, a_{j_1 j_2 l_1}) = \underbrace{(\text{UP}:\text{DP},\text{nF})}_{\text{hyperbolic-secant flow }(-)}. \tag{1.133}$$

The equilibrium of $(x_{j_1}^*, x_{j_2}^*) = (a_{j_1 j_1 1}, a_{j_1 j_2 l_1})$ *is a* (UP:DP,nF)-*negative hyperbolic-secant flow.*

- *For* $a_{j_1 j_1 0}(a_{j_1 j_2 l_1} - a_{j_1 j_2 l_2}) < 0$ *and* $a_{j_2 j_2 0}(a_{j_1 j_2 l_1} - a_{j_2 j_2 1}) < 0$,

$$(a_{j_1 j_1 1}, a_{j_1 j_2 l_1}) = \underbrace{(\text{DP}:\text{UP},\text{nF})}_{\text{hyperbolic flow }(-)}. \tag{1.134}$$

The equilibrium of $(x_{j_1}^*, x_{j_2}^*) = (a_{j_1 j_1 1}, a_{j_1 j_2 l_1})$ *is a* (DP:UP,nF)-*negative hyperbolic flow.*

(vii₂) The first integral manifold for $a_{j_2 j_2 1} = a_{j_1 j_2 l_1} (l_1 = 1, 2)$ *is*

$$\ln \frac{|x_{j_1} - a_{j_1 j_1 1}|}{|x_{j_1 0} - a_{j_1 j_1 1}|} = \frac{a_{j_1 j_1 0}}{a_{j_2 j_2 0}} \left[\frac{1}{2} \ln \frac{|(x_{j_2} - a_{j_2 j_2})^2 + b_{j_2 j_2}|}{|(x_{j_2 0} - a_{j_2 j_2})^2 + b_{j_2 j_2}|} \right.$$
$$\left. + \frac{a_{j_2 j_2} - a_{j_1 j_2 l_2}}{\sqrt{b_{j_2 j_2}}} \left(\arctan \frac{x_{j_2} - a_{j_2 j_2}}{\sqrt{b_{j_2 j_2}}} - \arctan \frac{x_{j_2 0} - a_{j_2 j_2}}{\sqrt{b_{j_2 j_2}}} \right) \right]. \tag{1.135}$$

(vii₂ₐ) The infinite-equilibrium of $x_{j_2}^* = a_{j_1 j_2 l_1} = a_{j_2 j_2 1}$ *with* $\overline{x}_{j_1} \neq a_{j_1 j_1 1}$ *has the following properties.*

- *For* $a_{j_1 j_1 0}(\overline{x}_{j_1} - a_{j_1 j_1 1})(a_{j_1 j_2 l_1} - a_{j_1 j_2 l_2}) > 0$ *and* $a_{j_2 j_2 0} > 0$,

$$(\overline{x}_{j_1}, a_{j_2 j_2 1}) = \underbrace{(\text{II},\text{SO})}_{\text{increasing-inflection source}}. \tag{1.136}$$

The infinite-equilibrium of $x_{j_2}^* = a_{j_1 j_2 l_1} = a_{j_2 j_2 1}$ *is an* (II,SO)-*increasing-inflection source.*

- *For* $a_{j_1 j_1 0}(x_{j_1} - a_{j_1 j_1 1})(a_{j_1 j_2 l_1} - a_{j_1 j_2 l_2}) < 0$ *and* $a_{j_2 j_2 0} > 0$,

$$(\overline{x}_{j_1}, a_{j_2 j_2 1}) = \underbrace{(\text{DI},\text{SO})}_{\text{decreasing-inflection source}}. \tag{1.137}$$

The infinite-equilibrium of $x_{j_2}^* = a_{j_1 j_2 l_1} = a_{j_2 j_2 1}$ *is a* (DI,SO)-*decreasing-inflection source.*

- *For* $a_{j_1j_1 0}(\overline{x}_{j_1} - a_{j_1j_1 1})(a_{j_1j_2l_1} - a_{j_1j_2l_2}) > 0$ *and* $a_{j_2j_2 0} < 0$,

$$(\overline{x}_{j_1}, a_{j_2j_2 1}) = \underbrace{(\text{DP, SI})}_{\text{down-parabola sink}}. \tag{1.138}$$

The infinite-equilibrium of $x^*_{j_2} = a_{j_1j_2l_1} = a_{j_2j_2 1}$ *is a* (DI,SI)-*decreasing-inflection sink.*

- *For* $a_{j_1j_1 0}(\overline{x}_{j_1} - a_{j_1j_1 1})(a_{j_1j_2l_1} - a_{j_1j_2l_2}) < 0$ *and* $a_{j_2j_2 0} < 0$,

$$(\overline{x}_{j_1}, a_{j_2j_2 1}) = \underbrace{(\text{II, SI})}_{\text{increasing-inflection sink}}. \tag{1.139}$$

The infinite-equilibrium of $x^*_{j_2} = a_{j_1j_2l_1} = a_{j_2j_2 1}$ *is an* (II,SI)-*increasing-inflection sink.*

(vii$_{2b}$) The equilibrium of $(x^*_{j_1}, x^*_{j_2}) = (a_{j_1j_1 1}, a_{j_2j_2l_1})$ *with* $a_{j_1j_2l_1} = a_{j_2j_2 1}$ *has the following properties.*

- *For* $a_{j_1j_1 0}(a_{j_1j_2l_1} - a_{j_1j_2l_2}) > 0$ *and* $a_{j_2j_2 0} > 0$,

$$(a_{j_1j_1 1}, a_{j_2j_2 1}) = \underbrace{(\text{DI : II, SO})}_{\text{inflection source}}. \tag{1.140}$$

The equilibrium of $(x^*_{j_1}, x^*_{j_2}) = (a_{j_1j_1 1}, a_{j_2j_2 1})$ *is a* (DI:II,SO)-*inflection source.*

- *For* $a_{j_1j_1 0}(a_{j_1j_2l_1} - a_{j_1j_2l_2}) < 0$ *and* $a_{j_2j_2 0} > 0$,

$$(a_{j_1j_1 1}, a_{j_2j_2 1}) = \underbrace{(\text{II : DI, SO})}_{\text{inflection source}}. \tag{1.141}$$

The equilibrium of $(x^*_{j_1}, x^*_{j_2}) = (a_{j_1j_1 1}, a_{j_2j_2 1})$ *is an* (II:DI,SO)-*inflection source.*

- *For* $a_{j_1j_1 0}(a_{j_1j_2l_1} - a_{j_1j_2l_2}) > 0$ *and* $a_{j_2j_2 0} < 0$,

$$(a_{j_1j_1 1}, a_{j_2j_2 1}) = \underbrace{(\text{II : DI, SI})}_{\text{inflection sink}}. \tag{1.142}$$

The equilibrium of $(x^*_{j_1}, x^*_{j_2}) = (a_{j_1j_1 1}, a_{j_2j_2 1})$ *is an* (II:DI,SI)-*inflection sink.*

- *For* $a_{j_1j_1 0}(a_{j_1j_2l_1} - a_{j_1j_2l_2}) < 0$ *and* $a_{j_2j_2 0} < 0$,

$$(a_{j_1j_1 1}, a_{j_2j_2 1}) = \underbrace{(\text{DI : II, SI})}_{\text{inflection sink}}. \tag{1.143}$$

The equilibrium of $(x^*_{j_1}, x^*_{j_2}) = (a_{j_1j_1 1}, a_{j_2j_2 1})$ *is a* (DI:II,SI)-*inflection sink.*

(viii) *For* $\Delta_{j_1j_2}>0$ and $\Delta_{j_2j_2}=0$, the standard form is

$$\begin{aligned}\dot{x}_{j_2}&=a_{j_2j_20}(x_{j_2}-a_{j_2j_2s_1})(x_{j_2}-a_{j_2j_2s_2})^2,\\ \dot{x}_{j_1}&=a_{j_1j_10}(x_{j_1}-a_{j_1j_11})(x_{j_2}-a_{j_1j_21})(x_{j_2}-a_{j_1j_22})\end{aligned}\tag{1.144}$$

where

$$\begin{aligned}&b_{j_1j_21},b_{j_1j_22}=-\frac{1}{2}(B_{j_1j_2}\pm\sqrt{\Delta_{j_1j_2}}),\\ &\{a_{j_1j_21},a_{j_1j_22}\}=\mathrm{sort}\{b_{j_1j_21},b_{j_1j_22}\},a_{j_1j_21}<a_{j_1j_22};\\ &a_{j_2j_2s_2}=-\frac{1}{2}B_{j_2j_2},a_{j_2j_2s_1}=b_{j_2j_21}.\end{aligned}\tag{1.145}$$

(viii$_1$) The first integral manifold for $a_{j_2j_2s}\neq a_{j_1j_21},a_{j_1j_22}$ $(s=s_1,s_2)$ *is*

$$\begin{aligned}\ln\frac{|x_{j_1}-a_{j_1j_11}|}{|x_{j_10}-a_{j_1j_11}|}=&\frac{a_{j_1j_10}}{a_{j_2j_20}}\Bigg\{\frac{\prod_{l_1=1}^{2}(a_{j_2j_2s_1}-a_{j_1j_2l_1})}{(a_{j_2j_2s_1}-a_{j_2j_2s_2})^2}\ln\frac{|x_{j_2}-a_{j_2j_2s_1}|}{|x_{j_20}-a_{j_2j_2s_1}|}\\ &+[\frac{\sum_{l_1=1}^{2}(a_{j_2j_2s_2}-a_{j_1j_2l_1})}{(a_{j_2j_2s_2}-a_{j_2j_2s_1})}-\frac{\prod_{l_1=1}^{2}(a_{j_2j_2s_2}-a_{j_1j_2l_1})}{(a_{j_2j_2s_2}-a_{j_2j_2s_1})^2}]\\ &\times\ln\frac{|x_{j_2}-a_{j_2j_2s_2}|}{|x_{j_20}-a_{j_2j_2s_2}|}-\frac{\prod_{l_1=1}^{2}(a_{j_2j_2s_2}-a_{j_1j_2l_1})}{(a_{j_2j_2s_2}-a_{j_2j_2s_1})}\\ &\times(\frac{1}{x_{j_2}-a_{j_2j_2s_2}}-\frac{1}{x_{j_20}-a_{j_2j_2s_2}})\Bigg\}.\end{aligned}\tag{1.146}$$

(viii$_{1a}$) The equilibrium of $(x_{j_1}^*,x_{j_2}^*)=(a_{j_1j_11},a_{j_2j_2s_1})$ *has the following properties.*

- *For* $a_{j_1j_10}\prod_{l_1=1}^{2}(a_{j_2j_2s_1}-a_{j_1j_2l_1})>0$ *and* $a_{j_2j_20}(a_{j_2j_2s_1}-a_{j_2j_2s_2})^2>0$,

$$(a_{j_1j_11},a_{j_2j_2s_1})=\underbrace{(\mathrm{SO},\mathrm{SO})}_{\text{source}}.\tag{1.147}$$

The equilibrium of $(x_{j_1}^*,x_{j_2}^*)=(a_{j_1j_11},a_{j_2j_2s_1})$ *is an* (SO,SO)*-source.*

- *For* $a_{j_1j_10}\prod_{l_1=1}^{2}(a_{j_2j_2s_1}-a_{j_1j_2l_1})<0$ *and* $a_{j_2j_20}(a_{j_2j_2s_1}-a_{j_2j_2s_2})^2>0$,

$$(a_{j_1j_11},a_{j_2j_2s_1})=\underbrace{(\mathrm{SI},\mathrm{SO})}_{\text{saddle}}.\tag{1.148}$$

The equilibrium of $(x_{j_1}^*,x_{j_2}^*)=(a_{j_1j_11},a_{j_2j_2s_1})$ *is an* (SI,SO)*-saddle.*

- *For* $a_{j_1j_10}\prod_{l_1=1}^{2}(a_{j_2j_2s_1}-a_{j_1j_2l_1})>0$ *and* $a_{j_2j_20}(a_{j_2j_2s_1}-a_{j_2j_2s_2})^2<0$,

$$(a_{j_1j_1 1},a_{j_2j_2s_1})=\underbrace{(\text{SO},\text{SI})}_{\text{saddle}}. \tag{1.149}$$

The equilibrium of $(x_{j_1}^*,x_{j_2}^*)=(a_{j_1j_1 1},a_{j_2j_2s_1})$ *is an* (SO,SI)-*saddle.*

- *For* $a_{j_1j_10}\prod_{l_1=1}^{2}(a_{j_2j_2s_1}-a_{j_1j_2l_1})<0$ *and* $a_{j_2j_20}(a_{j_2j_2s_1}-a_{j_2j_2s_2})^2<0$,

$$(a_{j_1j_1 1},a_{j_2j_2s_1})=\underbrace{(\text{SI},\text{SI})}_{\text{sink}}. \tag{1.150}$$

The equilibrium of $(x_{j_1}^*,x_{j_2}^*)=(a_{j_1j_1 1},a_{j_2j_2s_1})$ *is an* (SI,SI)-*sink.*

(viii$_{1b}$*) The equilibrium of* $(x_{j_1}^*,x_{j_2}^*)=(a_{j_1j_1 1},a_{j_2j_2s_2})$ *has the following properties.*

- *For* $a_{j_1j_10}\prod_{s_1=1}^{2}(a_{j_2j_2s_2}-a_{j_1j_2l_1})>0$ *and* $a_{j_2j_20}(a_{j_2j_2s_2}-a_{j_2j_2s_1})>0$,

$$(a_{j_1j_1 1},a_{j_2j_2s_2})=\underbrace{(\text{SO},\text{US})}_{\text{upper-saddle source}}. \tag{1.151}$$

The equilibrium of $(x_{j_1}^*,x_{j_2}^*)=(a_{j_1j_1 1},a_{j_2j_2s_2})$ *is an* (SO,US)-*upper-saddle source.*

- *For* $a_{j_1j_10}\prod_{s_1=1}^{2}(a_{j_2j_2s_2}-a_{j_1j_2l_1})<0$ *and* $a_{j_2j_20}(a_{j_2j_2s_2}-a_{j_2j_2s_1})>0$,

$$(a_{j_1j_1 1},a_{j_2j_2s_2})=\underbrace{(\text{SI},\text{US})}_{\text{upper-saddle-sink}}. \tag{1.152}$$

The equilibrium of $(x_{j_1}^*,x_{j_2}^*)=(a_{j_1j_1 1},a_{j_2j_2s_2})$ *is an* (SI,US)-*upper-saddle sink.*

- *For* $a_{j_1j_10}\prod_{s_1=1}^{2}(a_{j_2j_2s_2}-a_{j_1j_2l_1})>0$ *and* $a_{j_2j_20}(a_{j_2j_2s_2}-a_{j_2j_2s_1})<0$,

$$(a_{j_1j_1 1},a_{j_2j_2s_2})=\underbrace{(\text{SO},\text{LS})}_{\text{lower-saddle source}}. \tag{1.153}$$

The equilibrium of $(x_{j_1}^*,x_{j_2}^*)=(a_{j_1j_1 1},a_{j_2j_2s_2})$ *is an* (SO,LS)-*lower-saddle source.*

- *For* $a_{j_1j_10}\prod_{s_1=1}^{2}(a_{j_2j_2s_2}-a_{j_1j_2l_1})<0$ *and* $a_{j_2j_20}(a_{j_2j_2s_2}-a_{j_2j_2s_1})<0$,

$$(a_{j_1j_1 1},a_{j_2j_2s_2})=\underbrace{(\text{SI},\text{LS})}_{\text{lower-saddle source}}. \tag{1.154}$$

The equilibrium of $(x_{j_1}^*,x_{j_2}^*)=(a_{j_1j_1 1},a_{j_2j_2s_2})$ *is an* (SI,LS)-*lower-saddle sink.*

– *The saddle-source and saddle-sink are the appearing bifurcations of saddle-source and saddle-sink.*

(viii$_{1c}$*) The equilibrium of* $(x_{j_1}^*, x_{j_2}^*) = (a_{j_1j_1}1, a_{j_1j_2l_1})$ *with* $a_{j_1j_2l_1} \neq a_{j_2j_2s_1}, a_{j_2j_2s_2}$ $(l_1, l_2, s_1, s_2 \in \{1,2\}, l_1 \neq l_2, s_1 \neq s_2)$ *has the following properties.*

- *For* $a_{j_1j_1}0(a_{j_1j_2l_1} - a_{j_1j_2l_2}) > 0$ *and* $a_{j_2j_2}0(a_{j_1j_2l_1} - a_{j_2j_2s_1})(a_{j_1j_2l_1} - a_{j_2j_2s_2})^2 > 0$,

$$(a_{j_1j_1}1, a_{j_1j_2l_1}) = \underbrace{(\mathrm{DP:UP,pF})}_{\text{hyperbolic flow }(+)}. \tag{1.155}$$

The equilibrium of $(x_{j_1}^*, x_{j_2}^*) = (a_{j_1j_1}1, a_{j_1j_2l_1})$ *is a* (DP:UP, pF)*-positive hyperbolic flow.*

- *For* $a_{j_1j_1}0(a_{j_1j_2l_1} - a_{j_1j_2l_2}) < 0$ *and* $a_{j_2j_2}0(a_{j_1j_2l_1} - a_{j_2j_2s_1})(a_{j_1j_2l_1} - a_{j_2j_2s_2})^2 > 0$,

$$(a_{j_1j_1}1, a_{j_1j_2l_1}) = \underbrace{(\mathrm{UP:DP,pF})}_{\text{hyperbolic-secant flow }(+)}. \tag{1.156}$$

The equilibrium of $(x_{j_1}^*, x_{j_2}^*) = (a_{j_1j_1}1, a_{j_1j_2l_1})$ *is a* (UP:DP, pF)*-positive hyperbolic-secant flow.*

- *For* $a_{j_1j_1}0(a_{j_1j_2l_1} - a_{j_1j_2l_2}) > 0$ *and* $a_{j_2j_2}0(a_{j_1j_2l_1} - a_{j_2j_2s_1})(a_{j_1j_2l_1} - a_{j_2j_2s_2})^2 < 0$,

$$(a_{j_1j_1}1, a_{j_1j_2l_1}) = \underbrace{(\mathrm{UP:DP,nF})}_{\text{hyperbolic-secant flow }(-)}. \tag{1.157}$$

The equilibrium of $(x_{j_1}^*, x_{j_2}^*) = (a_{j_1j_1}1, a_{j_1j_2l_1})$ *is a* (UP:DP,nF)*-negative hyperbolic-secant flow.*

- *For* $a_{j_1j_1}0(a_{j_1j_2l_1} - a_{j_1j_2l_2}) < 0$ *and* $a_{j_2j_2}0(a_{j_1j_2l_1} - a_{j_2j_2s_1})(a_{j_1j_2l_1} - a_{j_2j_2s_2})^2 < 0$,

$$(a_{j_1j_1}1, a_{j_1j_2l_1}) = \underbrace{(\mathrm{DP:UP,nF})}_{\text{hyperbolic flow }(-)}. \tag{1.158}$$

The equilibrium of $(x_{j_1}^*, x_{j_2}^*) = (a_{j_1j_1}1, a_{j_1j_2l_1})$ *is a* (DP:UP,nF)*-negative hyperbolic flow.*

(viii$_2$*) For* $a_{j_2j_2s_1} = a_{j_2j_2s_2} = a_{j_2j_2}1$, *the standard form is*

$$\begin{aligned}\dot{x}_{j_2} &= a_{j_2j_2}0(x_{j_2} - a_{j_2j_2}1)^3,\\ \dot{x}_{j_1} &= a_{j_1j_1}0(x_{j_1} - a_{j_1j_1}1)(x_{j_2} - a_{j_1j_2}1)(x_{j_2} - a_{j_1j_2}2).\end{aligned} \tag{1.159}$$

The first integral manifold for $a_{j_2j_2}1 \neq a_{j_1j_2}1, a_{j_1j_2}2$ *is*

$$\ln\frac{|x_{j_1}-a_{j_1j_1 1}|}{|x_{j_1 0}-a_{j_1j_1 1}|}=\frac{a_{j_1j_1 0}}{a_{j_2j_2 0}}\left\{\ln\frac{|x_{j_2}-a_{j_2j_2 1}|}{|x_{j_2 0}-a_{j_2j_2 1}|}\right.$$
$$-\sum\nolimits_{l_1=1}^{2}(a_{j_2j_2 1}-a_{j_1j_2l_1})\left[\frac{1}{x_{j_2}-a_{j_2j_2 1}}-\frac{1}{x_{j_2 0}-a_{j_2j_2 1}}\right]$$
$$\left.-\frac{1}{2}\prod\nolimits_{l_1=1}^{2}(a_{j_2j_2 1}-a_{j_1j_2l_1})\left[\frac{1}{(x_{j_2}-a_{j_2j_2 1})^2}-\frac{1}{(x_{j_2 0}-a_{j_2j_2 1})^2}\right]\right\}. \tag{1.160}$$

The equilibrium of $(x_{j_1}^*, x_{j_2}^*)=(a_{j_1j_1 1}, a_{j_2j_2 1})$ *has the following properties.*

- *For* $a_{j_1j_1 0}\prod_{l_1=1}^{2}(a_{j_2j_2 1}-a_{j_1j_2l_1})>0$ *and* $a_{j_2j_2 0}>0$,

$$(a_{j_1j_1 1}, a_{j_2j_2 1})=\underbrace{(\mathrm{SO}, 3^{\mathrm{rd}}\mathrm{SO})}_{\text{third-order source}}. \tag{1.161}$$

  The equilibrium of $(x_{j_1}^*, x_{j_2}^*)=(a_{j_1j_1 1}, a_{j_2j_2 1})$ *is an* (SO,3^rd^SO)*-third-order source, which is for an appearing bifurcation from source to source, saddle, and source.*

- *For* $a_{j_1j_1 0}\prod_{l_1=1}^{2}(a_{j_2j_2 1}-a_{j_1j_2l_1})<0$ *and* $a_{j_2j_2 0}>0$,

$$(a_{j_1j_1 1}, a_{j_2j_2 1})=\underbrace{(\mathrm{SI}, 3^{\mathrm{rd}}\mathrm{SO})}_{\text{third-order saddle}}. \tag{1.162}$$

  The equilibrium of $(x_{j_1}^*, x_{j_2}^*)=(a_{j_1j_1 1}, a_{j_2j_2 1})$ *is an* (SI,3^rd^SO)*-third-order saddle, which is for an appearing bifurcation from saddle to saddle, sink, and saddle.*

- *For* $a_{j_1j_1 0}\prod_{l_1=1}^{2}(a_{j_2j_2 1}-a_{j_1j_2l_1})>0$ *and* $a_{j_2j_2 0}<0$,

$$(a_{j_1j_1 1}, a_{j_2j_2 1})=\underbrace{(\mathrm{SO}, 3^{\mathrm{rd}}\mathrm{SI})}_{\text{third-order saddle}}. \tag{1.163}$$

  The equilibrium of $(x_{j_1}^*, x_{j_2}^*)=(a_{j_1j_1 1}, a_{j_2j_2 1})$ *is an* (SO,3^rd^SI)*-third-order saddle, which is for an appearing bifurcation from saddle to saddle, source, and saddle.*

- *For* $a_{j_1j_1 0}\prod_{l_1=1}^{2}(a_{j_2j_2 1}-a_{j_1j_2l_1})<0$ *and* $a_{j_2j_2 0}<0$,

$$(a_{j_1j_1 1}, a_{j_2j_2 1})=\underbrace{(\mathrm{SI}, 3^{\mathrm{rd}}\mathrm{SI})}_{\text{third-order sink}}. \tag{1.164}$$

  The equilibrium of $(x_{j_1}^*, x_{j_2}^*)=(a_{j_1j_1 1}, a_{j_2j_2 1})$ *is an* (SI,3^rd^SI)*-third-order sink, which is for an appearing bifurcation from sink to sink, saddle, and sink.*

(viii$_3$) The first integral manifold for $a_{j_2j_2s_1} = a_{j_1j_2l_1}$ $(l_1, l_2 \in \{1,2\}, l_1 \neq l_2)$ is

$$\begin{aligned}\ln\frac{|x_{j_1} - a_{j_1j_1 1}|}{|x_{j_1 0} - a_{j_1j_1 1}|} = \frac{a_{j_1j_1 0}}{a_{j_2j_2 0}}\Bigg[\ln\frac{|x_{j_2} - a_{j_2j_2s_2}|}{|x_{j_2 0} - a_{j_2j_2s_2}|} \\ - (a_{j_2j_2s_2} - a_{j_1j_2l_2})(\frac{1}{x_{j_2} - a_{j_2j_2s_2}} - \frac{1}{x_{j_2 0} - a_{j_2j_2s_2}})\Bigg].\end{aligned} \tag{1.165}$$

(viii$_{3a}$) The infinite-equilibrium of $x_{j_2}^* = a_{j_1j_2l_1} = a_{j_2j_2s_1}$ with $\overline{x}_{j_1} \neq a_{j_1j_1 1}$ has the following properties.

- *For $a_{j_1j_1 0}(\overline{x}_{j_1} - a_{j_1j_1 1})(a_{j_1j_2l_1} - a_{j_1j_2l_2}) > 0$ and $a_{j_2j_2 0}(a_{j_2j_2s_1} - a_{j_2j_2s_2})^2 > 0$,*

$$(\overline{x}_{j_1}, a_{j_2j_2s_1}) = \underbrace{(\text{II, SO})}_{\text{increasing-inflection source}}. \tag{1.166}$$

The infinite-equilibrium of $x_{j_2}^* = a_{j_1j_2l_1} = a_{j_2j_2s_1}$ is an (II,SO)-increasing-inflection source.

- *For $a_{j_1j_1 0}(\overline{x}_{j_1} - a_{j_1j_1 1})(a_{j_1j_2l_1} - a_{j_1j_2l_2}) < 0$ and $a_{j_2j_2 0}(a_{j_2j_2s_1} - a_{j_2j_2s_2})^2 > 0$,*

$$(\overline{x}_{j_1}, a_{j_2j_2s_1}) = \underbrace{(\text{DI, SO})}_{\text{decreasing-inflection source}}. \tag{1.167}$$

The infinite-equilibrium of $x_{j_2}^* = a_{j_1j_2l_1} = a_{j_2j_2s_1}$ is a (DI,SO)-decreasing-inflection source.

- *For $a_{j_1j_1 0}(\overline{x}_{j_1} - a_{j_1j_1 1})(a_{j_1j_2l_1} - a_{j_1j_2l_2}) > 0$ and $a_{j_2j_2 0}(a_{j_2j_2s_1} - a_{j_2j_2s_2})^2 < 0$,*

$$(\overline{x}_{j_1}, a_{j_2j_2s_1}) = \underbrace{(\text{DI, SI})}_{\text{decreasing-inflection sink}}. \tag{1.168}$$

The infinite-equilibrium of $x_{j_2}^* = a_{j_1j_2l_1} = a_{j_2j_2s_1}$ is a (DI,SI)-decreasing-inflection sink.

- *For $a_{j_1j_1 0}(\overline{x}_{j_1} - a_{j_1j_1 1})(a_{j_1j_2l_1} - a_{j_1j_2l_2}) < 0$ and $a_{j_2j_2 0}(a_{j_2j_2s_1} - a_{j_2j_2s_2})^2 < 0$,*

$$(\overline{x}_{j_1}, a_{j_2j_2s_1}) = \underbrace{(\text{II, SI})}_{\text{increasing-inflection sink}}. \tag{1.169}$$

The infinite-equilibrium of $x_{j_2}^* = a_{j_1j_2l_1} = a_{j_2j_2s_1}$ is an (II,SI)-increasing-inflection sink.

(viii$_{3b}$) The equilibrium of $(x^*_{j_1}, x^*_{j_2}) = (a_{j_1j_1 1}, a_{j_2j_2s_1})$ *with* $a_{j_1j_2l_1} = a_{j_2j_2s_1}$ *has the following properties.*

- *For* $a_{j_1j_1 0}(a_{j_1j_2l_1} - a_{j_1j_2l_2}) > 0$ *and* $a_{j_2j_2 0}(a_{j_2j_2s_1} - a_{j_2j_2s_2})^2 > 0$,

$$(a_{j_1j_1 1}, a_{j_2j_2s_1}) = \underbrace{(\text{DI : II, SO})}_{\text{inflection source}}. \tag{1.170}$$

The equilibrium of $(x^*_{j_1}, x^*_{j_2}) = (a_{j_1j_1 1}, a_{j_2j_2s_1})$ *is a* (DI:II,SO)-*inflection source.*

- *For* $a_{j_1j_1 0}(a_{j_1j_2l_1} - a_{j_1j_2l_2}) < 0$ *and* $a_{j_2j_2 0}(a_{j_2j_2s_1} - a_{j_2j_2s_2})^2 > 0$,

$$(a_{j_1j_1 1}, a_{j_2j_2s_1}) = \underbrace{(\text{II : DI, SO})}_{\text{inflection source}}. \tag{1.171}$$

The equilibrium of $(x^*_{j_1}, x^*_{j_2}) = (a_{j_1j_1 1}, a_{j_2j_2s_1})$ *is an* (II:DI,SO)-*inflection source.*

- *For* $a_{j_1j_1 0}(a_{j_1j_2l_1} - a_{j_1j_2l_2}) > 0$ *and* $a_{j_2j_2 0}(a_{j_2j_2s_1} - a_{j_2j_2s_2})^2 < 0$,

$$(a_{j_1j_1 1}, a_{j_2j_2s_1}) = \underbrace{(\text{II : DI, SI})}_{\text{inflection sink}}. \tag{1.172}$$

The equilibrium of $(x^*_{j_1}, x^*_{j_2}) = (a_{j_1j_1 1}, a_{j_2j_2s_1})$ *is an* (II:DI,SI)-*inflection sink.*

- *For* $a_{j_1j_1 0}(a_{j_1j_2l_1} - a_{j_1j_2l_2}) < 0$ *and* $a_{j_2j_2 0}(a_{j_2j_2s_1} - a_{j_2j_2s_2})^2 < 0$,

$$(a_{j_1j_1 1}, a_{j_2j_2s_1}) = \underbrace{(\text{DI : II, SI})}_{\text{inflection sink}}. \tag{1.173}$$

The equilibrium of $(x^*_{j_1}, x^*_{j_2}) = (a_{j_1j_1 1}, a_{j_2j_2s_1})$ *is a* (DI:II,SI)-*inflection sink.*

(viii$_4$) The first integral manifold for $a_{j_2j_2s_2} = a_{j_1j_2l_1}$ $(l_1, l_2 \in \{1,2\}, l_1 \neq l_2)$ *is*

$$\begin{aligned} \ln\frac{|x_{j_1} - a_{j_1j_1 1}|}{|x_{j_1 0} - a_{j_1j_1 1}|} = \frac{a_{j_1j_1 0}}{a_{j_2j_2 0}}\bigg[&\frac{(a_{j_2j_2s_1} - a_{j_1j_2l_2})}{(a_{j_2j_2s_1} - a_{j_2j_2s_2})}\ln\frac{|x_{j_2} - a_{j_2j_2s_1}|}{|x_{j_2 0} - a_{j_2j_2s_1}|} \\ &+\frac{(a_{j_2j_2s_2} - a_{j_1j_2l_2})}{(a_{j_2j_2s_2} - a_{j_2j_2s_1})}\ln\frac{|x_{j_2} - a_{j_2j_2s_2}|}{|x_{j_2 0} - a_{j_2j_2s_2}|}\bigg]. \end{aligned} \tag{1.174}$$

(viii$_{4a}$) The infinite-equilibrium of $x^*_{j_2} = a_{j_1j_2l_1} = a_{j_2j_2s_2}$ *with* $\overline{x}_{j_1} \neq a_{j_1j_1 1}$ *has the following properties.*

- *For* $a_{j_1j_1 0}(\overline{x}_{j_1} - a_{j_1j_1 1})(a_{j_1j_2l_1} - a_{j_1j_2l_2}) > 0$ *and* $a_{j_2j_2 0}(a_{j_2j_2s_2} - a_{j_2j_2s_1}) > 0$,

$$(\overline{x}_{j_1}, a_{j_1 j_2 l_1}) = \underbrace{(\text{DU}, \text{US})}_{\text{down-up upper-saddle}}. \quad (1.175)$$

The infinite-equilibrium of $x_{j_2}^* = a_{j_1 j_2 l_1} = a_{j_2 j_2 s_2}$ *is a* (DU,US)-*down-up upper-saddle.*

- *For* $a_{j_1 j_1 0}(\overline{x}_{j_1} - a_{j_1 j_1 1})(a_{j_1 j_2 l_1} - a_{j_1 j_2 l_2}) < 0$ *and* $a_{j_2 j_2 0}(a_{j_2 j_2 s_2} - a_{j_2 j_2 s_1}) > 0$,

$$(\overline{x}_{j_1}, a_{j_1 j_2 l_1}) = \underbrace{(\text{UD}, \text{US})}_{\text{up-down upper-saddle}}. \quad (1.176)$$

The infinite-equilibrium of $x_{j_2}^* = a_{j_1 j_2 l_1} = a_{j_2 j_2 s_2}$ *is a* (UD,US)-*up-down upper-saddle.*

- *For* $a_{j_1 j_1 0}(\overline{x}_{j_1} - a_{j_1 j_1 1})(a_{j_1 j_2 l_1} - a_{j_1 j_2 l_2}) > 0$ *and* $a_{j_2 j_2 0}(a_{j_2 j_2 s_2} - a_{j_2 j_2 s_1}) < 0$,

$$(\overline{x}_{j_1}, a_{j_1 j_2 l_1}) = \underbrace{(\text{UD}, \text{LS})}_{\text{up-down lower-saddle}}. \quad (1.177)$$

The infinite-equilibrium of $x_{j_2}^* = a_{j_1 j_2 l_1} = a_{j_2 j_2 s_2}$ *is a* (UD,LS)-*up-down lower-saddle.*

- *For* $a_{j_1 j_1 0}(\overline{x}_{j_1} - a_{j_1 j_1 1})(a_{j_1 j_2 l_1} - a_{j_1 j_2 l_2}) < 0$ *and* $a_{j_2 j_2 0}(a_{j_2 j_2 s_2} - a_{j_2 j_2 s_1}) < 0$,

$$(\overline{x}_{j_1}, a_{j_1 j_2 l_1}) = \underbrace{(\text{DU}, \text{LS})}_{\text{down-up lower-saddle}}. \quad (1.178)$$

The infinite-equilibrium of $x_{j_2}^* = a_{j_1 j_2 l_1} = a_{j_2 j_2 s_2}$ *is a* (DU,LS)-*down-up lower-saddle.*

(viii$_{4b}$) The equilibrium of $(x_{j_1}^*, x_{j_2}^*) = (a_{j_1 j_1 1}, a_{j_2 j_2 s_2})$ *with* $a_{j_2 j_2 s_2} = a_{j_1 j_2 l_1}$ *has the following properties.*

- *For* $a_{j_1 j_1 0}(a_{j_1 j_2 l_1} - a_{j_1 j_2 l_2}) > 0$ *and* $a_{j_2 j_2 0}(a_{j_2 j_2 s_2} - a_{j_2 j_2 s_1}) > 0$,

$$(a_{j_1 j_1 1}, a_{j_2 j_2 s_2}) = \underbrace{(\text{DU} : \text{UD}, \text{US})}_{\text{down-up-to-up-down upper-saddle}}. \quad (1.189)$$

The equilibrium of $(x_{j_1}^*, x_{j_2}^*) = (a_{j_1 j_1 1}, a_{j_2 j_2 1})$ *is a* (DU:UD,US)-*down-up-to-up-down upper-saddle.*

- *For* $a_{j_1 j_1 0}(a_{j_1 j_2 l_1} - a_{j_1 j_2 l_2}) < 0$ *and* $a_{j_2 j_2 0}(a_{j_2 j_2 s_2} - a_{j_2 j_2 s_1}) > 0$,

$$(a_{j_1 j_1 1}, a_{j_2 j_2 s_2}) = \underbrace{(\text{UD} : \text{DU}, \text{US})}_{\text{up-down-to-down-up upper-saddle}}. \quad (1.180)$$

The equilibrium of $(x_{j_1}^*, x_{j_2}^*) = (a_{j_1 j_1 1}, a_{j_2 j_2 1})$ *is a* (UD:DU,US)-*up-down-to-down-up upper-saddle.*

- *For* $a_{j_1j_10}(a_{j_1j_2l_1}-a_{j_1j_2l_2})>0$ *and* $a_{j_2j_20}(a_{j_2j_2s_2}-a_{j_2j_2s_1})<0$,

$$(a_{j_1j_11},a_{j_2j_2s_2})=\underbrace{(\text{UD}:\text{DU},\text{LS})}_{\text{up-down-to-down-up lower-saddle}}. \tag{1.181}$$

The equilibrium of $(x^*_{j_1},x^*_{j_2})=(a_{j_1j_11},a_{j_2j_21})$ *is a* (UD:DU,LS)-*up-down-to-down-up lower-saddle.*

- *For* $a_{j_1j_10}(a_{j_1j_2l_1}-a_{j_1j_2l_2})<0$ *and* $a_{j_2j_20}(a_{j_2j_2s_2}-a_{j_2j_2s_1})<0$,

$$(a_{j_1j_11},a_{j_2j_2s_2})=\underbrace{(\text{DU}:\text{UD},\text{LS})}_{\text{down-up-to-up-down lower-saddle}}. \tag{1.182}$$

The equilibrium of $(x^*_{j_1},x^*_{j_2})=(a_{j_1j_11},a_{j_2j_21})$ *is a* (DU:UD,LS)-*down-up-to-up-down lower-saddle.*

(viii$_5$) For $a_{j_2j_2s_1}=a_{j_2j_2s_2}=a_{j_2j_21}=a_{j_1j_2l_1}$, *the first integral manifold for is*

$$\begin{aligned}\ln\frac{|x_{j_1}-a_{j_1j_11}|}{|x_{j_10}-a_{j_1j_11}|}=\frac{a_{j_1j_10}}{a_{j_2j_20}}\Bigg[&\ln\frac{|x_{j_2}-a_{j_2j_21}|}{|x_{j_20}-a_{j_2j_21}|}\\&-(a_{j_2j_21}-a_{j_1j_2l_2})(\frac{1}{x_{j_2}-a_{j_2j_21}}-\frac{1}{x_{j_20}-a_{j_2j_21}})\Bigg].\end{aligned} \tag{1.183}$$

(viii$_{5a}$) The infinite-equilibrium of $x^*_{j_2}=a_{j_1j_2l_1}=a_{j_2j_21}$ *with* $\bar{x}_{j_1}\neq a_{j_1j_11}$ *has the following properties.*

- *For* $a_{j_1j_10}(\bar{x}_{j_1}-a_{j_1j_11})(a_{j_2j_21}-a_{j_1j_2l_2})>0$ *and* $a_{j_2j_20}>0$,

$$(\bar{x}_{j_1},a_{j_2j_21})=\underbrace{(\text{IC},3^{\text{rd}}\text{SO})}_{\text{third-order increasing-concave source}}. \tag{1.184}$$

The infinite-equilibrium of $x^*_{j_2}=a_{j_1j_21}=a_{j_2j_21}$ *is an* (IC,3rdSO)-*third-order increasing-concave source.*

- *For* $a_{j_1j_10}(\bar{x}_{j_1}-a_{j_1j_11})(a_{j_2j_21}-a_{j_1j_2l_2})<0$ *and* $a_{j_2j_20}>0$,

$$(\bar{x}_{j_1},a_{j_2j_21})=\underbrace{(\text{DC},3^{\text{rd}}\text{SO})}_{\text{third-order decreasing-concave source}}. \tag{1.185}$$

The infinite-equilibrium of $x^*_{j_2}=a_{j_1j_2l_1}=a_{j_2j_21}$ *is a* (DC,3rdSO)-*third-order decreasing-concave source.*

- *For* $a_{j_1j_10}(\overline{x}_{j_1} - a_{j_1j_11})(a_{j_2j_21} - a_{j_1j_2l_2}) > 0$ *and* $a_{j_2j_20} < 0$,

$$(\overline{x}_{j_1}, a_{j_2j_21}) = \underbrace{(\text{DC}, 3^{\text{rd}}\text{SI})}_{\text{third-order decreasing-concave sink}}. \tag{1.186}$$

The infinite-equilibrium of $x^*_{j_2} = a_{j_1j_2l_1} = a_{j_2j_21}$ *is a* (DC,3rdSI)-*third-order decreasing-concave sink.*

- *For* $a_{j_1j_10}(\overline{x}_{j_1} - a_{j_1j_11})(a_{j_2j_21} - a_{j_1j_2l_2}) < 0$ *and* $a_{j_2j_20} < 0$,

$$(\overline{x}_{j_1}, a_{j_2j_21}) = \underbrace{(\text{IC}, 3^{\text{rd}}\text{SI})}_{\text{third-order increasing-concave sink}}. \tag{1.187}$$

The infinite-equilibrium of $x^*_{j_2} = a_{j_1j_2l_1} = a_{j_2j_21}$ *is an* (IC,3rdSO)-third-order increasing-concave sink.

(viii$_{5b}$) The equilibrium of $(x^*_{j_1}, x^*_{j_2}) = (a_{j_1j_11}, a_{j_2j_21})$ *with* $a_{j_1j_21} = a_{j_2j_21}$ *has the following properties.*

- *For* $a_{j_1j_10}(a_{j_2j_21} - a_{j_1j_2l_2}) > 0$ *and* $a_{j_2j_20} > 0$,

$$(a_{j_1j_11}, a_{j_2j_21}) = \underbrace{(\text{DC}:\text{IC}, 3^{\text{rd}}\text{SO})}_{\text{third-order decreasing-to-increasing concave source}}. \tag{1.188}$$

The equilibrium of $x^*_{j_2} = a_{j_1j_2l_1} = a_{j_2j_21}$ *is a* (DC:IC,3rdSO)-*third-order decreasing-to-increasing concave source.*

- *For* $a_{j_1j_10}(a_{j_2j_21} - a_{j_1j_2l_2}) < 0$ *and* $a_{j_2j_20} > 0$,

$$(a_{j_1j_11}, a_{j_2j_21}) = \underbrace{(\text{IC}:\text{DC}, 3^{\text{rd}}\text{SO})}_{\text{third-order increasing-to-decreasing concave source}}. \tag{1.189}$$

The equilibrium of $(x^*_{j_1}, x^*_{j_2}) = (a_{j_1j_11}, a_{j_2j_21})$ *is an* (IC:DC,3rdSO)-*third-order increasing-to-decreasing concave source.*

- *For* $a_{j_1j_10}(a_{j_2j_21} - a_{j_1j_2l_2}) > 0$ *and* $a_{j_2j_20} < 0$,

$$(a_{j_1j_11}, a_{j_2j_21}) = \underbrace{(\text{IC}:\text{DC}, 3^{\text{rd}}\text{SI})}_{\text{third-order increasing-to-decreasing concave sink}}. \tag{1.190}$$

The equilibrium of $(x^*_{j_1}, x^*_{j_2}) = (a_{j_1j_11}, a_{j_2j_21})$ *is an* (IC:DC,3rdSI)-*third-order increasing-to-deceasing concave sink.*

- *For* $a_{j_1j_10}(a_{j_2j_21}-a_{j_1j_2l_2})<0$ *and* $a_{j_2j_20}<0$,

$$(a_{j_1j_11},a_{j_2j_21})=\underbrace{(\mathrm{DC:IC},3^{\mathrm{rd}}\mathrm{SI})}_{\text{third-order decreasing-to-increasing concave sink}}. \tag{1.191}$$

The equilibrium of $(x_{j_1}^*,x_{j_2}^*)=(a_{j_1j_11},a_{j_2j_21})$ *is a* (DC:IC,3rdSI)-*third-order decreasing-to-increasing concave sink.*

(ix) *For* $\Delta_{j_1j_2}>0$ *and* $\Delta_{j_2j_2}>0$, *the standard form is*

$$\begin{aligned}\dot{x}_{j_2}&=a_{j_2j_20}(x_{j_2}-a_{j_2j_21})(x_{j_2}-a_{j_2j_22})(x_{j_2}-a_{j_2j_23}),\\ \dot{x}_{j_1}&=a_{j_1j_10}(x_{j_1}-a_{j_1j_11})(x_{j_2}-a_{j_1j_21})(x_{j_2}-a_{j_1j_22})\end{aligned} \tag{1.192}$$

where

$$\begin{aligned}&b_{j_1j_21},b_{j_1j_22}=-\frac{1}{2}(B_{j_1j_2}\pm\sqrt{\Delta_{j_1j_2}}),\\ &\{a_{j_1j_21},a_{j_1j_22}\}=\mathrm{sort}\{b_{j_1j_21},b_{j_1j_22}\},a_{j_1j_21}<a_{j_1j_22};\\ &b_{j_2j_22},b_{j_2j_23}=-\frac{1}{2}(B_{j_2j_2}\pm\sqrt{\Delta_{j_2j_2}}),\\ &\{a_{j_2j_21},a_{j_2j_22},a_{j_2j_23}\}=\mathrm{sort}\{b_{j_2j_21},b_{j_2j_22},b_{j_2j_23}\},\\ &a_{j_2j_2s_1}<a_{j_2j_2s_2};s_1,s_2\in\{1,2,3\},s_1<s_2.\end{aligned} \tag{1.193}$$

(ix$_1$) The first integral manifold $(a_{j_2j_2s_1}\neq a_{j_1j_21})$ *is*

$$\ln\frac{|x_{j_1}-a_{j_1j_11}|}{|x_{j_10}-a_{j_1j_11}|}=\frac{a_{j_1j_10}}{a_{j_2j_20}}\sum\nolimits_{s_2=1,s_2\neq s_1}^{3}\frac{\prod_{l_1=1}^{2}(a_{j_2j_2s_1}-a_{j_1j_2l_1})}{\prod_{s_2=1,s_2\neq s_1}^{3}(a_{j_2j_2s_1}-a_{j_2j_2s_2})}\ln\frac{|x_{j_2}-a_{j_2j_2s_2}|}{|x_{j_20}-a_{j_2j_2s_2}|}. \tag{1.194}$$

(ix$_{1a}$) The equilibrium of $(x_{j_1}^*,x_{j_2}^*)=(a_{j_1j_11},a_{j_2j_2s_1})$ $(s_1,\ s_2\in\{1,2\},\ s_1\neq s_2)$ *has the following properties.*

- *For* $a_{j_1j_10}\prod_{l_1=1}^{2}(a_{j_2j_2s_1}-a_{j_1j_2l_1})>0$ *and* $a_{j_2j_20}\prod_{s_2=1,s_2\neq s_1}^{3}(a_{j_2j_2s_1}-a_{j_2j_2s_2})>0$,

$$(a_{j_1j_11},a_{j_2j_2s_1})=\underbrace{(\mathrm{SO},\mathrm{SO})}_{\text{source}}. \tag{1.195}$$

The equilibrium of $(x_{j_1}^*,x_{j_2}^*)=(a_{j_1j_11},a_{j_2j_2s_1})$ *is an* (SO,SO)-*source.*

- *For* $a_{j_1j_10}\prod_{l_1=1}^{2}(a_{j_2j_2s_1}-a_{j_1j_2l_1})<0$ *and* $a_{j_2j_20}\prod_{s_2=1,s_2\neq s_1}^{3}(a_{j_2j_2s_1}-a_{j_2j_2s_2})>0$,

$$(a_{j_1j_11},a_{j_2j_2s_1})=\underbrace{(\mathrm{SI},\mathrm{SO})}_{\text{saddle}}. \tag{1.196}$$

The equilibrium of $(x_{j_1}^*,x_{j_2}^*)=(a_{j_1j_11},a_{j_2j_2s_1})$ *is an* (SI,SO)-*saddle.*

- *For* $a_{j_1j_10}\prod_{l_1=1}^{2}(a_{j_2j_2s_1}-a_{j_1j_2l_1})>0$ *and* $a_{j_2j_20}\prod_{s_2=1,s_2\neq s_1}^{3}(a_{j_2j_2s_1}-a_{j_2j_2s_2})<0$,

$$(a_{j_1j_1 1},a_{j_2j_2s_1})=\underbrace{(\text{SO},\text{SI})}_{\text{saddle}}. \tag{1.197}$$

The equilibrium of $(x_{j_1}^*,x_{j_2}^*)=(a_{j_1j_1 1},a_{j_2j_2s_1})$ *is an* (SO,SI)-*saddle.*

- *For* $a_{j_1j_10}\prod_{l_1=1}^{2}(a_{j_2j_2s_1}-a_{j_1j_2l_1})<0$ *and* $a_{j_2j_20}\prod_{s_2=1,s_2\neq s_1}^{3}(a_{j_2j_2s_1}-a_{j_2j_2s_2})<0$,

$$(a_{j_1j_1 1},a_{j_2j_2s_1})=\underbrace{(\text{SI},\text{SI})}_{\text{sink}}. \tag{1.198}$$

The equilibrium of $(x_{j_1}^*,x_{j_2}^*)=(a_{j_1j_1 1},a_{j_2j_2s_1})$ *is an* (SI,SI)-*sink.*

(ix$_{1b}$) The equilibrium of $(x_{j_1}^*,x_{j_2}^*)=(a_{j_1j_1 1},a_{j_1j_2l_1})$ $(l_1,\ l_2\in\{1,2\},\ l_1\neq l_2)$ *with* $a_{j_1j_2l_1}\neq a_{j_2j_2 1},a_{j_2j_2 2}$ *has the following properties.*

- *For* $a_{j_1j_10}(a_{j_1j_2l_1}-a_{j_1j_2l_2})>0$ *and* $a_{j_2j_20}\prod_{s_1=1}^{3}(a_{j_1j_2l_1}-a_{j_2j_2s_1})>0$,

$$(a_{j_1j_1 1},a_{j_1j_2l_1})=\underbrace{(\text{DP}:\text{UP},\text{pF})}_{\text{hyperbolic flow }(+)}. \tag{1.199}$$

The equilibrium of $(x_{j_1}^*,x_{j_2}^*)=(a_{j_1j_1 1},a_{j_1j_2l_1})$ *is a* (DP:UP,pF)-*positive hyperbolic flow.*

- *For* $a_{j_1j_10}(a_{j_1j_2l_1}-a_{j_1j_2l_2})<0$ *and* $a_{j_2j_20}\prod_{s_1=1}^{3}(a_{j_1j_2l_1}-a_{j_2j_2s_1})>0$,

$$(a_{j_1j_1 1},a_{j_1j_2l_1})=\underbrace{(\text{UP}:\text{DP},\text{pF})}_{\text{hyperbolic-secant flow }(+)}. \tag{1.200}$$

The equilibrium of $(x_{j_1}^*,x_{j_2}^*)=(a_{j_1j_1 1},a_{j_1j_2l_1})$ *is a* (UP:DP,pF)-*positive hyperbolic-secant flow.*

- *For* $a_{j_1j_10}(a_{j_1j_2l_1}-a_{j_1j_2l_2})>0$ *and* $a_{j_2j_20}\prod_{s_1=1}^{3}(a_{j_1j_2l_1}-a_{j_2j_2s_1})<0$,

$$(a_{j_1j_1 1},a_{j_1j_2l_1})=\underbrace{(\text{UP}:\text{DP},\text{nF})}_{\text{hyperbolic-secant flow }(-)}. \tag{1.201}$$

The equilibrium of $(x_{j_1}^*,x_{j_2}^*)=(a_{j_1j_1 1},a_{j_1j_2l_1})$ *is a* (UP:DP,nF)-*negative hyperbolic-secant flow.*

- *For* $a_{j_1j_10}(a_{j_1j_2l_1}-a_{j_1j_2l_2})<0$ *and* $a_{j_2j_20}\prod_{s_1=1}^{3}(a_{j_1j_2l_1}-a_{j_2j_2s_1})<0$,

$$(a_{j_1j_11},a_{j_1j_2l_1})=\underbrace{(\mathrm{DP}:\mathrm{UP},\mathrm{nF})}_{\text{hyperbolic flow }(-)}. \tag{1.202}$$

The equilibrium of $(x_{j_1}^*,x_{j_2}^*)=(a_{j_1j_11},a_{j_1j_2l_1})$ *is a* (DP:UP,nF)-*negative hyperbolic flow.*

(ix₂) The first integral manifold ($a_{j_2j_2s_1}=a_{j_1j_2l_1}$ and $a_{j_2j_2s_2}\neq a_{j_1j_2l_2}$, l_1, $l_2\in\{1,2\}$; $l_1\neq l_2$; s_1, s_2, $s_3\in\{1,2,3\}$; $s_1\neq s_2\neq s_3$) *is*

$$\ln\frac{|x_{j_1}-a_{j_1j_11}|}{|x_{j_10}-a_{j_1j_11}|}=\frac{a_{j_1j_10}}{a_{j_2j_20}}\sum_{s_2=1,s_2\neq s_1,s_3}^{3}\frac{(a_{j_2j_2s_2}-a_{j_1j_2l_2})}{(a_{j_2j_2s_2}-a_{j_2j_2s_3})}\ln\frac{|x_{j_2}-a_{j_2j_2s_2}|}{|x_{j_20}-a_{j_2j_2s_2}|} \tag{1.203}$$

(ix_{2a}) The infinite-equilibrium of $x_{j_2}^*=a_{j_1j_2l_1}=a_{j_2j_2s_1}$ *with* $\bar{x}_{j_1}\neq a_{j_1j_11}$ *has the following properties.*

- *For* $a_{j_1j_10}(\bar{x}_{j_1}-a_{j_1j_11})(a_{j_1j_2l_1}-a_{j_1j_2l_2})>0$ *and* $a_{j_2j_20}\prod_{s_2=1,s_2\neq s_1}^{3}(a_{j_2j_2s_1}-a_{j_2j_2s_2})>0$,

$$(\bar{x}_{j_1},a_{j_2j_2s_1})=\underbrace{(\mathrm{II},\mathrm{SO})}_{\text{increasing-inflection source}}. \tag{1.204}$$

The infinite-equilibrium of $x_{j_2}^*=a_{j_1j_2l_1}=a_{j_2j_2s_1}$ *is an* (II,SO)-*increasing-inflection source.*

- *For* $a_{j_1j_10}(\bar{x}_{j_1}-a_{j_1j_11})(a_{j_1j_2l_1}-a_{j_1j_2l_2})<0$ *and* $a_{j_2j_20}\prod_{s_2=1,s_2\neq s_1}^{3}(a_{j_2j_2s_1}-a_{j_2j_2s_2})>0$,

$$(\bar{x}_{j_1},a_{j_2j_2s_1})=\underbrace{(\mathrm{DI},\mathrm{SO})}_{\text{decreasing-inflection source}}. \tag{1.205}$$

The infinite-equilibrium of $x_{j_2}^*=a_{j_1j_2l_1}=a_{j_2j_2s_1}$ *is a* (DI,SO)-*decreasing-inflection source.*

- *For* $a_{j_1j_10}(\bar{x}_{j_1}-a_{j_1j_11})(a_{j_1j_2l_1}-a_{j_1j_2l_2})>0$ *and* $a_{j_2j_20}\prod_{s_2=1,s_2\neq s_1}^{3}(a_{j_2j_2s_1}-a_{j_2j_2s_2})<0$,

$$(\bar{x}_{j_1},a_{j_2j_2s_1})=\underbrace{(\mathrm{DI},\mathrm{SI})}_{\text{decreasing-inflection sink}}. \tag{1.206}$$

The infinite-equilibrium of $x_{j_2}^*=a_{j_1j_2l_1}=a_{j_2j_2s_1}$ *is a* (DI,SI)-*decreasing-inflection sink.*

- *For* $a_{j_1j_10}(\overline{x}_{j_1}-a_{j_1j_11})(a_{j_1j_2l_1}-a_{j_1j_2l_2})<0$ *and* $a_{j_2j_20}\prod_{s_2=1,s_2\neq s_1}^{3}(a_{j_2j_2s_1}-a_{j_2j_2s_2})<0$,

$$(\overline{x}_{j_1},a_{j_2j_2s_1})=\underbrace{(\text{II},\text{SI})}_{\text{increasing-inflection sink}}. \tag{1.207}$$

The infinite-equilibrium of $x_{j_2}^{*}=a_{j_1j_2l_1}=a_{j_2j_2s_1}$ *is an* (II,SI)*-increasing-inflection sink.*

(ix_{2b}) The equilibrium of $(x_{j_1}^{*},x_{j_2}^{*})=(a_{j_1j_11},a_{j_2j_2s_1})$ *with* $a_{j_2j_2s_1}=a_{j_1j_2l_1}$ *has the following properties.*

- *For* $a_{j_1j_10}(a_{j_1j_2l_1}-a_{j_1j_2l_2})>0$ *and* $a_{j_2j_20}\prod_{s_2=1,s_2\neq s_1}^{3}(a_{j_2j_2s_1}-a_{j_2j_2s_2})>0$,

$$(a_{j_1j_11},a_{j_2j_2s_1})=\underbrace{(\text{DI}:\text{II},\text{SO})}_{\text{inflection source}}. \tag{1.208}$$

The equilibrium of $(x_{j_1}^{*},x_{j_2}^{*})=(a_{j_1j_11},a_{j_2j_2s_1})$ *is a* (DI:II,SO)*-inflection source.*

- *For* $a_{j_1j_10}(a_{j_1j_2l_1}-a_{j_1j_2l_2})<0$ *and* $a_{j_2j_20}\prod_{s_2=1,s_2\neq s_1}^{3}(a_{j_2j_2s_1}-a_{j_2j_2s_2})>0$,

$$(a_{j_1j_11},a_{j_2j_2s_1})=\underbrace{(\text{II}:\text{DI},\text{SO})}_{\text{inflection source}}. \tag{1.209}$$

The equilibrium of $(x_{j_1}^{*},x_{j_2}^{*})=(a_{j_1j_11},a_{j_2j_2s_1})$ *is an* (II:DI,SO)*-inflection source.*

- *For* $a_{j_1j_10}(a_{j_1j_2l_1}-a_{j_1j_2l_2})>0$ *and* $a_{j_2j_20}\prod_{s_2=1,s_2\neq s_1}^{3}(a_{j_2j_2s_1}-a_{j_2j_2s_2})<0$,

$$(a_{j_1j_11},a_{j_2j_2s_1})=\underbrace{(\text{II}:\text{DI},\text{SI})}_{\text{inflection sink}}. \tag{1.210}$$

The equilibrium of $(x_{j_1}^{*},x_{j_2}^{*})=(a_{j_1j_11},a_{j_2j_2s_1})$ *is an* (II:DI,SI)*-inflection sink.*

- *For* $a_{j_1j_10}(a_{j_1j_2l_1}-a_{j_1j_2l_2})<0$ *and* $a_{j_2j_20}\prod_{s_2=1,s_2\neq s_1}^{3}(a_{j_2j_2s_1}-a_{j_2j_2s_2})<0$,

$$(a_{j_1j_11},a_{j_2j_2s_1})=\underbrace{(\text{DI}:\text{II},\text{SI})}_{\text{inflection sink}}. \tag{1.211}$$

The equilibrium of $(x_{j_1}^{*},x_{j_2}^{*})=(a_{j_1j_11},a_{j_2j_2s_1})$ *is a* (DI:II,SI)*-inflection sink.*

(ix_3) The first integral manifold ($a_{j_2j_2s_1}=a_{j_1j_2l_1}$ and $a_{j_2j_2s_2}=a_{j_1j_2l_2}$, $l_1,\ l_2\in\{1,2\}$; $l_1\neq l_2$; $s_1,\ s_2,\ s_3\in\{1,2,3\}$; $s_1\neq s_2\neq s_3$) *is*

$$\ln\frac{|x_{j_1}-a_{j_1j_11}|}{|x_{j_10}-a_{j_1j_11}|}=\frac{a_{j_1j_10}}{a_{j_2j_20}}\ln\frac{|x_{j_2}-a_{j_2j_2s_3}|}{|x_{j_20}-a_{j_2j_2s_3}|}. \tag{1.212}$$

(ix_{3a}) The infinite-equilibrium of $x^*_{j_2} = a_{j_1j_2l_\alpha} = a_{j_2j_2s_\alpha}$ ($\alpha, \beta \in \{1,2\}, \gamma \in \{1,2,3\}, \beta, \gamma \neq \alpha$) *with* $\bar{x}_{j_1} \neq a_{j_1j_11}$ *has the following properties.*

- *For* $a_{j_1j_10}(\bar{x}_{j_1} - a_{j_1j_11})(a_{j_1j_2l_\alpha} - a_{j_1j_2l_\beta}) > 0$ *and* $a_{j_2j_20}\prod^3_{s_\gamma=1,s_\gamma\neq s_\alpha}(a_{j_2j_2s_\alpha} - a_{j_2j_2s_\gamma}) > 0$,

$$(\bar{x}_{j_1}, a_{j_2j_2s_\alpha}) = \underbrace{(\text{II, SO})}_{\text{increasing-inflection source}}. \tag{1.213}$$

The infinite-equilibrium of $x^*_{j_2} = a_{j_1j_2l_\alpha} = a_{j_2j_2s_\alpha}$ *is an* (II,SO)-*increasing-inflection source.*

- *For* $a_{j_1j_10}(\bar{x}_{j_1} - a_{j_1j_11})(a_{j_1j_2l_\alpha} - a_{j_1j_2l_\beta}) < 0$ *and* $a_{j_2j_20}\prod^3_{s_\gamma=1,s_\gamma\neq s_\alpha}(a_{j_2j_2s_\alpha} - a_{j_2j_2s_\gamma}) > 0$,

$$(\bar{x}_{j_1}, a_{j_2j_2s_\alpha}) = \underbrace{(\text{DI, SO})}_{\text{decreasing-inflection source}}. \tag{1.214}$$

The infinite-equilibrium of $x^*_{j_2} = a_{j_1j_2l_\alpha} = a_{j_2j_2s_\alpha}$ *is a* (DI,SO)-*decreasing-inflection source.*

- *For* $a_{j_1j_10}(\bar{x}_{j_1} - a_{j_1j_11})(a_{j_1j_2l_\alpha} - a_{j_1j_2l_\beta}) > 0$ *and* $a_{j_2j_20}\prod^3_{s_\gamma=1,s_\gamma\neq s_\alpha}(a_{j_2j_2s_\alpha} - a_{j_2j_2s_\gamma}) < 0$,

$$(\bar{x}_{j_1}, a_{j_2j_2s_\alpha}) = \underbrace{(\text{DI, SI})}_{\text{decreasing-inflection sink}}. \tag{1.215}$$

The infinite-equilibrium of $x^*_{j_2} = a_{j_1j_2l_\alpha} = a_{j_2j_2s_\alpha}$ *is a* (DI,SI)-*decreasing-inflection sink.*

- *For* $a_{j_1j_10}(\bar{x}_{j_1} - a_{j_1j_11})(a_{j_1j_2l_\alpha} - a_{j_1j_2l_\beta}) < 0$ *and* $a_{j_2j_20}\prod^3_{s_\gamma=1,s_\gamma\neq s_\alpha}(a_{j_2j_2s_\alpha} - a_{j_2j_2s_\gamma}) < 0$,

$$(\bar{x}_{j_1}, a_{j_2j_2s_\alpha}) = \underbrace{(\text{II, SI})}_{\text{increasing-inflection sink}}. \tag{1.216}$$

The infinite-equilibrium of $x^*_{j_2} = a_{j_1j_2l_\alpha} = a_{j_2j_2s_\alpha}$ *is an* (II,SI)-*increasing-inflection sink.*

(ix_{3b}) The equilibrium of $(x^*_{j_1}, x^*_{j_2}) = (a_{j_1j_11}, a_{j_2j_2s_\alpha})$ *with* $a_{j_2j_2s_\alpha} = a_{j_1j_2l_\alpha}$ *has the following properties.*

- *For* $a_{j_1j_10}(a_{j_1j_2l_\alpha} - a_{j_1j_2l_\beta}) > 0$ *and* $a_{j_2j_20}\prod^3_{s_\gamma=1,s_\gamma\neq s_\alpha}(a_{j_2j_2s_\alpha} - a_{j_2j_2s_\gamma}) > 0$,

$$(a_{j_1j_11}, a_{j_2j_2s_\alpha}) = \underbrace{(\text{DI : II, SO})}_{\text{inflection source}}. \tag{1.217}$$

The equilibrium of $(x^*_{j_1}, x^*_{j_2}) = (a_{j_1j_11}, a_{j_2j_2s_\alpha})$ *is a* (DI:II,SO)-*inflection source.*

- *For* $a_{j_1j_10}(a_{j_1j_2l_\alpha} - a_{j_1j_2l_\beta}) < 0$ *and* $a_{j_2j_20}\prod_{s_\gamma=1, s_\gamma\neq s_\alpha}^{3}(a_{j_2j_2s_\alpha} - a_{j_2j_2s_\gamma}) > 0$,

$$(a_{j_1j_11}, a_{j_2j_2s_\alpha}) = \underbrace{(\text{II} : \text{DI}, \text{SO})}_{\text{inflection source}}. \tag{1.218}$$

The equilibrium of $(x_{j_1}^*, x_{j_2}^*) = (a_{j_1j_11}, a_{j_2j_2s_\alpha})$ *is an* (II:DI,SO)-*inflection source.*

- *For* $a_{j_1j_10}(a_{j_1j_2l_\alpha} - a_{j_1j_2l_\beta}) > 0$ *and* $a_{j_2j_20}\prod_{s_\gamma=1, s_\gamma\neq s_\alpha}^{3}(a_{j_2j_2s_\alpha} - a_{j_2j_2s_\gamma}) < 0$,

$$(a_{j_1j_11}, a_{j_2j_2s_\alpha}) = \underbrace{(\text{II} : \text{DI}, \text{SI})}_{\text{inflection sink}}. \tag{1.219}$$

The equilibrium of $(x_{j_1}^*, x_{j_2}^*) = (a_{j_1j_11}, a_{j_2j_2s_\alpha})$ *is an* (II:DI,SI)-*inflection sink.*

- *For* $a_{j_1j_10}(a_{j_1j_2l_\alpha} - a_{j_1j_2l_\beta}) < 0$ *and* $a_{j_2j_20}\prod_{s_\gamma=1, s_\gamma\neq s_\alpha}^{3}(a_{j_2j_2s_\alpha} - a_{j_2j_2s_\gamma}) < 0$,

$$(a_{j_1j_11}, a_{j_2j_2s_\alpha}) = \underbrace{(\text{DI} : \text{II}, \text{SI})}_{\text{inflection sink}}. \tag{1.220}$$

The equilibrium of $(x_{j_1}^*, x_{j_2}^*) = (a_{j_1j_11}, a_{j_2j_2s_\alpha})$ *is a* (DI:II,SI)-*inflection sink.*

1.2 Proof of Theorem 1.1

Consider a two-dimensional, nonlinear dynamical system as

$$\begin{aligned}\dot{x}_{j_2} &= a_{j_2j_20}(x_{j_2} - b_{j_2j_21})(x_{j_2}^2 + B_{j_2j_2}x_{j_2} + C_{j_2j_2}),\\ \dot{x}_{j_1} &= a_{j_1j_10}(x_{j_1} - a_{j_1j_11})(x_{j_2}^2 + B_{j_1j_2}x_{j_2} + C_{j_1j_2})\end{aligned}$$

with

$$\begin{aligned}\Delta_{j_2j_2} &= B_{j_2j_2}^2 - 4C_{j_2j_2},\\ \Delta_{j_1j_2} &= B_{j_1j_2}^2 - 4C_{j_1j_2}.\end{aligned}$$

(i) For $\Delta_{j_1j_2} < 0$ and $\Delta_{j_2j_2} < 0$, the standard form is

$$\begin{aligned}\dot{x}_{j_2} &= a_{j_2j_20}(x_{j_2} - a_{j_2j_21})\left[(x_{j_2} - a_{j_2j_2})^2 + b_{j_2j_2}\right],\\ \dot{x}_{j_1} &= a_{j_1j_10}(x_{j_1} - a_{j_1j_11})\left[(x_{j_2} - a_{j_1j_2})^2 + b_{j_1j_2}\right]\end{aligned}$$

where

$$
\begin{aligned}
&a_{j_2j_21} = b_{j_2j_21}, a_{j_2j_2} = -\frac{1}{2}B_{j_2j_2}, b_{j_2j_2} = \frac{1}{4}(-\Delta_{j_1j_2});\\
&a_{j_1j_2} = -\frac{1}{2}B_{j_1j_2}, b_{j_1j_2} = \frac{1}{4}(-\Delta_{j_1j_2}).
\end{aligned}
$$

In phase space,

$$
\frac{dx_{j_1}}{dx_{j_2}} = \frac{a_{j_1j_10}}{a_{j_2j_20}} \frac{(x_{j_1} - a_{j_1j_11})\left[(x_{j_2} - a_{j_1j_2})^2 + b_{j_1j_2}\right]}{(x_{j_2} - a_{j_2j_21})\left[(x_{j_2} - a_{j_2j_2})^2 + b_{j_2j_2}\right]},
$$

and the deformation of the foregoing equation is

$$
\begin{aligned}
\frac{dx_{j_1}}{x_{j_1} - a_{j_1j_11}} = \frac{a_{j_1j_10}}{a_{j_2j_20}} \Big\{ &\frac{x_{j_2} - a_{j_2j_21}}{(x_{j_2} - a_{j_2j_2})^2 + b_{j_2j_2}} + \frac{2(a_{j_2j_21} - a_{j_1j_2})}{(x_{j_2} - a_{j_2j_2})^2 + b_{j_2j_2}}\\
&+ \frac{(a_{j_2j_21} - a_{j_1j_2})^2 + b_{j_1j_2}}{(x_{j_2} - a_{j_2j_21})\left[(x_{j_2} - a_{j_2j_2})^2 + b_{j_2j_2}\right]} \Big\} dx_{j_2}.
\end{aligned}
$$

With

$$
\begin{aligned}
&\frac{(a_{j_2j_21} - a_{j_1j_2})^2 + b_{j_1j_2}}{(x_{j_2} - a_{j_2j_21})\left[(x_{j_2} - a_{j_2j_2})^2 + b_{j_2j_2}\right]}\\
&= \frac{(a_{j_2j_21} - a_{j_1j_2})^2 + b_{j_1j_2}}{(a_{j_2j_21} - a_{j_2j_2})^2 + b_{j_2j_2}} \left[\frac{1}{x_{j_2} - a_{j_2j_21}} - \frac{x_2 - a_{j_2j_21} + 2(a_{j_2j_21} - a_{j_2j_2})}{(x_{j_2} - a_{j_2j_2})^2 + b_{j_2j_2}} \right],
\end{aligned}
$$

we have

$$
\begin{aligned}
\frac{dx_{j_1}}{x_{j_1} - a_{j_1j_11}} = \frac{a_{j_1j_10}}{a_{j_2j_20}} \Big\{ &\frac{(a_{j_2j_21} - a_{j_1j_2})^2 + b_{j_1j_2}}{(a_{j_2j_21} - a_{j_2j_2})^2 + b_{j_2j_2}} \frac{1}{(x_{j_2} - a_{j_2j_21})}\\
&+ \left[1 - \frac{(a_{j_2j_21} - a_{j_1j_2})^2 + b_{j_1j_2}}{(a_{j_2j_21} - a_{j_2j_2})^2 + b_{j_2j_2}} \right] \frac{x_{j_2} - a_{j_2j_2}}{(x_{j_2} - a_{j_2j_2})^2 + b_{j_2j_2}}\\
&+ \left[a_{j_2j_2} + a_{j_2j_21} - 2a_{j_1j_2} - (a_{j_2j_21} - a_{j_2j_2}) \frac{(a_{j_2j_21} - a_{j_1j_2})^2 + b_{j_1j_2}}{(a_{j_2j_21} - a_{j_2j_2})^2 + b_{j_2j_2}} \right]\\
&\times \frac{1}{(x_{j_2} - a_{j_2j_2})^2 + b_{j_2j_2}} \Big\} dx_{j_2}.
\end{aligned}
$$

With an initial condition of (x_{j_10}, x_{j_20}) at $t = t_0$, the integration of the above equation gives the first integral manifold as

$$\ln\frac{|x_{j_1}-a_{j_1j_11}|}{|x_{j_10}-a_{j_1j_11}|}=\frac{a_{j_1j_10}}{a_{j_2j_20}}\left[A\ln\frac{|x_{j_2}-a_{j_2j_21}|}{|x_{j_20}-a_{j_2j_21}|}+\frac{1}{2}(1-A)\ln\frac{|(x_{j_2}-a_{j_2j_2})^2+b_{j_2j_2}|}{|(x_{j_20}-a_{j_2j_2})^2+b_{j_2j_2}|}\right.$$
$$\left.+\frac{B}{\sqrt{b_{j_2j_2}}}(\arctan\frac{x_{j_2}-a_{j_2j_2}}{\sqrt{b_{j_2j_2}}}-\arctan\frac{x_{j_20}-a_{j_2j_2}}{\sqrt{b_{j_2j_2}}})\right]$$

where

$$A=\frac{(a_{j_2j_21}-a_{j_1j_2})^2+b_{j_1j_2}}{(a_{j_2j_21}-a_{j_2j_2})^2+b_{j_2j_2}},B=a_{j_2j_2}+a_{j_2j_21}-2a_{j_1j_2}-(a_{j_2j_21}-a_{j_2j_2})A.$$

At $(x_{j_1}^*,x_{j_2}^*)=(a_{j_1j_11},a_{j_2j_21})$, the variational equations are

$$\Delta\dot{x}_{j_2}=a_{j_2j_20}\Delta x_{j_2}\left[(a_{j_2j_21}-a_{j_2j_2})^2+b_{j_2j_2}\right],$$
$$\Delta\dot{x}_{j_1}=a_{j_1j_10}\Delta x_{j_1}\left[(a_{j_2j_21}-a_{j_1j_2})^2+b_{j_1j_2}\right].$$

Because of

$$(a_{j_2j_21}-a_{j_2j_2})^2+b_{j_2j_2}>0 \text{ and } (a_{j_2j_21}-a_{j_1j_2})^2+b_{j_1j_2}>0,$$

the flows at $(x_{j_1}^*,x_{j_2}^*)=(a_{j_1j_11},a_{j_2j_21})$ in the x_{j_1}-direction are source and sink for $a_{j_1j_10}>0$ and $a_{j_1j_10}<0$, respectively; and the flows at $(x_{j_1}^*,x_{j_2}^*)=(a_{j_1j_11},a_{j_2j_21})$ in the x_{j_2}-direction are source and sink for $a_{j_2j_20}>0$ and $a_{j_2j_20}<0$, respectively.

Therefore, the equilibrium of $(x_{j_1}^*,x_{j_2}^*)=(a_{j_1j_11},a_{j_2j_21})$ has the following properties as in Eqs. (1.7)–(1.10).

- For $a_{j_1j_10}>0$ and $a_{j_2j_20}>0$,

$$(a_{j_1j_11},a_{j_2j_21})=\underbrace{(\text{SO},\text{SO})}_{\text{source}}.$$

The equilibrium of $(x_{j_1}^*,x_{j_2}^*)=(a_{j_1j_11},a_{j_2j_21})$ is an (SO,SO)-source.

- For $a_{j_1j_10}<0$ and $a_{j_2j_20}>0$,

$$(a_{j_1j_11},a_{j_2j_21})=\underbrace{(\text{SI},\text{SO})}_{\text{saddle}}.$$

The equilibrium of $(x_{j_1}^*,x_{j_2}^*)=(a_{j_1j_11},a_{j_2j_21})$ is an (SI,SO)-saddle.

- For $a_{j_1j_10} > 0$ and $a_{j_2j_20} < 0$,

$$(a_{j_1j_11}, a_{j_2j_21}) = \underbrace{(\text{SO, SI})}_{\text{saddle}}.$$

The equilibrium of $(x^*_{j_1}, x^*_{j_2}) = (a_{j_1j_11}, a_{j_2j_21})$ is an (SO,SI)-saddle.

- For $a_{j_1j_10} < 0$ and $a_{j_2j_20} < 0$,

$$(a_{j_1j_11}, a_{j_2j_21}) = \underbrace{(\text{SI, SI})}_{\text{sink}}.$$

The equilibrium of $(x^*_{j_1}, x^*_{j_2}) = (a_{j_1j_11}, a_{j_2j_21})$ is an (SI,SI)-sink.

(ii) For $\Delta_{j_1j_2} < 0$ and $\Delta_{j_2j_2} = 0$, the standard form is

$$\begin{aligned}\dot{x}_{j_2} &= a_{j_2j_20}(x_{j_2} - a_{j_2j_2s_1})(x_{j_2} - a_{j_2j_2s_2})^2,\\ \dot{x}_{j_1} &= a_{j_1j_10}(x_{j_1} - a_{j_1j_11})\left[(x_{j_2} - a_{j_1j_2})^2 + b_{j_1j_2}\right]\end{aligned}$$

where

$$\begin{aligned}&a_{j_1j_2} = -\frac{1}{2}B_{j_1j_2}, b_{j_1j_2} = \frac{1}{4}(-\Delta_{j_1j_2}),\\ &a_{j_2j_2s_1} = b_{j_2j_21}, a_{j_2j_2s_2} = -\frac{1}{2}B_{j_2j_2};\\ &\{a_{j_2j_21}, a_{j_2j_22}\} = \text{sort}\{a_{j_2j_2s_1}, a_{j_2j_2s_2}\}, a_{j_2j_21} < a_{j_2j_22}.\end{aligned}$$

(ii$_1$) For $a_{j_2j_2s_1} \neq a_{j_2j_2s_2}$, in phase space,

$$\frac{dx_{j_1}}{dx_{j_2}} = \frac{a_{j_1j_10}}{a_{j_2j_20}}\frac{(x_{j_1} - a_{j_1j_11})\left[(x_{j_2} - a_{j_1j_2})^2 + b_{j_1j_2}\right]}{(x_{j_2} - a_{j_2j_2s_1})(x_{j_2} - a_{j_2j_2s_2})^2},$$

and the deformation of the foregoing equation is

$$\begin{aligned}\frac{dx_{j_1}}{x_{j_1} - a_{j_1j_11}} = \frac{a_{j_1j_10}}{a_{j_2j_20}}\bigg\{&\frac{x_{j_2} - a_{j_2j_2s_1}}{(x_{j_2} - a_{j_2j_2s_2})^2} + \frac{2(a_{j_2j_2s_1} - a_{j_1j_2})}{(x_{j_2} - a_{j_2j_2s_2})^2}\\ &+ \frac{(a_{j_2j_2s_1} - a_{j_1j_2})^2 + b_{j_1j_2}}{(x_{j_2} - a_{j_2j_2s_1})(x_{j_2} - a_{j_2j_2s_2})^2}\bigg\}dx_{j_2}.\end{aligned}$$

With

$$\frac{(a_{j_2j_2s_1}-a_{j_1j_2})^2+b_{j_1j_2}}{(x_{j_2}-a_{j_2j_2s_1})\left[(x_{j_2}-a_{j_2j_2s_2})^2+b_{j_2j_2}\right]}$$
$$=\frac{(a_{j_2j_2s_1}-a_{j_1j_2})^2+b_{j_1j_2}}{(a_{j_2j_2s_1}-a_{j_2j_2s_2})^2}\left[\frac{1}{x_{j_2}-a_{j_2j_2s_1}}-\frac{x_2-a_{j_2j_2s_2}+(a_{j_2j_2s_1}-a_{j_2j_2s_2})}{(x_{j_2}-a_{j_2j_2s_2})^2}\right],$$

we have

$$\begin{aligned}\frac{dx_{j_1}}{x_{j_1}-a_{j_1j_1 1}}=&\frac{a_{j_1j_1 0}}{a_{j_2j_2 0}}\left\{\frac{(a_{j_2j_2s_1}-a_{j_1j_2})^2+b_{j_1j_2}}{(a_{j_2j_2s_1}-a_{j_2j_2s_2})^2}\frac{1}{(x_{j_2}-a_{j_2j_2s_1})}\right.\\&+\left[\frac{2(a_{j_2j_2s_2}-a_{j_1j_2})}{a_{j_2j_2s_2}-a_{j_2j_2s_1}}-\frac{(a_{j_2j_2s_2}-a_{j_1j_2})^2+b_{j_1j_2}}{(a_{j_2j_2s_1}-a_{j_2j_2s_2})^2}\right]\frac{1}{x_{j_2}-a_{j_2j_2s_2}}\\&\left.+\frac{(a_{j_2j_2s_2}-a_{j_1j_2})^2+b_{j_1j_2}}{(a_{j_2j_2s_2}-a_{j_2j_2s_2})}\frac{1}{(x_{j_2}-a_{j_2j_2s_2})^2}\right\}dx_{j_2}.\end{aligned}$$

With an initial condition of $(x_{j_1 0},x_{j_2 0})$ at $t=t_0$, the integration of the above equation gives the first integral manifold as

$$\begin{aligned}\ln\frac{|x_{j_1}-a_{j_1j_1 1}|}{|x_{j_1 0}-a_{j_1j_1 1}|}=&\frac{a_{j_1j_1 0}}{a_{j_2j_2 0}}\left\{\frac{(a_{j_2j_2s_1}-a_{j_1j_2})^2+b_{j_1j_2}}{(a_{j_2j_2s_1}-a_{j_2j_2s_2})^2}\ln\frac{|x_{j_2}-a_{j_2j_2s_1}|}{|x_{j_2 0}-a_{j_2j_2s_1}|}\right.\\&+\left[\frac{2(a_{j_2j_2s_2}-a_{j_1j_2})}{a_{j_2j_2s_2}-a_{j_2j_2s_1}}-\frac{(a_{j_2j_2s_1}-a_{j_1j_2})^2+b_{j_1j_2}}{(a_{j_2j_2s_1}-a_{j_2j_2s_2})^2}\right]\ln\frac{|x_{j_2}-a_{j_2j_2s_2}|}{|x_{j_2 0}-a_{j_2j_2s_2}|}\\&\left.-\frac{(a_{j_2j_2s_2}-a_{j_1j_2})^2+b_{j_1j_2}}{a_{j_2j_2s_2}-a_{j_2j_2s_2}}\left(\frac{1}{x_{j_2}-a_{j_2j_2s_2}}-\frac{1}{x_{j_2 0}-a_{j_2j_2s_2}}\right)\right\}.\end{aligned}$$

(ii$_{1a}$) For $a_{j_2j_2s_1}\neq a_{j_2j_2s_2}$,at $(x_{j_1}^*,x_{j_2}^*)=(a_{j_1j_1 1},a_{j_2j_2s_1})$, the variational equations are

$$\begin{aligned}\Delta\dot{x}_{j_2}&=a_{j_2j_2 0}\Delta x_{j_2}(a_{j_2j_2s_1}-a_{j_2j_2s_2})^2,\\\Delta\dot{x}_{j_1}&=a_{j_1j_1 0}\Delta x_{j_1}\left[(a_{j_2j_2s_1}-a_{j_1j_2})^2+b_{j_1j_2}\right].\end{aligned}$$

Because of

$$(a_{j_2j_2s_1}-a_{j_2j_2s_2})^2>0 \text{ and } (a_{j_2j_2 1}-a_{j_1j_2})^2+b_{j_1j_2}>0,$$

the flows at $(x_{j_1}^*,x_{j_2}^*)=(a_{j_1j_1 1},a_{j_2j_2s_1})$ in the x_{j_1} -direction are source and sink for $a_{j_1j_1 0}>0$ and $a_{j_1j_1 0}<0$, respectively; and the flows at $(x_{j_1}^*,x_{j_2}^*)=(a_{j_1j_1 1},a_{j_2j_2s_1})$ in the x_{j_2}-direction are source and sink for $a_{j_2j_2 0}(a_{j_2j_2s_1}-a_{j_2j_2s_2})^2>0$ and $a_{j_2j_2 0}(a_{j_2j_2s_1}-a_{j_2j_2s_2})^2<0$, respectively.

Therefore, the equilibrium of $(x_{j_1}^*,x_{j_2}^*)=(a_{j_1j_1 1},a_{j_2j_2s_1})$ has the following properties as in Eqs. (1.14)–(1.17).

- For $a_{j_1 j_1 0} > 0$ and $a_{j_2 j_2 0}(a_{j_2 j_2 s_1} - a_{j_2 j_2 s_2})^2 > 0$,

$$(a_{j_1 j_1 1}, a_{j_2 j_2 s_1}) = \underbrace{(\text{SO}, \text{SO})}_{\text{source}}.$$

The equilibrium of $(x_{j_1}^*, x_{j_2}^*) = (a_{j_1 j_1 1}, a_{j_2 j_2 s_1})$ is an (SO,SO)-source.

- For $a_{j_1 j_1 0} < 0$ and $a_{j_2 j_2 0}(a_{j_2 j_2 s_1} - a_{j_2 j_2 s_2})^2 > 0$,

$$(a_{j_1 j_1 1}, a_{j_2 j_2 s_1}) = \underbrace{(\text{SI}, \text{SO})}_{\text{saddle}}.$$

The equilibrium of $(x_{j_1}^*, x_{j_2}^*) = (a_{j_1 j_1 1}, a_{j_2 j_2 s_1})$ is an (SI,SO)-saddle.

- For $a_{j_1 j_1 0} > 0$ and $a_{j_2 j_2 0}(a_{j_2 j_2 s_1} - a_{j_2 j_2 s_2})^2 < 0$,

$$(a_{j_1 j_1 1}, a_{j_2 j_2 s_1}) = \underbrace{(\text{SO}, \text{SI})}_{\text{saddle}}.$$

The equilibrium of $(x_{j_1}^*, x_{j_2}^*) = (a_{j_1 j_1 1}, a_{j_2 j_2 1})$ is an (SO,SI)-saddle.

- For $a_{j_1 j_1 0} < 0$ and $a_{j_2 j_2 0}(a_{j_2 j_2 s_1} - a_{j_2 j_2 s_2})^2 < 0$,

$$(a_{j_1 j_1 1}, a_{j_2 j_2 s_1}) = \underbrace{(\text{SI}, \text{SI})}_{\text{sink}}.$$

The equilibrium of $(x_{j_1}^*, x_{j_2}^*) = (a_{j_1 j_1 1}, a_{j_2 j_2 s_1})$ is an (SI,SI)-sink.

($\text{ii}_{1\text{b}}$) At $(x_{j_1}^*, x_{j_2}^*) = (a_{j_1 j_1 1}, a_{j_2 j_2 s_2})$, the variational equations are

$$\begin{aligned}
\Delta\dot{x}_{j_2} &= a_{j_2 j_2 0}(a_{j_2 j_2 s_2} - a_{j_2 j_2 s_1})(\Delta x_{j_2})^2,\\
\Delta\dot{x}_{j_1} &= a_{j_1 j_1 0}\left[(a_{j_2 j_2 s_2} - a_{j_1 j_2})^2 + b_{j_1 j_2}\right]\Delta x_{j_1}.
\end{aligned}$$

Because of

$$(a_{j_2 j_2 s_2} - a_{j_1 j_2})^2 + b_{j_1 j_2} > 0,$$

the flows at $(x_{j_1}^*, x_{j_2}^*) = (a_{j_1 j_1 1}, a_{j_2 j_2 s_2})$ in the x_{j_1}-direction are source and sink for $a_{j_1 j_1 0} > 0$ and $a_{j_1 j_1 0} < 0$, respectively. The flows at $(x_{j_1}^*, x_{j_2}^*) = (a_{j_1 j_1 1}, a_{j_2 j_2 s_2})$ in the x_{j_2}-direction are upper-saddle and lower-saddle for $a_{j_2 j_2 0}(a_{j_2 j_2 s_2} - a_{j_2 j_2 s_1}) > 0$ and $a_{j_2 j_2 0}(a_{j_2 j_2 s_2} - a_{j_2 j_2 s_1}) < 0$, respectively.

Therefore, the equilibrium of $(x_{j_1}^*, x_{j_2}^*) = (a_{j_1 j_1 1}, a_{j_2 j_2 s_2})$ has the following properties as in Eqs. (1.18)–(1.21).

- For $a_{j_1j_10}>0$ and $a_{j_2j_20}(a_{j_2j_2s_2}-a_{j_2j_2s_1})>0$,

$$(a_{j_1j_11}, a_{j_2j_21}) = \underbrace{(\text{SO}, \text{US})}_{\text{upper-saddle source}}.$$

The equilibrium of $(x^*_{j_1}, x^*_{j_2}) = (a_{j_1j_11}, a_{j_2j_2s_2})$ is an (SO,US)-upper-saddle source.

- For $a_{j_1j_10}<0$ and $a_{j_2j_20}(a_{j_2j_2s_2}-a_{j_2j_2s_1})>0$,

$$(a_{j_1j_11}, a_{j_2j_21}) = \underbrace{(\text{SI}, \text{US})}_{\text{upper-saddle sink}}.$$

The equilibrium of $(x^*_{j_1}, x^*_{j_2}) = (a_{j_1j_11}, a_{j_2j_2s_2})$ is an (SI,US)-upper-saddle sink.

- For $a_{j_1j_10}>0$ and $a_{j_2j_20}(a_{j_2j_2s_2}-a_{j_2j_2s_1})<0$,

$$(a_{j_1j_11}, a_{j_2j_21}) = \underbrace{(\text{SO}, \text{LS})}_{\text{lower-saddle source}}.$$

The equilibrium of $(x^*_{j_1}, x^*_{j_2}) = (a_{j_1j_11}, a_{j_2j_2s_2})$ is an (SO,LS)-lower-saddle source.

- For $a_{j_1j_10}<0$ and $a_{j_2j_20}(a_{j_2j_2s_2}-a_{j_2j_2s_1})<0$,

$$(a_{j_1j_11}, a_{j_2j_21}) = \underbrace{(\text{SI}, \text{LS})}_{\text{lower-saddle sink}}.$$

The equilibrium of $(x^*_{j_1}, x^*_{j_2}) = (a_{j_1j_11}, a_{j_2j_2s_2})$ is an (SI,LS)-lower-saddle sink.

– The saddle-source and saddle-sink are the appearing bifurcations of saddle-source and saddle-sink.

(ii$_2$) For $a_{j_2j_2s_1}=a_{j_2j_2s_2}=a_{j_2j_21}$, the equation becomes

$$\begin{aligned}\dot{x}_{j_2} &= a_{j_2j_20}(x_{j_2}-a_{j_2j_21})^3,\\ \dot{x}_{j_1} &= a_{j_1j_10}(x_{j_1}-a_{j_1j_11})\left[(x_{j_2}-a_{j_1j_2})^2+b_{j_1j_2}\right].\end{aligned}$$

In phase space,

$$\frac{dx_{j_1}}{dx_{j_2}} = \frac{a_{j_1j_10}}{a_{j_2j_20}} \frac{(x_{j_1}-a_{j_1j_11})\left[(x_{j_2}-a_{j_1j_2})^2+b_{j_1j_2}\right]}{(x_{j_2}-a_{j_2j_21})^3},$$

and the deformation of the foregoing equation is

$$\frac{dx_{j_1}}{x_{j_1}-a_{j_1j_11}}=\frac{a_{j_1j_10}}{a_{j_2j_20}}\left[\frac{1}{x_{j_2}-a_{j_2j_21}}+\frac{2(a_{j_2j_21}-a_{j_1j_2})}{(x_{j_2}-a_{j_2j_21})^2}\right.$$
$$\left.+\frac{(a_{j_2j_21}-a_{j_1j_2})^2+b_{j_1j_2}}{(x_{j_2}-a_{j_2j_21})^3}\right]dx_{j_2}.$$

With an initial condition of (x_{j_10},x_{j_20}) at $t=t_0$, the integration of the above equation gives the first integral manifold as

$$\ln\frac{|x_{j_1}-a_{j_1j_11}|}{|x_{j_10}-a_{j_1j_11}|}=\frac{a_{j_1j_10}}{a_{j_2j_20}}\left\{\ln\frac{|x_{j_2}-a_{j_2j_21}|}{|x_{j_20}-a_{j_2j_21}|}\right.$$
$$-2(a_{j_2j_21}-a_{j_1j_2})\left[\frac{1}{x_{j_2}-a_{j_2j_21}}-\frac{1}{x_{j_20}-a_{j_2j_21}}\right]$$
$$\left.-\frac{1}{2}\left[(a_{j_2j_21}-a_{j_1j_2})^2+b_{j_1j_2}\right]\left[\frac{1}{(x_{j_2}-a_{j_2j_21})^2}-\frac{1}{(x_{j_20}-a_{j_2j_21})^2}\right]\right\}.$$

For $a_{j_2j_2s_1}=a_{j_2j_2s_2}=a_{j_2j_21}$,at $(x^*_{j_1},x^*_{j_2})=(a_{j_1j_11},a_{j_2j_21})$, the variational equations are

$$\Delta\dot{x}_{j_2}=a_{j_2j_20}(\Delta x_{j_2})^3,$$
$$\Delta\dot{x}_{j_1}=a_{j_1j_10}\Delta x_{j_1}\left[(a_{j_2j_21}-a_{j_1j_2})^2+b_{j_1j_2}\right].$$

Because of

$$(a_{j_2j_21}-a_{j_1j_2})^2+b_{j_1j_2}>0,$$

the flows at $(x^*_{j_1},x^*_{j_2})=(a_{j_1j_11},a_{j_2j_2s_1})$ in the x_{j_1} -direction are source and sink for $a_{j_1j_10}>0$ and $a_{j_1j_10}<0$, respectively; and the flows at $(x^*_{j_1},x^*_{j_2})=(a_{j_1j_11},a_{j_2j_2s_1})$ in the x_{j_2}-direction are third-order source and sink for $a_{j_2j_20}>0$ and $a_{j_2j_20}<0$, respectively.

Therefore, the equilibrium of $(x^*_{j_1},x^*_{j_2})=(a_{j_1j_11},a_{j_2j_21})$ has the following properties as in Eqs. (1.23)–(1.26).

- For $a_{j_1j_10}>0$ and $a_{j_2j_20}>0$,

$$(a_{j_1j_11},a_{j_2j_21})=\underbrace{(\mathrm{SO},3^{\mathrm{rd}}\mathrm{SO})}_{\text{third-order source}}.$$

  The equilibrium of $(x^*_{j_1},x^*_{j_2})=(a_{j_1j_11},a_{j_2j_21})$ is an (SO,3rdSO)-third-order source, which is for an appearing bifurcation from source to source, saddle, and source.

- For $a_{j_1j_10}<0$ and $a_{j_2j_20}>0$,

$$(a_{j_1j_11},a_{j_2j_21})=\underbrace{(\mathrm{SI},3^{\mathrm{rd}}\mathrm{SO})}_{\text{third-order saddle}}.$$

The equilibrium of $(x_{j_1}^*,x_{j_2}^*)=(a_{j_1j_11},a_{j_2j_21})$ is an (SI,3rdSO)-third-order saddle, which is for an appearing bifurcation from saddle to saddle, sink, and saddle.

- For $a_{j_1j_10}>0$ and $a_{j_2j_20}<0$,

$$(a_{j_1j_11},a_{j_2j_21})=\underbrace{(\mathrm{SO},3^{\mathrm{rd}}\mathrm{SI})}_{\text{third-order saddle}}.$$

The equilibrium of $(x_{j_1}^*,x_{j_2}^*)=(a_{j_1j_11},a_{j_2j_21})$ is an (SO,3rdSI)-third-order saddle, which is for an appearing bifurcation from saddle to saddle, source, and saddle.

- For $a_{j_1j_10}<0$ and $a_{j_2j_20}<0$,

$$(a_{j_1j_11},a_{j_2j_21})=\underbrace{(\mathrm{SI},3^{\mathrm{rd}}\mathrm{SI})}_{\text{third-order sink}}.$$

The equilibrium of $(x_{j_1}^*,x_{j_2}^*)=(a_{j_1j_11},a_{j_2j_21})$ is an (SI,3rdSI)-third-order sink, which is for an appearing bifurcation from sink to sink, saddle, and sink.

(iii) For $\Delta_{j_1j_2}<0$ and $\Delta_{j_2j_2}>0$, the standard form is

$$\begin{aligned}
\dot{x}_{j_2}&=a_{j_2j_20}(x_{j_2}-a_{j_2j_21})(x_{j_2}-a_{j_2j_22})(x_{j_2}-a_{j_2j_23}),\\
\dot{x}_{j_1}&=a_{j_1j_10}(x_{j_1}-a_{j_1j_11})\left[(x_{j_2}-a_{j_1j_2})^2+b_{j_1j_2}\right]
\end{aligned}$$

where

$$\begin{aligned}
&a_{j_1j_2}=-\frac{1}{2}B_{j_1j_2},\,b_{j_1j_2}=\frac{1}{4}(-\Delta_{j_1j_2}),\\
&b_{j_2j_22},b_{j_2j_23}=-\frac{1}{2}(B_{j_2j_2}\pm\sqrt{\Delta_{j_2j_2}}),\\
&\{a_{j_2j_21},a_{j_2j_22},a_{j_2j_23}\}=\mathrm{sort}\{b_{j_2j_21},b_{j_2j_22},b_{j_2j_23}\},\\
&a_{j_2j_2s_1}<a_{j_2j_2s_2},s_1,s_2\in\{1,2,3\},s_1<s_2.
\end{aligned}$$

In phase space,

$$\frac{dx_{j_1}}{dx_{j_2}}=\frac{a_{j_1j_10}}{a_{j_2j_20}}\frac{(x_{j_1}-a_{j_1j_11})\left[(x_{j_2}-a_{j_1j_2})^2+b_{j_1j_2}\right]}{(x_{j_2}-a_{j_2j_21})(x_{j_2}-a_{j_2j_22})(x_{j_2}-a_{j_2j_23})},$$

and the deformation of the foregoing equation is

$$\frac{1}{(x_{j_1} - a_{j_1 j_1 1})} dx_{j_1} = \frac{a_{j_1 j_1 0}}{a_{j_2 j_2 0}} \sum\nolimits_{s_1=1}^{2} \frac{(a_{j_2 j_2 s_1} - a_{j_1 j_2}) + b_{j_1 j_2}}{\prod\nolimits_{s_1=1, s_1 \neq s_2}^{3} (a_{j_2 j_2 s_1} - a_{j_2 j_2 s_2})} \frac{1}{x_{j_2} - a_{j_2 j_2 s_1}} dx_{j_2}.$$

With an initial condition of $(x_{j_1 0}, x_{j_2 0})$ at $t = t_0$, the integration of the above equation gives

$$\ln \frac{|x_{j_1} - a_{j_1 j_1 1}|}{|x_{j_1 0} - a_{j_1 j_1 1}|} = \frac{a_{j_1 j_1 0}}{a_{j_2 j_2 0}} \sum\nolimits_{s_1=1}^{3} \frac{(a_{j_2 j_2 s_1} - a_{j_1 j_2})^2 + b_{j_1 j_2}}{\prod\nolimits_{s_2=1, s_2 \neq s_1}^{3} (a_{j_2 j_2 s_1} - a_{j_1 j_2 s_2})} \ln \frac{|x_{j_2} - a_{j_2 j_2 s_1}|}{|x_{j_2 0} - a_{j_2 j_2 s_1}|}.$$

At $(x^*_{j_1}, x^*_{j_2}) = (a_{j_1 j_1 1}, a_{j_2 j_2 s_1})$, the variational equations are

$$\begin{aligned}
\Delta \dot{x}_{j_2} &= a_{j_2 j_2 0} \prod\nolimits_{s_1=1, s_1 \neq s_2}^{3} (a_{j_2 j_2 s_1} - a_{j_2 j_2 s_2}) \Delta x_{j_2}, \\
\Delta \dot{x}_{j_1} &= a_{j_1 j_1 0} \left[(a_{j_2 j_2 s_1} - a_{j_1 j_2})^2 + b_{j_1 j_2} \right] \Delta x_{j_1}.
\end{aligned}$$

Because of

$$(a_{j_2 j_2 s_1} - a_{j_1 j_2})^2 + b_{j_1 j_2} > 0,$$

the flows at $(x^*_{j_1}, x^*_{j_2}) = (a_{j_1 j_1 1}, a_{j_2 j_2 s_1})$ in the x_{j_1}-direction are source and sink for $a_{j_1 j_1 0} > 0$ and $a_{j_1 j_1 0} < 0$, respectively. The flows at $(x^*_{j_1}, x^*_{j_2}) = (a_{j_1 j_1 1}, a_{j_2 j_2 s_1})$ in the x_{j_2}-direction are source and sink for $a_{j_2 j_2 0} \prod\nolimits_{s_2=1, s_2 \neq s_1}^{3} (a_{j_2 j_2 s_1} - a_{j_1 j_2 s_2}) > 0$ and $a_{j_2 j_2 0} \prod\nolimits_{s_2=1, s_2 \neq s_1}^{3} (a_{j_2 j_2 s_1} - a_{j_1 j_2 s_2}) < 0$, respectively.

Therefore, the equilibrium of $(x^*_{j_1}, x^*_{j_2}) = (a_{j_1 j_1 1}, a_{j_2 j_2 s_1})$ (s_1, $s_2 \in \{1, 2\}$, $s_1 \neq s_2$) has the following properties as in Eqs. (1.30)–(1.33).

- For $a_{j_1 j_1 0} > 0$ and $a_{j_2 j_2 0} \prod\nolimits_{s_2=1, s_2 \neq s_1}^{3} (a_{j_2 j_2 s_1} - a_{j_1 j_2 s_2}) > 0$,

$$(a_{j_1 j_1 1}, a_{j_2 j_2 s_1}) = \underbrace{(\mathrm{SO}, \mathrm{SO})}_{\text{source}}.$$

  The equilibrium of $(x^*_{j_1}, x^*_{j_2}) = (a_{j_1 j_1 1}, a_{j_2 j_2 s_1})$ is an (SO,SO)-source.
- For $a_{j_1 j_1 0} < 0$ and $a_{j_2 j_2 0} \prod\nolimits_{s_2=1, s_2 \neq s_1}^{3} (a_{j_2 j_2 s_1} - a_{j_1 j_2 s_2}) > 0$,

$$(a_{j_1 j_1 1}, a_{j_2 j_2 s_1}) = \underbrace{(\mathrm{SI}, \mathrm{SO})}_{\text{saddle}}.$$

  The equilibrium of $(x^*_{j_1}, x^*_{j_2}) = (a_{j_1 j_1 1}, a_{j_2 j_2 s_1})$ is an (SI,SO)-saddle.
- For $a_{j_1 j_1 0} > 0$ and $a_{j_2 j_2 0} \prod\nolimits_{s_2=1, s_2 \neq s_1}^{3} (a_{j_2 j_2 s_1} - a_{j_1 j_2 s_2}) < 0$,

$$(a_{j_1 j_1 1}, a_{j_2 j_2 s_1}) = \underbrace{(\mathrm{SO}, \mathrm{SI})}_{\text{saddle}}.$$

  The equilibrium of $(x^*_{j_1}, x^*_{j_2}) = (a_{j_1 j_1 1}, a_{j_2 j_2 s_1})$ is an (SO,SI)-saddle.

- For $a_{j_1j_10}<0$ and $a_{j_2j_20}\prod_{s_2=1,s_2\neq s_1}^{3}(a_{j_2j_2s_1}-a_{j_1j_2s_2})<0$,

$$(a_{j_1j_11},a_{j_2j_2s_1})=\underbrace{(\mathrm{SI},\mathrm{SI})}_{\text{sink}}.$$

The equilibrium of $(x_{j_1}^*,x_{j_2}^*)=(a_{j_1j_11},a_{j_2j_2s_1})$ is an (SI,SI)-sink.

(iv) For $\Delta_{j_1j_2}=0$ and $\Delta_{j_2j_2}<0$, the standard form is

$$\begin{aligned}\dot{x}_{j_2}&=a_{j_2j_20}(x_{j_2}-a_{j_2j_21})\left[(x_{j_2}-a_{j_2j_2})^2+b_{j_2j_2}\right],\\ \dot{x}_{j_1}&=a_{j_1j_10}(x_{j_1}-a_{j_1j_11})(x_{j_2}-a_{j_1j_21})^2\end{aligned}$$

where

$$\begin{aligned}a_{j_1j_21}&=-\frac{1}{2}B_{j_1j_2},a_{j_2j_21}=b_{j_2j_21},\\ a_{j_2j_2}&=-\frac{1}{2}B_{j_2j_2},b_{j_2j_2}=\frac{1}{4}(-\Delta_{j_2j_2}).\end{aligned}$$

(iv$_1$) For $a_{j_2j_21}\neq a_{j_1j_21}$, in phase space,

$$\frac{dx_{j_1}}{dx_{j_2}}=\frac{a_{j_1j_10}}{a_{j_2j_20}}\frac{(x_{j_1}-a_{j_1j_11})(x_{j_2}-a_{j_1j_21})^2}{(x_{j_2}-a_{j_2j_21})\left[(x_{j_2}-a_{j_2j_2})^2+b_{j_2j_2}\right]},$$

and the deformation of the foregoing equation is

$$\begin{aligned}\frac{dx_{j_1}}{x_{j_1}-a_{j_1j_11}}=\frac{a_{j_1j_10}}{a_{j_2j_20}}&\left\{\frac{x_{j_2}-a_{j_2j_21}}{(x_{j_2}-a_{j_2j_2})^2+b_{j_2j_2}}+\frac{2(a_{j_2j_21}-a_{j_1j_21})}{(x_{j_2}-a_{j_2j_2})^2+b_{j_2j_2}}\right.\\ &\left.+\frac{(a_{j_2j_21}-a_{j_1j_21})^2}{(x_{j_2}-a_{j_2j_21})\left[(x_{j_2}-a_{j_2j_2})^2+b_{j_2j_2}\right]}\right\}dx_{j_2}.\end{aligned}$$

With

$$\begin{aligned}&\frac{(a_{j_2j_21}-a_{j_1j_21})^2}{(x_{j_2}-a_{j_2j_21})\left[(x_{j_2}-a_{j_2j_2})^2+b_{j_2j_2}\right]}\\ &=\frac{(a_{j_2j_21}-a_{j_1j_21})^2}{(a_{j_2j_21}-a_{j_2j_2})^2+b_{j_2j_2}}\left[\frac{1}{x_{j_2}-a_{j_2j_21}}-\frac{x_2-a_{j_2j_21}+2(a_{j_2j_21}-a_{j_2j_2})}{(x_{j_2}-a_{j_2j_2})^2+b_{j_2j_2}}\right],\end{aligned}$$

we have

$$
\begin{aligned}
\frac{dx_{j_1}}{x_{j_1}-a_{j_1j_11}} &= \frac{a_{j_1j_10}}{a_{j_2j_20}}\left\{\frac{(a_{j_2j_21}-a_{j_1j_21})^2}{(a_{j_2j_21}-a_{j_2j_2})^2+b_{j_2j_2}}\frac{1}{(x_{j_2}-a_{j_2j_21})}\right.\\
&+\left[1-\frac{(a_{j_2j_21}-a_{j_1j_21})^2}{(a_{j_2j_21}-a_{j_2j_2})^2+b_{j_2j_2}}\right]\frac{x_{j_2}-a_{j_2j_2}}{(x_{j_2}-a_{j_2j_2})^2+b_{j_2j_2}}\\
&+\left[a_{j_2j_2}+a_{j_2j_21}-2a_{j_1j_21}-(a_{j_2j_21}-a_{j_2j_2})\frac{(a_{j_2j_21}-a_{j_1j_21})^2}{(a_{j_2j_21}-a_{j_2j_2})^2+b_{j_2j_2}}\right]\\
&\left.\times\frac{1}{(x_{j_2}-a_{j_2j_2})^2+b_{j_2j_2}}\right\}dx_{j_2}.
\end{aligned}
$$

With an initial condition of (x_{j_10},x_{j_20}) at $t=t_0$, the integration of the above equation gives the first integral manifold as

$$
\begin{aligned}
\ln\frac{|x_{j_1}-a_{j_1j_11}|}{|x_{j_10}-a_{j_1j_11}|} &= \frac{a_{j_1j_10}}{a_{j_2j_20}}\left[A\ln\frac{|x_{j_2}-a_{j_2j_21}|}{|x_{j_20}-a_{j_2j_21}|}\right.\\
&+\frac{1}{2}(1-A)\ln\frac{|(x_{j_2}-a_{j_2j_2})^2+b_{j_2j_2}|}{|(x_{j_20}-a_{j_2j_2})^2+b_{j_2j_2}|}\\
&\left.+\frac{B}{\sqrt{b_{j_2j_2}}}\left(\arctan\frac{x_{j_2}-a_{j_2j_2}}{\sqrt{b_{j_2j_2}}}-\arctan\frac{x_{j_20}-a_{j_2j_2}}{\sqrt{b_{j_2j_2}}}\right)\right]
\end{aligned}
$$

where

$$
A=\frac{(a_{j_2j_21}-a_{j_1j_21})^2}{(a_{j_2j_21}-a_{j_2j_2})^2+b_{j_2j_2}},\; B=a_{j_2j_2}+a_{j_2j_21}-2a_{j_1j_21}-(a_{j_2j_21}-a_{j_2j_2})A.
$$

(iv$_{1a}$) For $a_{j_2j_21}\neq a_{j_1j_21}$, at $(x^*_{j_1},x^*_{j_2})=(a_{j_1j_11},a_{j_2j_21})$, the variational equations are

$$
\begin{aligned}
\Delta\dot{x}_{j_2} &= a_{j_2j_20}\Delta x_{j_2}\left[(a_{j_2j_21}-a_{j_2j_2})^2+b_{j_2j_2}\right],\\
\Delta\dot{x}_{j_1} &= a_{j_1j_10}\Delta x_{j_1}(a_{j_2j_21}-a_{j_1j_21})^2.
\end{aligned}
$$

Because of

$$
(a_{j_2j_21}-a_{j_2j_2})^2+b_{j_2j_2}>0 \text{ and } (a_{j_2j_21}-a_{j_1j_21})^2>0,
$$

the flows at $(x^*_{j_1},x^*_{j_2})=(a_{j_1j_11},a_{j_2j_21})$ in the x_{j_1}-direction are source and sink for $a_{j_1j_10}(a_{j_2j_21}-a_{j_1j_21})^2>0$ and $a_{j_1j_10}(a_{j_2j_21}-a_{j_1j_21})^2<0$, respectively; and the flows at $(x^*_{j_1},x^*_{j_2})=(a_{j_1j_11},a_{j_2j_21})$ in the x_{j_2}-direction are source and sink for $a_{j_2j_20}>0$ and $a_{j_2j_20}<0$, respectively.

Therefore, the equilibrium of $(x^*_{j_1},x^*_{j_2})=(a_{j_1j_11},a_{j_2j_21})$ has the following properties as in Eqs. (1.38)–(1.41).

- For $a_{j_1j_10}(a_{j_2j_21} - a_{j_1j_21})^2 > 0$ and $a_{j_2j_20} > 0$,

$$(a_{j_1j_11}, a_{j_2j_21}) = \underbrace{(\text{SO}, \text{SO})}_{\text{source}}.$$

The equilibrium of $(x^*_{j_1}, x^*_{j_2}) = (a_{j_1j_11}, a_{j_2j_21})$ is an (SO,SO)-source.

- For $a_{j_1j_10}(a_{j_2j_21} - a_{j_1j_21})^2 < 0$ and $a_{j_2j_20} > 0$,

$$(a_{j_1j_11}, a_{j_2j_21}) = \underbrace{(\text{SI}, \text{SO})}_{\text{saddle}}.$$

The equilibrium of $(x^*_{j_1}, x^*_{j_2}) = (a_{j_1j_11}, a_{j_2j_21})$ is an (SI,SO)-saddle.

- For $a_{j_1j_10}(a_{j_2j_21} - a_{j_1j_21})^2 > 0$ and $a_{j_2j_20} < 0$,

$$(a_{j_1j_11}, a_{j_2j_21}) = \underbrace{(\text{SO}, \text{SI})}_{\text{saddle}}.$$

The equilibrium of $(x^*_{j_1}, x^*_{j_2}) = (a_{j_1j_11}, a_{j_2j_21})$ is an (SO,SI)-saddle.

- For $a_{j_1j_10}(a_{j_2j_21} - a_{j_1j_21})^2 < 0$ and $a_{j_2j_20} < 0$,

$$(a_{j_1j_11}, a_{j_2j_21}) = \underbrace{(\text{SI}, \text{SI})}_{\text{sink}}.$$

The equilibrium of $(x^*_{j_1}, x^*_{j_2}) = (a_{j_1j_11}, a_{j_2j_21})$ is an (SI,SI)-sink.

($\text{iv}_{1\text{b}}$) In phase space, at $x^*_{j_2} = a_{j_1j_21}$ with $\overline{x}_{j_1} \neq a_{j_1j_11}$,

$$\left.\frac{dx_{j_1}}{dx_{j_2}}\right|_{x^*_{j_2} = a_{j_1j_21}} = \frac{a_{j_1j_10}}{a_{j_2j_20}} \left.\frac{(\overline{x}_{j_1} - a_{j_1j_11})(x_{j_2} - a_{j_1j_21})^2}{(x_{j_2} - a_{j_2j_21})\left[(x_{j_2} - a_{j_2j_2})^2 + b_{j_2j_2}\right]}\right|_{x^*_{j_2} = a_{j_1j_21}} = 0,$$

$$\left.\frac{d^2x_{j_1}}{dx^2_{j_2}}\right|_{x^*_{j_2} = a_{j_1j_21}} = 0.$$

If

$$\left.\frac{d^3x_{j_1}}{dx^3_{j_2}}\right|_{x^*_{j_2} = a_{j_1j_21}} = \frac{a_{j_1j_10}}{a_{j_2j_20}} \frac{2(\overline{x}_{j_1} - a_{j_1j_11})}{(a_{j_1j_21} - a_{j_2j_21})\left[(a_{j_1j_21} - a_{j_2j_2})^2 + b_{j_2j_2}\right]} > 0,$$

there is a third-order increasing-inflection flow at $x^*_{j_2} = a_{j_1j_21}$ with $\overline{x}_{j_1} \neq a_{j_1j_11}$. If

$$\left.\frac{d^3x_{j_1}}{dx_{j_2}^3}\right|_{x_{j_2}^*=a_{j_1j_21}} = \frac{a_{j_1j_10}}{a_{j_2j_20}}\frac{2(\overline{x}_{j_1}-a_{j_1j_11})}{(a_{j_1j_21}-a_{j_2j_21})\left[(a_{j_1j_21}-a_{j_2j_2})^2+b_{j_2j_2}\right]}<0,$$

there is a third-order decreasing-inflection flow at $x_{j_2}^*=a_{j_1j_21}$ with $\overline{x}_{j_1}\neq a_{j_1j_11}$. Let

$$\dot{x}_{j_2}=a_{j_2j_20}(a_{j_1j_21}-a_{j_2j_21})\left[(a_{j_1j_21}-a_{j_2j_2})^2+b_{j_2j_2}\right],$$

with

$$(a_{j_1j_21}-a_{j_2j_2})^2+b_{j_2j_2}>0.$$

The third-order inflection flows at $x_{j_2}^*=a_{j_1j_21}$ with $\overline{x}_{j_1}\neq a_{j_1j_11}$ are positive and negative for $a_{j_2j_20}(a_{j_1j_21}-a_{j_2j_21})>0$ and $a_{j_2j_20}(a_{j_1j_21}-a_{j_2j_21})<0$, respectively.

Thus, the equilibrium of $x_{j_2}^*=a_{j_1j_21}$ with $\overline{x}_{j_1}\neq a_{j_1j_11}$ has the following properties.

- For $a_{j_1j_10}(\overline{x}_{j_1}-a_{j_1j_11})>0$ and $a_{j_2j_20}(a_{j_1j_21}-a_{j_2j_21})>0$,
$$(a_{j_1j_21},\dot{x}_{j_2})=\underbrace{(\text{II, pF})}_{\text{increasing-inflection flow }(+)}.$$
- For $a_{j_1j_10}(\overline{x}_{j_1}-a_{j_1j_11})<0$ and $a_{j_2j_20}(a_{j_1j_21}-a_{j_2j_21})>0$,
$$(a_{j_1j_21},\dot{x}_{j_2})=\underbrace{(\text{DI, pF})}_{\text{decreasing-inflection flow }(+)}.$$
- For $a_{j_1j_10}(\overline{x}_{j_1}-a_{j_1j_11})>0$ and $a_{j_2j_20}(a_{j_1j_21}-a_{j_2j_21})<0$,
$$(a_{j_1j_21},\dot{x}_{j_2})=\underbrace{(\text{DI, nF})}_{\text{decreasing-inflection flow }(-)}.$$
- For $a_{j_1j_10}(\overline{x}_{j_1}-a_{j_1j_11})<0$ and $a_{j_2j_20}(a_{j_1j_21}-a_{j_2j_21})<0$,
$$(a_{j_1j_21},\dot{x}_{j_2})=\underbrace{(\text{II, nF})}_{\text{increasing-inflection flow }(-)}.$$

The variational equation at $x_{j_1}^*=a_{j_1j_11}$ with $\overline{x}_{j_2}\neq a_{j_1j_21}$ is

$$\Delta\dot{x}_{j_1}=a_{j_1j_10}(\overline{x}_{j_2}-a_{j_1j_21})^2\Delta x_{j_1}$$

and

$$\dot{x}_{j_2} = a_{j_2j_20}(\overline{x}_{j_2} - a_{j_2j_21})\left[(\overline{x}_{j_2} - a_{j_2j_2})^2 + b_{j_2j_2}\right].$$

Thus, the equilibrium of $x^*_{j_2} = a_{j_1j_21}$ with $\overline{x}_{j_1} \neq a_{j_1j_11}$ has the following properties.

- For $a_{j_1j_10}(\overline{x}_{j_2} - a_{j_1j_21})^2 > 0$ and $a_{j_2j_20}(\overline{x}_{j_2} - a_{j_2j_21}) > 0$,

$$(a_{j_1j_11}, \dot{x}_{j_2}) = \underbrace{(\text{SO}, \text{pF})}_{\text{source flow }(+)}.$$

- For $a_{j_1j_10}(\overline{x}_{j_2} - a_{j_1j_21})^2 < 0$ and $a_{j_2j_20}(\overline{x}_{j_2} - a_{j_2j_21}) > 0$,

$$(a_{j_1j_11}, \dot{x}_{j_2}) = \underbrace{(\text{SI}, \text{pF})}_{\text{sink flow }(+)}.$$

- For $a_{j_1j_10}(\overline{x}_{j_2} - a_{j_1j_21})^2 > 0$ and $a_{j_2j_20}(\overline{x}_{j_2} - a_{j_2j_21}) < 0$,

$$(a_{j_1j_11}, \dot{x}_{j_2}) = \underbrace{(\text{SO}, \text{nF})}_{\text{source flow }(-)}.$$

- For $a_{j_1j_10}(\overline{x}_{j_2} - a_{j_1j_21})^2 < 0$ and $a_{j_2j_20}(\overline{x}_{j_2} - a_{j_2j_21}) < 0$,

$$(a_{j_1j_11}, \dot{x}_{j_2}) = \underbrace{(\text{SI}, \text{nF})}_{\text{sink flow }(-)}.$$

Therefore, the equilibrium of $(x^*_{j_1}, x^*_{j_2}) = (a_{j_1j_11}, a_{j_1j_21})$ has the following properties as in Eqs. (1.42)–(1.45).

- For $a_{j_1j_10} > 0$ and $a_{j_2j_20}(a_{j_1j_21} - a_{j_2j_21}) > 0$,

$$(a_{j_1j_11}, a_{j_1j_21}) = \underbrace{({}_{\text{DI:II}}\text{SO}, \text{pF})}_{\text{inflection-source flow }(+)}.$$

  The equilibrium of $(x^*_{j_1}, x^*_{j_2}) = (a_{j_1j_11}, a_{j_1j_21})$ is a $({}_{\text{DI:II}}\text{SO}, \text{pF})$-positive inflection-source flow.

- For $a_{j_1j_10} < 0$ and $a_{j_2j_20}(a_{j_1j_21} - a_{j_2j_21}) > 0$,

$$(a_{j_1j_11}, a_{j_1j_21}) = \underbrace{({}_{\text{II:DI}}\text{SI}, \text{pF})}_{\text{inflection-sink flow }(+)}.$$

The equilibrium of $(x^*_{j_1}, x^*_{j_2}) = (a_{j_1j_11}, a_{j_1j_21})$ is an $({}_{\mathrm{II:DI}}\mathrm{SI}, \mathrm{pF})$-positive inflection-sink flow.

- For $a_{j_1j_10} > 0$ and $a_{j_2j_20}(a_{j_1j_21} - a_{j_2j_21}) < 0$,

$$(a_{j_1j_11}, a_{j_1j_21}) = \underbrace{({}_{\mathrm{II:DI}}\mathrm{SO}, \mathrm{nF})}_{\text{inflection-source flow }(-)}.$$

The equilibrium of $(x^*_{j_1}, x^*_{j_2}) = (a_{j_1j_11}, a_{j_1j_21})$ is an $({}_{\mathrm{II:DI}}\mathrm{SO}, \mathrm{nF})$-negative inflection-source flow.

- For $a_{j_1j_10} < 0$ and $a_{j_2j_20}(a_{j_1j_21} - a_{j_2j_21}) < 0$,

$$(a_{j_1j_11}, a_{j_1j_21}) = \underbrace{({}_{\mathrm{DI:II}}\mathrm{SI}, \mathrm{nF})}_{\text{inflection-sink flow }(-)}.$$

The equilibrium of $(x^*_{j_1}, x^*_{j_2}) = (a_{j_1j_11}, a_{j_1j_21})$ is a $({}_{\mathrm{DI:II}}\mathrm{SI}, \mathrm{nF})$-negative inflection-sink flow.

(iv$_2$) For $a_{j_2j_21} = a_{j_1j_21}$, in phase space,

$$\frac{dx_{j_1}}{dx_{j_2}} = \frac{a_{j_1j_10}}{a_{j_2j_20}} \frac{(x_{j_1} - a_{j_1j_11})(x_{j_2} - a_{j_1j_21})}{(x_{j_2} - a_{j_2j_2})^2 + b_{j_2j_2}},$$

and the deformation of the foregoing equation is

$$\frac{dx_{j_1}}{x_{j_1} - a_{j_1j_11}} = \frac{a_{j_1j_10}}{a_{j_2j_20}} \frac{x_{j_2} - a_{j_2j_2} + a_{j_2j_2} - a_{j_1j_21}}{(x_{j_2} - a_{j_2j_2})^2 + b_{j_2j_2}} dx_{j_2}.$$

With an initial condition of (x_{j_10}, x_{j_20}) at $t = t_0$, the integration of the above equation gives the first integral manifold as

$$\ln \frac{|x_{j_1} - a_{j_1j_11}|}{|x_{j_10} - a_{j_1j_11}|} = \frac{a_{j_1j_10}}{a_{j_2j_20}} \left[\frac{1}{2} \ln \frac{|(x_{j_2} - a_{j_2j_2})^2 + b_{j_2j_2}|}{|(x_{j_20} - a_{j_2j_2})^2 + b_{j_2j_2}|} + \frac{(a_{j_2j_2} - a_{j_1j_21})}{\sqrt{b_{j_2j_2}}} \left(\arctan \frac{x_{j_2} - a_{j_2j_2}}{\sqrt{b_{j_2j_2}}}, - \arctan \frac{x_{j_20} - a_{j_2j_2}}{\sqrt{b_{j_2j_2}}} \right) \right].$$

(iv$_{2a}$) In phase space, at $x^*_{j_2} = a_{j_1j_21}$ with $\bar{x}_{j_1} \neq a_{j_1j_11}$,

$$\left.\frac{dx_{j_1}}{dx_{j_2}}\right|_{x^*_{j_2} = a_{j_1j_21}} = \left.\frac{a_{j_1j_10}}{a_{j_2j_20}} \frac{(\bar{x}_{j_1} - a_{j_1j_11})(x_{j_2} - a_{j_1j_21})}{\left[(x_{j_2} - a_{j_2j_2})^2 + b_{j_2j_2}\right]}\right|_{x^*_{j_2} = a_{j_1j_21}} = 0.$$

If

$$\left.\frac{d^2x_{j_1}}{dx_{j_2}^2}\right|_{x_{j_2}^*=a_{j_1j_21}} = \frac{a_{j_1j_10}}{a_{j_2j_20}}\frac{(\bar{x}_{j_1}-a_{j_1j_11})}{(a_{j_1j_21}-a_{j_2j_2})^2+b_{j_2j_2}} > 0,$$

there is an up-parabola flow at $x_{j_2}^* = a_{j_1j_21}$ with $\bar{x}_{j_1} \neq a_{j_1j_11}$. If

$$\left.\frac{d^2x_{j_1}}{dx_{j_2}^2}\right|_{x_{j_2}^*=a_{j_1j_21}} = \frac{a_{j_1j_10}}{a_{j_2j_20}}\frac{(\bar{x}_{j_1}-a_{j_1j_11})}{(a_{j_1j_21}-a_{j_2j_2})^2+b_{j_2j_2}} < 0,$$

there is a down-parabola flow at $x_{j_2}^* = a_{j_1j_21}$ with $\bar{x}_{j_1} \neq a_{j_1j_11}$.

The variational equation is

$$\Delta\dot{x}_{j_2} = a_{j_2j_20}\left[(a_{j_2j_21}-a_{j_2j_2})^2+b_{j_2j_2}\right]\Delta x_{j_2}$$

and

$$(a_{j_2j_21}-a_{j_2j_2})^2+b_{j_2j_2} > 0.$$

The parabola flows for $x_{j_2}^* = a_{j_1j_21}$ with $\bar{x}_{j_1} \neq a_{j_1j_11}$ are source and sink, respectively.

Therefore, the infinite-equilibrium of $x_{j_2}^* = a_{j_1j_21} = a_{j_2j_21}$ with $\bar{x}_{j_1} \neq a_{j_1j_11}$ has the following properties as in Eqs. (1.47)–(1.50).

- For $a_{j_1j_10}(\bar{x}_{j_1}-a_{j_1j_11}) > 0$ and $a_{j_2j_20} > 0$,

$$(\bar{x}_{j_1}, a_{j_2j_21}) = \underbrace{(\mathrm{UP}, \mathrm{SO})}_{\text{up-parabola source}}.$$

  The infinite-equilibrium of $x_{j_2}^* = a_{j_1j_21} = a_{j_2j_21}$ is a (UP,SO)-up-parabola source.

- For $a_{j_1j_10}(\bar{x}_{j_1}-a_{j_1j_11}) < 0$ and $a_{j_2j_20} > 0$,

$$(\bar{x}_{j_1}, a_{j_2j_21}) = \underbrace{(\mathrm{DP}, \mathrm{SO})}_{\text{down-parabola source}}.$$

  The infinite-equilibrium of $x_{j_2}^* = a_{j_1j_21} = a_{j_2j_21}$ is a (DP,SO)-down-parabola source.

- For $a_{j_1j_10}(\bar{x}_{j_1}-a_{j_1j_11}) > 0$ and $a_{j_2j_20} < 0$,

$$(\bar{x}_{j_1}, a_{j_2j_21}) = \underbrace{(\mathrm{DP}, \mathrm{SI})}_{\text{down-parabola sink}}.$$

  The infinite-equilibrium of $x_{j_2}^* = a_{j_1j_21} = a_{j_2j_21}$ is a (DP,SI)-down-parabola sink.

- For $a_{j_1j_10}(\bar{x}_{j_1}-a_{j_1j_11})<0$ and $a_{j_2j_20}<0$,

$$(\bar{x}_{j_1},a_{j_2j_21})=\underbrace{(\text{UP},\text{SI})}_{\text{up-parabola sink}}.$$

The infinite-equilibrium of $x^*_{j_2}=a_{j_1j_21}=a_{j_2j_21}$ is a (UP,SI)-up-parabola sink.

(iv$_{2b}$) From (iv$_{2a}$), for $\bar{x}_{j_1}=a_{j_1j_11}$, the equilibrium of $(x^*_{j_1},x^*_{j_2})=(a_{j_1j_11},a_{j_2j_21})$ with $a_{j_1j_21}=a_{j_2j_21}$ has the following properties as in Eqs. (1.51)–(1.54).

- For $a_{j_1j_10}>0$ and $a_{j_2j_20}>0$,

$$(a_{j_1j_11},a_{j_2j_21})=\underbrace{(\text{DP}:\text{UP},\text{SO})}_{\text{hyperbolic source}}.$$

The equilibrium of $(x^*_{j_1},x^*_{j_2})=(a_{j_1j_11},a_{j_2j_21})$ is a (DP:UP,SO)-hyperbolic source.

- For $a_{j_1j_10}<0$ and $a_{j_2j_20}>0$,

$$(a_{j_1j_11},a_{j_2j_21})=\underbrace{(\text{UP}:\text{DP},\text{SO})}_{\text{hyperbolic-secant source}}.$$

The equilibrium of $(x^*_{j_1},x^*_{j_2})=(a_{j_1j_11},a_{j_2j_21})$ is a (UP:DP,SO)-hyperbolic-secant source.

- For $a_{j_1j_10}>0$ and $a_{j_2j_20}<0$,

$$(a_{j_1j_11},a_{j_2j_21})=\underbrace{(\text{UP}:\text{DP},\text{SI})}_{\text{hyperbolic-secant sink}}.$$

The equilibrium of $(x^*_{j_1},x^*_{j_2})=(a_{j_1j_11},a_{j_2j_21})$ is a (UP:DP,SI)-hyperbolic-secant sink.

- For $a_{j_1j_10}<0$ and $a_{j_2j_20}<0$,

$$(a_{j_1j_11},a_{j_2j_21})=\underbrace{(\text{DP}:\text{UP},\text{SI})}_{\text{hyperbolic sink}}.$$

The equilibrium of $(x^*_{j_1},x^*_{j_2})=(a_{j_1j_11},a_{j_2j_21})$ is a (DP:UP,SI)-hyperbolic sink.

(v) For $\Delta_{j_1j_2}=0$ and $\Delta_{j_2j_2}=0$, the standard form is

$$\begin{aligned}\dot{x}_{j_2}&=a_{j_2j_20}(x_{j_2}-a_{j_2j_2s_1})(x_{j_2}-a_{j_2j_2s_2})^2,\\ \dot{x}_{j_1}&=a_{j_1j_10}(x_{j_1}-a_{j_1j_11})(x_{j_2}-a_{j_1j_21})^2\end{aligned}$$

where

$$a_{j_1j_21} = -\frac{1}{2}B_{j_1j_2}, a_{j_2j_2s_2} = -\frac{1}{2}B_{j_2j_2}, a_{j_2j_2s_1} = b_{j_2j_21}$$
$$s_1, s_2 \in \{1,2\}; s_1 \neq s_2.$$

(v_1) For $a_{j_2j_2s} \neq a_{j_1j_21} (s = s_1,\ s_2)$, in phase space,

$$\frac{dx_{j_1}}{dx_{j_2}} = \frac{a_{j_1j_10}}{a_{j_2j_20}} \frac{(x_{j_1} - a_{j_1j_11})(x_{j_2} - a_{j_1j_21})^2}{(x_{j_2} - a_{j_2j_2s_1})(x_{j_2} - a_{j_2j_2s_2})^2},$$

and the deformation of the foregoing equation is

$$\begin{aligned}\frac{1}{(x_{j_1} - a_{j_1j_11})}dx_{j_1} = & \frac{a_{j_1j_10}}{a_{j_2j_20}}\Big\{\frac{(a_{j_2j_2s_1} - a_{j_1j_21})^2}{(a_{j_2j_2s_1} - a_{j_2j_2s_2})^2}\frac{1}{(x_{j_2} - a_{j_2j_2s_1})} \\ & + \Big[\frac{2(a_{j_2j_2s_2} - a_{j_1j_21})}{(a_{j_2j_2s_2} - a_{j_2j_2s_1})} - \frac{(a_{j_2j_2s_2} - a_{j_1j_21})^2}{(a_{j_2j_2s_2} - a_{j_2j_2s_1})^2}\Big]\frac{1}{(x_{j_2} - a_{j_2j_2s_2})} \\ & + \frac{(a_{j_2j_21} - a_{j_1j_21})^2}{(a_{j_2j_2s_2} - a_{j_2j_2s_1})}\frac{1}{(x_{j_2} - a_{j_2j_2s_2})^2}\Big\}dx_{j_2}.\end{aligned}$$

With an initial condition of (x_{j_10}, x_{j_20}) at $t = t_0$, the integration of the above equation gives the first integral manifold for $a_{j_2j_2s} \neq a_{j_1j_21} (s = s_1,\ s_2)$ as

$$\begin{aligned}\ln\frac{|x_{j_1} - a_{j_1j_11}|}{|x_{j_10} - a_{j_1j_11}|} = & \frac{a_{j_1j_10}}{a_{j_2j_20}}\Big\{\frac{(a_{j_2j_2s_1} - a_{j_1j_21})^2}{(a_{j_2j_2s_1} - a_{j_2j_2s_2})^2}\ln\frac{|x_{j_2} - a_{j_2j_2s_1}|}{|x_{j_20} - a_{j_2j_2s_1}|} \\ & + \Big[\frac{2(a_{j_2j_2s_2} - a_{j_1j_21})}{(a_{j_2j_2s_2} - a_{j_2j_2s_1})} - \frac{(a_{j_2j_2s_2} - a_{j_1j_21})^2}{(a_{j_2j_2s_2} - a_{j_2j_2s_1})^2}\Big]\ln\frac{|x_{j_2} - a_{j_2j_2s_1}|}{|x_{j_20} - a_{j_2j_2s_1}|} \\ & - \frac{(a_{j_2j_2s_2} - a_{j_1j_21})^2}{(a_{j_2j_2s_2} - a_{j_2j_2s_1})}\Big(\frac{1}{x_{j_2} - a_{j_2j_2s_2}} - \frac{1}{x_{j_20} - a_{j_2j_2s_2}}\Big)\Big\}.\end{aligned}$$

(v_{1a}) For $a_{j_2j_2s_1} \neq a_{j_2j_2s_2} \neq a_{j_1j_21}$,at $(x_{j_1}^*, x_{j_2}^*) = (a_{j_1j_11}, a_{j_2j_2s_1})$, the variational equations are

$$\begin{aligned}\Delta\dot{x}_{j_2} &= a_{j_2j_20}\Delta x_{j_2}(a_{j_2j_2s_1} - a_{j_2j_2s_2})^2, \\ \Delta\dot{x}_{j_1} &= a_{j_1j_10}\Delta x_{j_1}(a_{j_2j_2s_1} - a_{j_1j_21})^2.\end{aligned}$$

Because of

$$(a_{j_2j_2s_1} - a_{j_2j_2s_2})^2 > 0 \text{ and } (a_{j_2j_21} - a_{j_1j_21})^2 > 0,$$

the flows at $(x^*_{j_1}, x^*_{j_2}) = (a_{j_1j_1}1, a_{j_2j_2s_1})$ in the x_{j_1}-direction are source and sink for $a_{j_1j_1}0(a_{j_2j_2s_1} - a_{j_1j_2}1)^2 > 0$ and $a_{j_1j_1}0(a_{j_2j_2s_1} - a_{j_1j_2}1)^2 < 0$, respectively; and the flows at $(x^*_{j_1}, x^*_{j_2}) = (a_{j_1j_1}1, a_{j_2j_2s_1})$ are source and sink for $a_{j_2j_2}0(a_{j_2j_2s_1} - a_{j_2j_2s_2})^2 > 0$ and $a_{j_2j_2}0(a_{j_2j_2s_1} - a_{j_2j_2s_2})^2 < 0$ in the x_{j_2}-direction, respectively.

Therefore, the equilibrium of $(x^*_{j_1}, x^*_{j_2}) = (a_{j_1j_1}1, a_{j_2j_2s_1})$ has the following properties as in Eqs. (1.58)–(1.61).

- For $a_{j_1j_1}0(a_{j_2j_2s_1} - a_{j_1j_2}1)^2 > 0$ and $a_{j_2j_2}0(a_{j_2j_2s_1} - a_{j_2j_2s_2})^2 > 0$,

$$(a_{j_1j_1}1, a_{j_2j_2s_1}) = \underbrace{(\text{SO, SO})}_{\text{source}}.$$

The equilibrium of $(x^*_{j_1}, x^*_{j_2}) = (a_{j_1j_1}1, a_{j_2j_2s_1})$ is an (SO,SO)-source.

- For $a_{j_1j_1}0(a_{j_2j_2s_1} - a_{j_1j_2}1)^2 < 0$ and $a_{j_2j_2}0(a_{j_2j_2s_1} - a_{j_2j_2s_2})^2 > 0$,

$$(a_{j_1j_1}1, a_{j_2j_2s_1}) = \underbrace{(\text{SI, SO})}_{\text{saddle}}.$$

The equilibrium of $(x^*_{j_1}, x^*_{j_2}) = (a_{j_1j_1}1, a_{j_2j_2s_1})$ is an (SI,SO)-saddle.

- For $a_{j_1j_1}0(a_{j_2j_2s_1} - a_{j_1j_2}1)^2 > 0$ and $a_{j_2j_2}0(a_{j_2j_2s_1} - a_{j_2j_2s_2})^2 < 0$,

$$(a_{j_1j_1}1, a_{j_2j_2s_1}) = \underbrace{(\text{SO, SI})}_{\text{saddle}}.$$

The equilibrium of $(x^*_{j_1}, x^*_{j_2}) = (a_{j_1j_1}1, a_{j_2j_2}1)$ is an (SO,SI)-saddle.

- For $a_{j_1j_1}0(a_{j_2j_2s_1} - a_{j_1j_2}1)^2 < 0$ and $a_{j_2j_2}0(a_{j_2j_2s_1} - a_{j_2j_2s_2})^2 < 0$,

$$(a_{j_1j_1}1, a_{j_2j_2s_1}) = \underbrace{(\text{SI, SI})}_{\text{sink}}.$$

The equilibrium of $(x^*_{j_1}, x^*_{j_2}) = (a_{j_1j_1}1, a_{j_2j_2s_1})$ is an (SI,SI)-sink.

($\mathrm{v}_{1\mathrm{b}}$) For $(x^*_{j_1}, x^*_{j_2}) = (a_{j_1j_1}1, a_{j_2j_2}1)$, the variational equations are

$$\begin{aligned}
\Delta\dot{x}_{j_2} &= a_{j_2j_2}0(a_{j_2j_2s_2} - a_{j_2j_2s_1})(\Delta x_{j_2})^2,\\
\Delta\dot{x}_{j_1} &= a_{j_1j_1}0(a_{j_2j_2s_2} - a_{j_1j_2}1)^2 \Delta x_{j_1}.
\end{aligned}$$

the flows at $(x^*_{j_1}, x^*_{j_2}) = (a_{j_1j_1}1, a_{j_2j_2}1)$ in the x_{j_1}-direction are source and sink for $a_{j_1j_1}0(a_{j_2j_2s_2} - a_{j_1j_2}1)^2 > 0$ and $a_{j_1j_1}0(a_{j_2j_2s_2} - a_{j_1j_2}1)^2 < 0$, respectively. The flows at $(x^*_{j_1}, x^*_{j_2}) = (a_{j_1j_1}1, a_{j_2j_2}1)$ in the x_{j_2}-direction are upper-saddle and lower-saddle for $a_{j_2j_2}0(a_{j_2j_2s_2} - a_{j_2j_2s_1}) > 0$ and $a_{j_2j_2}0(a_{j_2j_2s_2} - a_{j_2j_2s_1}) < 0$, respectively.

Therefore, the equilibrium of $(x^*_{j_1}, x^*_{j_2}) = (a_{j_1j_1}1, a_{j_2j_2s_2})$ with $a_{j_2j_2}1 \neq a_{j_1j_2}1$ has the following properties as in Eqs. (1.62)–(1.65).

- For $a_{j_1j_1 0}(a_{j_2j_2s_2} - a_{j_1j_2 1})^2 > 0$ and $a_{j_2j_2 0}(a_{j_2j_2s_2} - a_{j_2j_2s_1}) > 0$,

$$(a_{j_1j_1 1}, a_{j_2j_2 1}) = \underbrace{(\mathrm{SO}, \mathrm{US})}_{\text{upper-saddle source}}.$$

The equilibrium of $(x^*_{j_1}, x^*_{j_2}) = (a_{j_1j_1 1}, a_{j_2j_2s_2})$ is an (SO,US)-upper-saddle source.

- For $a_{j_1j_1 0}(a_{j_2j_2s_2} - a_{j_1j_2 1})^2 < 0$ and $a_{j_2j_2 0}(a_{j_2j_2s_2} - a_{j_2j_2s_1}) > 0$,

$$(a_{j_1j_1 1}, a_{j_2j_2 1}) = \underbrace{(\mathrm{SI}, \mathrm{US})}_{\text{upper-saddle sink}}.$$

The equilibrium of $(x^*_{j_1}, x^*_{j_2}) = (a_{j_1j_1 1}, a_{j_2j_2s_2})$ is an (SI,US)-upper-saddle sink.

- For $a_{j_1j_1 0}(a_{j_2j_2s_2} - a_{j_1j_2 1})^2 > 0$ and $a_{j_2j_2 0}(a_{j_2j_2s_2} - a_{j_2j_2s_1}) < 0$,

$$(a_{j_1j_1 1}, a_{j_2j_2 1}) = \underbrace{(\mathrm{SO}, \mathrm{LS})}_{\text{lower-saddle source}}.$$

The equilibrium of $(x^*_{j_1}, x^*_{j_2}) = (a_{j_1j_1 1}, a_{j_2j_2s_2})$ is an (SO,LS)-lower-saddle source.

- For $a_{j_1j_1 0}(a_{j_2j_2s_2} - a_{j_1j_2 1})^2 < 0$ and $a_{j_2j_2 0}(a_{j_2j_2s_2} - a_{j_2j_2s_1}) < 0$,

$$(a_{j_1j_1 1}, a_{j_2j_2 1}) = \underbrace{(\mathrm{SI}, \mathrm{LS})}_{\text{lower-saddle sink}}.$$

The equilibrium of $(x^*_{j_1}, x^*_{j_2}) = (a_{j_1j_1 1}, a_{j_2j_2s_2})$ is an (SI,LS)-lower-saddle sink.

– The saddle-source and saddle-sink are the appearing bifurcations of saddle-source and saddle-sink.

(v_{1c}) In phase space, at $x^*_{j_2} = a_{j_1j_2 1} \neq a_{j_2j_2s_1} \neq a_{j_2j_2s_2}$ with $\bar{x}_{j_1} \neq a_{j_1j_1 1}$,

$$\left.\frac{dx_{j_1}}{dx_{j_2}}\right|_{x^*_{j_2} = a_{j_1j_2 1}} = \left.\frac{a_{j_1j_1 0}}{a_{j_2j_2 0}} \frac{(\bar{x}_{j_1} - a_{j_1j_1 1})(x_{j_2} - a_{j_1j_2 1})^2}{(x_{j_2} - a_{j_2j_2s_1})(x_{j_2} - a_{j_2j_2s_2})^2}\right|_{x^*_{j_2} = a_{j_1j_2 1}} = 0,$$

$$\left.\frac{d^2x_{j_1}}{dx^2_{j_2}}\right|_{x^*_{j_2} = a_{j_1j_2 1}} = 0.$$

If

$$\left.\frac{d^3x_{j_1}}{dx_{j_2}^3}\right|_{x_{j_2}^*=a_{j_1j_2}1}=\frac{a_{j_1j_1}0}{a_{j_2j_2}0}\frac{2(\overline{x}_{j_1}-a_{j_1j_1}1)}{(a_{j_1j_2}1-a_{j_2j_2s_1})(a_{j_1j_2}1-a_{j_2j_2s_2})^2}>0,$$

there is an increasing-inflection flow at $x_{j_2}^*=a_{j_1j_2}1$ with $\overline{x}_{j_1}\neq a_{j_1j_1}1$. If

$$\left.\frac{d^3x_{j_1}}{dx_{j_2}^3}\right|_{x_{j_2}^*=a_{j_1j_2}1}=\frac{a_{j_1j_1}0}{a_{j_2j_2}0}\frac{2(\overline{x}_{j_1}-a_{j_1j_1}1)}{(a_{j_1j_2}1-a_{j_2j_2s_1})(a_{j_1j_2}1-a_{j_2j_2s_2})^2}<0,$$

there is a decreasing-inflection flow at $x_{j_2}^*=a_{j_1j_2}1$ with $\overline{x}_{j_1}\neq a_{j_1j_1}1$. Let

$$\dot{x}_{j_2}=a_{j_2j_2}0(a_{j_1j_2}1-a_{j_2j_2s_1})(a_{j_1j_2}1-a_{j_2j_2s_2})^2\neq 0.$$

The inflection flows at $x_{j_2}^*=a_{j_1j_2}1$ with $\overline{x}_{j_1}\neq a_{j_1j_1}1$ are positive and negative for $\dot{x}_{j_2}>0$ and $\dot{x}_{j_2}<0$, respectively.

Thus, the equilibrium of $x_{j_2}^*=a_{j_1j_2}1$ with $\overline{x}_{j_1}\neq a_{j_1j_1}1$ has the following properties.

- For $a_{j_1j_1}0(\overline{x}_{j_1}-a_{j_1j_1}1)>0$ and $a_{j_2j_2}0(a_{j_1j_2}1-a_{j_2j_2s_1})(a_{j_1j_2}1-a_{j_2j_2s_2})^2>0$,

$$(a_{j_1j_2}1,\dot{x}_{j_2})=\underbrace{(\text{II},\text{pF})}_{\text{increasing-inflection flow }(+)}.$$

- For $a_{j_1j_1}0(\overline{x}_{j_1}-a_{j_1j_1}1)<0$ and $a_{j_2j_2}0(a_{j_1j_2}1-a_{j_2j_2s_1})(a_{j_1j_2}1-a_{j_2j_2s_2})^2>0$,

$$(a_{j_1j_2}1,\dot{x}_{j_2})=\underbrace{(\text{DI},\text{pF})}_{\text{decreasing-inflection flow }(+)}.$$

- For $a_{j_1j_1}0(\overline{x}_{j_1}-a_{j_1j_1}1)>0$ and $a_{j_2j_2}0(a_{j_1j_2}1-a_{j_2j_2s_1})(a_{j_1j_2}1-a_{j_2j_2s_2})^2<0$,

$$(a_{j_1j_2}1,\dot{x}_{j_2})=\underbrace{(\text{DI},\text{nF})}_{\text{decreasing-inflection flow }(-)}.$$

- For $a_{j_1j_1}0(\overline{x}_{j_1}-a_{j_1j_1}1)<0$ and $a_{j_2j_2}0(a_{j_1j_2}1-a_{j_2j_2s_1})(a_{j_1j_2}1-a_{j_2j_2s_2})^2<0$,

$$(a_{j_1j_2}1,\dot{x}_{j_2})=\underbrace{(\text{II},\text{nF})}_{\text{increasing-inflection flow }(-)}.$$

The variational equation at $x_{j_1}^*=a_{j_1j_1}1$ with $\overline{x}_{j_2}\neq a_{j_1j_2}1$ is

$$\Delta\dot{x}_{j_1} = a_{j_1j_10}(\overline{x}_{j_2} - a_{j_1j_21})^2 \Delta x_{j_1}$$

and

$$\dot{x}_{j_2} = a_{j_2j_20}(\overline{x}_{j_2} - a_{j_2j_2s_1})(\overline{x}_{j_2} - a_{j_2j_2s_2})^2.$$

Thus, the equilibrium of $x^*_{j_2} = a_{j_1j_21}$ with $\overline{x}_{j_1} \neq a_{j_1j_11}$ has the following properties.

- For $a_{j_1j_10}(\overline{x}_{j_2} - a_{j_1j_21})^2 > 0$ and $a_{j_2j_20}(\overline{x}_{j_2} - a_{j_2j_2s_1})(\overline{x}_{j_2} - a_{j_2j_2s_2})^2 > 0$,

$$(a_{j_1j_11}, \dot{x}_{j_2}) = \underbrace{(\text{SO}, \text{pF})}_{\text{source flow }(+)}.$$

- For $a_{j_1j_10}(\overline{x}_{j_2} - a_{j_1j_21})^2 < 0$ and $a_{j_2j_20}(\overline{x}_{j_2} - a_{j_2j_2s_1})(\overline{x}_{j_2} - a_{j_2j_2s_2})^2 > 0$,

$$(a_{j_1j_11}, \dot{x}_{j_2}) = \underbrace{(\text{SI}, \text{pF})}_{\text{sink flow }(+)}.$$

- For $a_{j_1j_10}(\overline{x}_{j_2} - a_{j_1j_21})^2 > 0$ and $a_{j_2j_20}(\overline{x}_{j_2} - a_{j_2j_2s_1})(\overline{x}_{j_2} - a_{j_2j_2s_2})^2 < 0$,

$$(a_{j_1j_11}, \dot{x}_{j_2}) = \underbrace{(\text{SO}, \text{nF})}_{\text{source flow }(-)}.$$

- For $a_{j_1j_10}(\overline{x}_{j_2} - a_{j_1j_21})^2 < 0$ and $a_{j_2j_20}(\overline{x}_{j_2} - a_{j_2j_2s_1})(\overline{x}_{j_2} - a_{j_2j_2s_2})^2 < 0$,

$$(a_{j_1j_11}, \dot{x}_{j_2}) = \underbrace{(\text{SI}, \text{nF})}_{\text{sink flow }(-)}.$$

From the foregoing analysis, the equilibrium of $(x^*_{j_1}, x^*_{j_2}) = (a_{j_1j_11}, a_{j_1j_21})$ with $a_{j_2j_21} \neq a_{j_1j_21}$ has the following properties as in Eqs. (1.66)–(1.69).

- For $a_{j_1j_10} > 0$ and $a_{j_2j_20}(a_{j_1j_21} - a_{j_2j_2s_1})(a_{j_1j_21} - a_{j_2j_2s_2})^2 > 0$,

$$(a_{j_1j_11}, a_{j_1j_21}) = \underbrace{({}_{\text{DI:II}}\text{SO}, \text{pF})}_{\text{inflection-source flow}(+)}.$$

The equilibrium of $(x^*_{j_1}, x^*_{j_2}) = (a_{j_1j_11}, a_{j_1j_21})$ is a $({}_{\text{DI:II}}\text{SO}, \text{pF})$-positive inflection-source flow.

- For $a_{j_1j_10} < 0$ and $a_{j_2j_20}(a_{j_1j_21} - a_{j_2j_2s_1})(a_{j_1j_21} - a_{j_2j_2s_2})^2 > 0$,

$$(a_{j_1j_11}, a_{j_1j_21}) = \underbrace{({}_{\text{II:DI}}\text{SO, pF})}_{\text{inflection-sink flow }(+)}.$$

The equilibrium of $(x_{j_1}^*, x_{j_2}^*) = (a_{j_1j_11}, a_{j_1j_21})$ is an (${}_{\text{II:DI}}$SI, pF)-positive inflection-sink flow.

- For $a_{j_1j_10} > 0$ and $a_{j_2j_20}(a_{j_1j_21} - a_{j_2j_2s_1})(a_{j_1j_21} - a_{j_2j_2s_2})^2 < 0$,

$$(a_{j_1j_11}, a_{j_1j_21}) = \underbrace{({}_{\text{II:DI}}\text{SO, nF})}_{\text{inflection-source flow }(-)}.$$

The equilibrium of $(x_{j_1}^*, x_{j_2}^*) = (a_{j_1j_11}, a_{j_1j_21})$ is an (${}_{\text{II:DI}}$SO, nF)-negative inflection-source flow.

- For $a_{j_1j_10} < 0$ and $a_{j_2j_20}(a_{j_1j_21} - a_{j_2j_2s_1})(a_{j_1j_21} - a_{j_2j_2s_2})^2 < 0$,

$$(a_{j_1j_11}, a_{j_1j_21}) = \underbrace{({}_{\text{DI:II}}\text{SI, nF})}_{\text{inflection-sink flow }(-)}.$$

The equilibrium of $(x_{j_1}^*, x_{j_2}^*) = (a_{j_1j_11}, a_{j_1j_21})$ is a (${}_{\text{DI:II}}$SI, nF)-negative inflection-sink flow.

(v$_2$) For $a_{j_2j_2s_1} = a_{j_2j_2s_2} = a_{j_2j_21}$, the equation becomes

$$\begin{aligned}\dot{x}_{j_2} &= a_{j_2j_20}(x_{j_2} - a_{j_2j_21})^3, \\ \dot{x}_{j_1} &= a_{j_1j_10}(x_{j_1} - a_{j_1j_11})(x_{j_2} - a_{j_1j_21})^2.\end{aligned}$$

In phase space,

$$\frac{dx_{j_1}}{dx_{j_2}} = \frac{a_{j_1j_10}}{a_{j_2j_20}} \frac{(x_{j_1} - a_{j_1j_11})(x_{j_2} - a_{j_1j_21})^2}{(x_{j_2} - a_{j_2j_21})^3},$$

and the deformation of the foregoing equation is

$$\frac{dx_{j_1}}{x_{j_1} - a_{j_1j_11}} = \frac{a_{j_1j_10}}{a_{j_2j_20}} \left[\frac{1}{x_{j_2} - a_{j_2j_21}} + \frac{2(a_{j_2j_21} - a_{j_1j_21})}{(x_{j_2} - a_{j_2j_21})^2} + \frac{(a_{j_2j_21} - a_{j_1j_21})^2}{(x_{j_2} - a_{j_2j_21})^3}\right] dx_{j_2}.$$

With an initial condition of (x_{j_10}, x_{j_20}) at $t = t_0$, the integration of the above equation gives the first integral manifold as

$$\ln\frac{|x_{j_1}-a_{j_1j_11}|}{|x_{j_10}-a_{j_1j_11}|}=\frac{a_{j_1j_10}}{a_{j_2j_20}}\left\{\ln\frac{|x_{j_2}-a_{j_2j_21}|}{|x_{j_20}-a_{j_2j_21}|}\right.$$
$$-2(a_{j_2j_21}-a_{j_1j_21})(\frac{1}{x_{j_2}-a_{j_2j_21}}-\frac{1}{x_{j_20}-a_{j_2j_21}})$$
$$\left.-\frac{1}{2}(a_{j_2j_21}-a_{j_1j_21})^2\left[\frac{1}{(x_{j_2}-a_{j_2j_21})^2}-\frac{1}{(x_{j_20}-a_{j_2j_21})^2}\right]\right\}.$$

For $a_{j_2j_2s_1}=a_{j_2j_2s_2}=a_{j_2j_21}$, at $(x^*_{j_1},x^*_{j_2})=(a_{j_1j_11},a_{j_2j_21})$, the variational equations are

$$\Delta\dot{x}_{j_2}=a_{j_2j_20}(\Delta x_{j_2})^3,$$
$$\Delta\dot{x}_{j_1}=a_{j_1j_10}\Delta x_{j_1}(a_{j_2j_21}-a_{j_1j_21})^2.$$

Because of

$$(a_{j_2j_21}-a_{j_1j_21})^2>0,$$

the flows at $(x^*_{j_1},x^*_{j_2})=(a_{j_1j_11},a_{j_2j_2s_1})$ in the x_{j_1}-direction are source and sink for $a_{j_1j_10}(a_{j_2j_21}-a_{j_1j_21})^2>0$ and $a_{j_1j_10}(a_{j_2j_21}-a_{j_1j_21})^2<0$, respectively; and the flows at $(x^*_{j_1},x^*_{j_2})=(a_{j_1j_11},a_{j_2j_2s_1})$ in the x_{j_2}-direction are third-order source and sink for $a_{j_2j_20}>0$ and $a_{j_2j_20}<0$, respectively.

Therefore, the equilibrium of $(x^*_{j_1},x^*_{j_2})=(a_{j_1j_11},a_{j_2j_21})$ has the following properties as in Eqs. (1.72)–(1.75).

- For $a_{j_1j_10}(a_{j_2j_21}-a_{j_1j_21})^2>0$ and $a_{j_2j_20}>0$,

$$(a_{j_1j_11},a_{j_2j_21})=\underbrace{(\text{SO},3^{\text{rd}}\text{SO})}_{\text{third-order source}}.$$

  The equilibrium of $(x^*_{j_1},x^*_{j_2})=(a_{j_1j_11},a_{j_2j_21})$ is an (SO,3rdSO)-third-order source, which is for an appearing bifurcation from source to source, saddle, and source.
- For $a_{j_1j_10}(a_{j_2j_21}-a_{j_1j_21})^2<0$ and $a_{j_2j_20}>0$,

$$(a_{j_1j_11},a_{j_2j_21})=\underbrace{(\text{SI},3^{\text{rd}}\text{SO})}_{\text{third-order saddle}}.$$

  The equilibrium of $(x^*_{j_1},x^*_{j_2})=(a_{j_1j_11},a_{j_2j_21})$ is an (SI,3rdSO)-third-order saddle, which is for an appearing bifurcation from saddle to saddle, sink, and saddle.

- For $a_{j_1j_10}(a_{j_2j_21}-a_{j_1j_21})^2>0$ and $a_{j_2j_20}<0$,

$$(a_{j_1j_11},a_{j_2j_21})=\underbrace{(\mathrm{SO},3^{\mathrm{rd}}\mathrm{SI})}_{\text{third-order saddle}}.$$

The equilibrium of $(x_{j_1}^*,x_{j_2}^*)=(a_{j_1j_11},a_{j_2j_21})$ is an (SO,3rdSI)-third-order saddle, which is for an appearing bifurcation from saddle to saddle, source, and saddle.

- For $a_{j_1j_10}(a_{j_2j_21}-a_{j_1j_21})^2<0$ and $a_{j_2j_20}<0$,

$$(a_{j_1j_11},a_{j_2j_21})=\underbrace{(\mathrm{SI},3^{\mathrm{rd}}\mathrm{SI})}_{\text{third-order sink}}.$$

The equilibrium of $(x_{j_1}^*,x_{j_2}^*)=(a_{j_1j_11},a_{j_2j_21})$ is an (SI,3rdSI)-third-order sink, which is for an appearing bifurcation from sink to sink, saddle, and sink.

(iv$_3$) For $a_{j_2j_2s_1}=a_{j_1j_21}$, in phase space,

$$\frac{dx_{j_1}}{dx_{j_2}}=\frac{a_{j_1j_10}}{a_{j_2j_20}}\frac{(x_{j_1}-a_{j_1j_11})(x_{j_2}-a_{j_1j_21})}{(x_{j_2}-a_{j_2j_2s_2})^2},$$

and the deformation of the foregoing equation is

$$\frac{dx_{j_1}}{x_{j_1}-a_{j_1j_11}}=\frac{a_{j_1j_10}}{a_{j_2j_20}}\frac{x_{j_2}-a_{j_2j_2s_2}+a_{j_2j_2s_2}-a_{j_1j_21}}{(x_{j_2}-a_{j_2j_2s_2})^2}dx_{j_2}.$$

With an initial condition of (x_{j_10},x_{j_20}) at $t=t_0$, the integration of the above equation gives the first integral manifold as

$$\begin{aligned}\ln\frac{|x_{j_1}-a_{j_1j_11}|}{|x_{j_10}-a_{j_1j_11}|}=\frac{a_{j_1j_10}}{a_{j_2j_20}}\Bigg[&\ln\frac{|x_{j_2}-a_{j_2j_2s_2}|}{|x_{j_20}-a_{j_2j_2s_2}|}\\&-(a_{j_2j_2s_2}-a_{j_1j_21})(\frac{1}{x_{j_2}-a_{j_2j_2s_2}}-\frac{1}{x_{j_20}-a_{j_2j_2s_2}})\Bigg].\end{aligned}$$

(v$_{3a}$) In phase space, at $x_{j_2}^*=a_{j_1j_21}$ with $\overline{x}_{j_1}\neq a_{j_1j_11}$,

$$\left.\frac{dx_{j_1}}{dx_{j_2}}\right|_{x_{j_2}^*=a_{j_1j_21}}=\left.\frac{a_{j_1j_10}}{a_{j_2j_20}}\frac{(\overline{x}_{j_1}-a_{j_1j_11})(x_{j_2}-a_{j_1j_21})}{(x_{j_2}-a_{j_2j_2s_2})^2}\right|_{x_{j_2}^*=a_{j_1j_21}}=0.$$

If

$$\left.\frac{d^2x_{j_1}}{dx_{j_2}^2}\right|_{x_{j_2}^*=a_{j_1j_2}1} = \frac{a_{j_1j_1}0}{a_{j_2j_2}0}\frac{(\overline{x}_{j_1}-a_{j_1j_1}1)}{(a_{j_1j_2}1-a_{j_2j_2s_2})^2} > 0,$$

there is an up-parabola flow at $x_{j_2}^* = a_{j_1j_2}1$ with $\overline{x}_{j_1} \neq a_{j_1j_1}1$. If

$$\left.\frac{d^2x_{j_1}}{dx_{j_2}^2}\right|_{x_{j_2}^*=a_{j_1j_2}1} = \frac{a_{j_1j_1}0}{a_{j_2j_2}0}\frac{(\overline{x}_{j_1}-a_{j_1j_1}1)}{(a_{j_1j_2}1-a_{j_2j_2s_2})^2} < 0,$$

there is a down-parabola flow at $x_{j_2}^* = a_{j_1j_2}1$ with $\overline{x}_{j_1} \neq a_{j_1j_1}1$.

The variational equation for $a_{j_2j_2s_1} \neq a_{j_2j_2s_2}$ is

$$\Delta\dot{x}_{j_2} = a_{j_2j_2}0(a_{j_2j_2s_1} - a_{j_2j_2s_2})^2\Delta x_{j_2}$$

and

$$(a_{j_2j_2s_1} - a_{j_2j_2s_2})^2 > 0.$$

The parabola flows for $x_{j_2}^* = a_{j_1j_2}1$ with $\overline{x}_{j_1} \neq a_{j_1j_1}1$ in the x_{j_2}-direction are source and sink for $a_{j_2j_2}0(a_{j_2j_2s_1} - a_{j_2j_2s_2})^2 > 0$ and $a_{j_2j_2}0(a_{j_2j_2s_1} - a_{j_2j_2s_2})^2 < 0$, respectively.

Therefore, the infinite-equilibrium of $x_{j_2}^* = a_{j_1j_2}1 = a_{j_2j_2s_1}$ with $\overline{x}_{j_1} \neq a_{j_1j_1}1$ has the following properties as in Eqs. (1.77)–(1.80).

- For $a_{j_1j_1}0(\overline{x}_{j_1} - a_{j_1j_1}1) > 0$ and $a_{j_2j_2}0(a_{j_2j_2s_1} - a_{j_2j_2s_2})^2 > 0$,

$$(\overline{x}_{j_1}, a_{j_2j_2s_1}) = \underbrace{(\text{UP, SO})}_{\text{up-parabola source}}.$$

  The infinite-equilibrium of $x_{j_2}^* = a_{j_1j_2}1 = a_{j_2j_2s_1}$ is a (UP,SO)-up-parabola source.
- For $a_{j_1j_1}0(\overline{x}_{j_1} - a_{j_1j_1}1) < 0$ and $a_{j_2j_2}0(a_{j_2j_2s_1} - a_{j_2j_2s_2})^2 > 0$,

$$(\overline{x}_{j_1}, a_{j_2j_2s_1}) = \underbrace{(\text{DP, SO})}_{\text{down-parabols source}}.$$

  The infinite-equilibrium of $x_{j_2}^* = a_{j_1j_2}1 = a_{j_2j_2s_1}$ is a (DP,SO)-down-parabola source.
- For $a_{j_1j_1}0(\overline{x}_{j_1} - a_{j_1j_1}1) > 0$ and $a_{j_2j_2}0(a_{j_2j_2s_1} - a_{j_2j_2s_2})^2 < 0$,

$$(\overline{x}_{j_1}, a_{j_2j_2s_1}) = \underbrace{(\text{DP, SI})}_{\text{down-parabola sink}}.$$

The infinite-equilibrium of $x^*_{j_2} = a_{j_1 j_2 1} = a_{j_2 j_2 s_1}$ is a (DP,SI)-down-parabola sink.

- For $a_{j_1 j_1 0}(\bar{x}_{j_1} - a_{j_1 j_1 1}) < 0$ and $a_{j_2 j_2 0}(a_{j_2 j_2 s_1} - a_{j_2 j_2 s_2})^2 < 0$,

$$(\bar{x}_{j_1}, a_{j_2 j_2 s_1}) = \underbrace{(\text{UP}, \text{SI})}_{\text{up-parabola sink}}.$$

The infinite-equilibrium of $x^*_{j_2} = a_{j_1 j_2 1} = a_{j_2 j_2 s_1}$ is a (UP,SI)-up-parabola sink.

(v_{3b}) From (iv_{2a}), for $\bar{x}_{j_1} = a_{j_1 j_1 1}$, the equilibrium of $(x^*_{j_1}, x^*_{j_2}) = (a_{j_1 j_1 1}, a_{j_2 j_2 s_1})$ with $a_{j_1 j_2 1} = a_{j_2 j_2 1}$ has the following properties as in Eqs. (1.81)–(1.84).

- For $a_{j_1 j_1 0} > 0$ and $a_{j_2 j_2 0}(a_{j_2 j_2 s_1} - a_{j_2 j_2 s_2})^2 > 0$,

$$(a_{j_1 j_1 1}, a_{j_2 j_2 s_1}) = \underbrace{(\text{DP} : \text{UP}, \text{SO})}_{\text{hyperbolic source}}.$$

The equilibrium of $(x^*_{j_1}, x^*_{j_2}) = (a_{j_1 j_1 1}, a_{j_2 j_2 1})$ is a (DP:UP,SO)-hyperbolic source.

- For $a_{j_1 j_1 0} < 0$ and $a_{j_2 j_2 0}(a_{j_2 j_2 s_1} - a_{j_2 j_2 s_2})^2 > 0$,

$$(a_{j_1 j_1 1}, a_{j_2 j_2 1}) = \underbrace{(\text{UP} : \text{DP}, \text{SO})}_{\text{hyperbolic-secant source}}.$$

The equilibrium of $(x^*_{j_1}, x^*_{j_2}) = (a_{j_1 j_1 1}, a_{j_2 j_2 1})$ is a (UP:DP,SO)-hyperbolic-secant source.

- For $a_{j_1 j_1 0} > 0$ and $a_{j_2 j_2 0}(a_{j_2 j_2 s_1} - a_{j_2 j_2 s_2})^2 < 0$,

$$(a_{j_1 j_1 1}, a_{j_2 j_2 1}) = \underbrace{(\text{UP} : \text{DP}, \text{SI})}_{\text{hyperbolic-secant sink}}.$$

The equilibrium of $(x^*_{j_1}, x^*_{j_2}) = (a_{j_1 j_1 1}, a_{j_2 j_2 1})$ is a (UP:DP,SI)-hyperbolic-secant sink.

- For $a_{j_1 j_1 0} < 0$ and $a_{j_2 j_2 0}(a_{j_2 j_2 s_1} - a_{j_2 j_2 s_2})^2 < 0$,

$$(a_{j_1 j_1 1}, a_{j_2 j_2 1}) = \underbrace{(\text{DP} : \text{UP}, \text{SI})}_{\text{hyperbolic sink}}.$$

The equilibrium of $(x^*_{j_1}, x^*_{j_2}) = (a_{j_1 j_1 1}, a_{j_2 j_2 1})$ is a (DP:UP,SI)-hyperbolic sink.

(v_4) For $a_{j_2 j_2 s_2} = a_{j_1 j_2 1}$, we have

$$\frac{dx_{j_1}}{dx_{j_2}} = \frac{a_{j_1 j_1 0}}{a_{j_2 j_2 0}} \frac{(x_{j_1} - a_{j_1 j_1 1})}{(x_{j_2} - a_{j_2 j_2 s_1})},$$

and the deformation of the foregoing equation is

$$\frac{1}{(x_{j_1}-a_{j_1j_11})}dx_{j_1}=\frac{a_{j_1j_10}}{a_{j_2j_20}}\frac{1}{(x_{j_2}-a_{j_2j_2s_1})}dx_{j_2}.$$

With an initial condition of (x_{j_10},x_{j_20}) at $t=t_0$, the integration of the above equation gives the first integral manifold for $a_{j_2j_2s_2}=a_{j_1j_21}$ as

$$\ln\frac{|x_{j_1}-a_{j_1j_11}|}{|x_{j_10}-a_{j_1j_11}|}=\frac{a_{j_1j_10}}{a_{j_2j_20}}\ln\frac{|x_{j_2}-a_{j_2j_2s_1}|}{|x_{j_20}-a_{j_2j_2s_1}|}.$$

(v_{2a}) In phase space, at $x_{j_2}^*=a_{j_2j_2s_2}=a_{j_1j_21}$ with $\bar{x}_{j_1}\neq a_{j_1j_11}$, if

$$\left.\frac{dx_{j_1}}{dx_{j_2}}\right|_{x_{j_2}^*=a_{j_1j_21}}=\frac{a_{j_1j_10}}{a_{j_2j_20}}\frac{(\bar{x}_{j_1}-a_{j_1j_11})}{(a_{j_2j_2s_2}-a_{j_2j_2s_1})}>0,$$

there is an increasing-inflection flow at $x_{j_2}^*=a_{j_2j_2s_2}=a_{j_1j_21}$ with $\bar{x}_{j_1}\neq a_{j_1j_11}$. If

$$\left.\frac{dx_{j_1}}{dx_{j_2}}\right|_{x_{j_2}^*=a_{j_1j_21}}=\frac{a_{j_1j_10}}{a_{j_2j_20}}\frac{(\bar{x}_{j_1}-a_{j_1j_11})}{(a_{j_2j_2s_2}-a_{j_2j_2s_1})}<0,$$

there is a decreasing-inflection flow at $x_{j_2}^*=a_{j_1j_21}$ with $\bar{x}_{j_1}\neq a_{j_1j_11}$. the variational equation is

$$\Delta\dot{x}_{j_2}=a_{j_2j_20}(a_{j_2j_2s_2}-a_{j_2j_2s_1})(\Delta x_{j_2})^2.$$

The inflection flows at $x_{j_2}^*=a_{j_1j_21}$ with $\bar{x}_{j_1}\neq a_{j_1j_11}$ are upper-saddle and lower-saddle for $a_{j_2j_20}(a_{j_2j_2s_2}-a_{j_2j_2s_1})>0$ and $a_{j_2j_20}(a_{j_2j_2s_2}-a_{j_2j_2s_1})<0$, respectively.

Therefore, the infinite-equilibrium of $x_{j_2}^*=a_{j_1j_21}=a_{j_2j_2s_2}$ with $\bar{x}_{j_1}\neq a_{j_1j_11}$ has the following properties as in Eqs. (1.86)–(1.89).

- For $a_{j_1j_10}(\bar{x}_{j_1}-a_{j_1j_11})>0$ and $a_{j_2j_20}(a_{j_2j_2s_2}-a_{j_2j_2s_1})>0$,

$$(\bar{x}_{j_1},a_{j_1j_21})=\underbrace{(\text{II, US})}_{\text{increasing-inflection upper-saddle}}.$$

The infinite-equilibrium of $x_{j_2}^*=a_{j_1j_21}=a_{j_2j_2s_2}$ is an (II,US)-increasing-inflection upper-saddle.

- For $a_{j_1j_10}(\bar{x}_{j_1}-a_{j_1j_11})<0$ and $a_{j_2j_20}(a_{j_2j_2s_2}-a_{j_2j_2s_1})>0$,

$$(\bar{x}_{j_1},a_{j_1j_21})=\underbrace{(\text{DI, US})}_{\text{decreasing-inflection upper-saddle}}.$$

The infinite-equilibrium of $x_{j_2}^* = a_{j_1j_2 1} = a_{j_2j_2s_2}$ is a (DI,US)-decreasing-inflection upper-saddle.

- For $a_{j_1j_1 0}(\overline{x}_{j_1} - a_{j_1j_1 1}) > 0$ and $a_{j_2j_2 0}(a_{j_2j_2s_2} - a_{j_2j_2s_1}) < 0$,

$$(\overline{x}_{j_1}, a_{j_1j_2 1}) = \underbrace{(\text{DI}, \text{LS})}_{\text{decreasing-inflection lower-saddle}} .$$

The infinite-equilibrium of $x_{j_2}^* = a_{j_1j_2 1} = a_{j_2j_2s_2}$ is a (DI,LS)-decreasing-inflection lower-saddle.

- For $a_{j_1j_1 0}(\overline{x}_{j_1} - a_{j_1j_1 1}) < 0$ and $a_{j_2j_2 0}(a_{j_2j_2s_2} - a_{j_2j_2s_1}) < 0$,

$$(\overline{x}_{j_1}, a_{j_1j_2 1}) = \underbrace{(\text{II}, \text{LS})}_{\text{increasing-inflection lower-saddle}} .$$

The infinite-equilibrium of $x_{j_2}^* = a_{j_1j_2 1} = a_{j_2j_2s_2}$ is an (II,LS)-increasing-inflection lower-saddle.

(v_{4b}) From (v_{4a}), for $\overline{x}_{j_1} = a_{j_1j_1 1}$, the equilibrium of $(x_{j_1}^*, x_{j_2}^*) = (a_{j_1j_1 1}, a_{j_2j_2s_2})$ with $a_{j_1j_2 1} = a_{j_2j_2s_2}$ has the following properties as in Eqs. (1.90)–(1.94).

- For $a_{j_1j_1 0} > 0$ and $a_{j_2j_2 0}(a_{j_2j_2s_2} - a_{j_2j_2s_1}) > 0$,

$$(a_{j_1j_1 1}, a_{j_2j_2s_2}) = \underbrace{(\text{DI} : \text{II}, \text{US})}_{\text{inflection upper-saddle}} .$$

The equilibrium of $(x_{j_1}^*, x_{j_2}^*) = (a_{j_1j_1 1}, a_{j_2j_2s_2})$ is a (DI:II,US)-inflection upper-saddle.

- For $a_{j_1j_1 0} < 0$ and $a_{j_2j_2 0}(a_{j_2j_2s_2} - a_{j_2j_2s_1}) > 0$,

$$(a_{j_1j_1 1}, a_{j_2j_2s_2}) = \underbrace{(\text{II} : \text{DI}, \text{US})}_{\text{inflection upper-saddle}} .$$

The equilibrium of $(x_{j_1}^*, x_{j_2}^*) = (a_{j_1j_1 1}, a_{j_2j_2s_2})$ is an (II:DI,US)-inflection upper-saddle.

- For $a_{j_1j_1 0} > 0$ and $a_{j_2j_2 0}(a_{j_2j_2s_2} - a_{j_2j_2s_1}) < 0$,

$$(a_{j_1j_1 1}, a_{j_2j_2s_2}) = \underbrace{(\text{II} : \text{DI}, \text{LS})}_{\text{inflection lower-saddle}} .$$

The equilibrium of $(x_{j_1}^*, x_{j_2}^*) = (a_{j_1j_1 1}, a_{j_2j_2s_2})$ is an (II:DI,LS)-inflection lower-saddle.

- For $a_{j_1j_10}<0$ and $a_{j_2j_20}(a_{j_2j_2s_2}-a_{j_2j_2s_1})<0$,

$$(a_{j_1j_11},a_{j_2j_2s_2})=\underbrace{(\mathrm{DI:II,LS})}_{\text{inflection lower-saddle}}.$$

The equilibrium of $(x_{j_1}^*,x_{j_2}^*)=(a_{j_1j_11},a_{j_2j_2s_2})$ is a (DI:II,LS)-inflection lower-saddle.

(v_5) For $a_{j_2j_2s_1}=a_{j_2j_2s_2}=a_{j_2j_21}=a_{j_1j_21}$, the equation becomes

$$\begin{aligned}\dot{x}_{j_2}&=a_{j_2j_20}(x_{j_2}-a_{j_2j_21})^3,\\ \dot{x}_{j_1}&=a_{j_1j_10}(x_{j_1}-a_{j_1j_11})(x_{j_2}-a_{j_1j_21})^2.\end{aligned}$$

In phase space,

$$\frac{dx_{j_1}}{dx_{j_2}}=\frac{a_{j_1j_10}}{a_{j_2j_20}}\frac{(x_{j_1}-a_{j_1j_11})}{(x_{j_2}-a_{j_2j_21})},$$

and the deformation of the foregoing equation is

$$\frac{dx_{j_1}}{x_{j_1}-a_{j_1j_11}}=\frac{a_{j_1j_10}}{a_{j_2j_20}}\frac{1}{x_{j_2}-a_{j_2j_21}}dx_{j_2}.$$

With an initial condition of (x_{j_10},x_{j_20}) at $t=t_0$, the integration of the above equation gives the first integral manifold as

$$\ln\frac{|x_{j_1}-a_{j_1j_11}|}{|x_{j_10}-a_{j_1j_11}|}=\frac{a_{j_1j_10}}{a_{j_2j_20}}\ln\frac{|x_{j_2}-a_{j_2j_21}|}{|x_{j_20}-a_{j_2j_21}|}.$$

(v_{5a}) For $x_{j_2}^*=a_{j_1j_21}=a_{j_2j_21}$ with $\overline{x}_{j_1}\neq a_{j_1j_11}$, in phase space,

$$\left.\frac{dx_{j_1}}{dx_{j_2}}\right|_{x_{j_2}^*=a_{j_2j_21}\pm\varepsilon}=\pm\frac{a_{j_1j_10}}{a_{j_2j_20}}\frac{(\overline{x}_{j_1}-a_{j_1j_11})}{\varepsilon}.$$

If

$$\frac{a_{j_1j_10}(\overline{x}_{j_1}-a_{j_1j_11})}{a_{j_2j_20}}>0,$$

there is an up-trumpet asymptotic flow at $x_{j_2}^*=a_{j_1j_21}=a_{j_2j_21}$ with $\overline{x}_{j_1}\neq a_{j_1j_11}$. If

$$\frac{a_{j_1 j_1 0}(\overline{x}_{j_1} - a_{j_1 j_1 1})}{a_{j_2 j_2 0}} < 0,$$

there is a down-trumpet asymptotic flow at $x_{j_2}^* = a_{j_1 j_2 1} = a_{j_2 j_2 1}$ with $\overline{x}_{j_1} \neq a_{j_1 j_1 1}$.

The variational equation at $x_{j_2}^* = a_{j_1 j_2 1} = a_{j_2 j_2 1}$ is

$$\Delta \dot{x}_{j_2} = a_{j_2 j_2 0}(\Delta x_{j_2})^3.$$

The trumpet asymptotic flows are the third-order source and sink for $a_{j_2 j_2 0} > 0$ and $a_{j_2 j_2 0} < 0$, respectively.

Therefore, the infinite-equilibrium of $x_{j_2}^* = a_{j_1 j_2 1} = a_{j_2 j_2 1}$ with $\overline{x}_{j_1} \neq a_{j_1 j_1 1}$ has the following properties as in Eqs. (1.95)–(1.98).

- For $a_{j_1 j_1 0}(\overline{x}_{j_1} - a_{j_1 j_1 1}) > 0$ and $a_{j_2 j_2 0} > 0$,

$$(\overline{x}_{j_1}, a_{j_2 j_2 1}) = \underbrace{(\text{UT}, 3^{\text{rd}}\text{SO})}_{\text{up-trumpet source}}.$$

  The infinite-equilibrium of $x_{j_2}^* = a_{j_1 j_2 1} = a_{j_2 j_2 1}$ is a (UT,3rdSO)-third-order up-trumpet source.
- For $a_{j_1 j_1 0}(\overline{x}_{j_1} - a_{j_1 j_1 1}) < 0$ and $a_{j_2 j_2 0} > 0$,

$$(\overline{x}_{j_1}, a_{j_2 j_2 1}) = \underbrace{(\text{DT}, 3^{\text{rd}}\text{SO})}_{\text{down-trumpet source}}.$$

  The infinite-equilibrium of $x_{j_2}^* = a_{j_1 j_2 1} = a_{j_2 j_2 s_2}$ is a (DT,3rdSO)-third-order down-trumpet source.
- For $a_{j_1 j_1 0}(\overline{x}_{j_1} - a_{j_1 j_1 1}) > 0$ and $a_{j_2 j_2 0} < 0$,

$$(\overline{x}_{j_1}, a_{j_2 j_2 1}) = \underbrace{(\text{DT}, 3^{\text{rd}}\text{SI})}_{\text{up-trumpet sink}}.$$

  The infinite-equilibrium of $x_{j_2}^* = a_{j_1 j_2 1} = a_{j_2 j_2 s_2}$ is a (DT,3rdSI)-third-order down-trumpet sink.
- For $a_{j_1 j_1 0}(\overline{x}_{j_1} - a_{j_1 j_1 1}) < 0$ and $a_{j_2 j_2 0} < 0$,

$$(\overline{x}_{j_1}, a_{j_2 j_2 1}) = \underbrace{(\text{UT}, 3^{\text{rd}}\text{SI})}_{\text{up-trumpet sink}}.$$

  The infinite-equilibrium of $x_{j_2}^* = a_{j_1 j_2 1} = a_{j_2 j_2 s_2}$ is a (UT,3rdSO)-third-order up-trumpet sink.

(v_{5b}) From (v_{5a}), for $\bar{x}_{j_1} = a_{j_1 j_1 1}$, the equilibrium of $(x^*_{j_1}, x^*_{j_2}) = (a_{j_1 j_1 1}, a_{j_2 j_2 1})$ with $a_{j_1 j_2 1} = a_{j_2 j_2 1}$ has the following properties as in Eqs. (1.99)–(1.102).

- For $a_{j_1 j_1 0} > 0$ and $a_{j_2 j_2 0} > 0$,

$$(a_{j_1 j_1 1}, a_{j_2 j_2 1}) = \underbrace{(\mathrm{DT}:\mathrm{UT}, 3^{\mathrm{rd}}\mathrm{SO})}_{\text{down-up trumpet source}}.$$

The equilibrium of $(x^*_{j_1}, x^*_{j_2}) = (a_{j_1 j_1 1}, a_{j_2 j_2 1})$ is a (DT:UT,3rdSO)-third-order down-up trumpet source.

- For $a_{j_1 j_1 0} < 0$ and $a_{j_2 j_2 0} > 0$,

$$(a_{j_1 j_1 1}, a_{j_2 j_2 1}) = \underbrace{(\mathrm{UT}:\mathrm{DT}, 3^{\mathrm{rd}}\mathrm{SO})}_{\text{up-down trumpet source}}.$$

The equilibrium of $(x^*_{j_1}, x^*_{j_2}) = (a_{j_1 j_1 1}, a_{j_2 j_2 1})$ is a (UT:DT,3rdSO)-third-order up-down trumpet source.

- For $a_{j_1 j_1 0} > 0$ and $a_{j_2 j_2 0} < 0$,

$$(a_{j_1 j_1 1}, a_{j_2 j_2 1}) = \underbrace{(\mathrm{UT}:\mathrm{DT}, 3^{\mathrm{rd}}\mathrm{SI})}_{\text{up-down trumpet sink}}.$$

The equilibrium of $(x^*_{j_1}, x^*_{j_2}) = (a_{j_1 j_1 1}, a_{j_2 j_2 1})$ is a (UT:DT,3rdSI)-third-order up-down trumpet sink.

- For $a_{j_1 j_1 0} < 0$ and $a_{j_2 j_2 0} < 0$,

$$(a_{j_1 j_1 1}, a_{j_2 j_2 1}) = \underbrace{(\mathrm{DT}:\mathrm{UT}, 3^{\mathrm{rd}}\mathrm{SI})}_{\text{down-up trumpet sink}}.$$

The equilibrium of $(x^*_{j_1}, x^*_{j_2}) = (a_{j_1 j_1 1}, a_{j_2 j_2 1})$ is a (DP:UP,3rdSI)-third-order down-up trumpet sink.

(vi) For $\Delta_{j_1 j_2} = 0$ and $\Delta_{j_2 j_2} > 0$, the standard form is

$$\begin{aligned}\dot{x}_{j_2} &= a_{j_2 j_2 0}(x_{j_2} - a_{j_2 j_2 1})(x_{j_2} - a_{j_2 j_2 2})(x_{j_2} - a_{j_2 j_2 3}),\\ \dot{x}_{j_1} &= a_{j_1 j_1 0}(x_{j_1} - a_{j_1 j_1 1})(x_{j_2} - a_{j_1 j_2 1})^2\end{aligned}$$

where

$$\begin{aligned}
a_{j_1j_2 1} &= -\frac{1}{2}B_{j_1j_2},\\
b_{j_2j_2 2}, b_{j_2j_2 3} &= -\frac{1}{2}(B_{j_2j_2} \pm \sqrt{\Delta_{j_2j_2}}),\\
\{a_{j_2j_2 1}, a_{j_2j_2 2}, a_{j_2j_2 3}\} &= \mathrm{sort}\{b_{j_2j_2 1}, b_{j_2j_2 2}, b_{j_2j_2 3}\},\\
a_{j_2j_2 s_1} &< a_{j_2j_2 s_2}; s_1, s_2 \in \{1,2,3\}, s_1 < s_2.
\end{aligned}$$

(vi_1) In phase space, $a_{j_2j_2s_1} \neq a_{j_1j_2 1}$, we have

$$\frac{dx_{j_1}}{dx_{j_2}} = \frac{a_{j_1j_1 0}}{a_{j_2j_2 0}}\frac{(x_{j_1} - a_{j_1j_1 1})(x_{j_2} - a_{j_1j_2 1})^2}{\prod_{s_1=1}^{3}(x_{j_2} - a_{j_2j_2s_1})},$$

and the deformation of the foregoing equation is

$$\frac{1}{(x_{j_1} - a_{j_1j_1 1})}dx_{j_1} = \frac{a_{j_1j_1 0}}{a_{j_2j_2 0}}\sum_{s_1=1}^{3}\frac{(a_{j_2j_2s_1} - a_{j_1j_2 1})^2}{\prod_{s_2=1,s_2\neq s_1}^{3}(a_{j_2j_2s_1} - a_{j_2j_2s_2})}\frac{dx_{j_2}}{x_{j_2} - a_{j_2j_2s_1}}.$$

With an initial condition of $(x_{j_1 0}, x_{j_2 0})$ at $t = t_0$, the integration of the above equation gives the first integral manifold ($a_{j_2j_2s_1} \neq a_{j_1j_2 1}$) as

$$\ln\frac{|x_{j_1} - a_{j_1j_1 1}|}{|x_{j_1 0} - a_{j_1j_1 1}|} = \frac{a_{j_1j_1 0}}{a_{j_2j_2 0}}\sum_{s_1=1}^{3}\frac{(a_{j_2j_2s_1} - a_{j_1j_2 1})^2}{\prod_{s_2=1,s_2\neq s_1}^{3}(a_{j_2j_2s_1} - a_{j_2j_2s_2})}\ln\frac{|x_{j_2} - a_{j_2j_2s_1}|}{|x_{j_2 0} - a_{j_2j_2s_1}|}.$$

(vi_{1a}) At $(x^*_{j_1}, x^*_{j_2}) = (a_{j_1j_1 1}, a_{j_2j_2s_1})$, the variational equations are

$$\begin{aligned}
\Delta\dot{x}_{j_2} &= a_{j_2j_2 0}\prod_{s_2=1,s_2\neq s_1}^{3}(a_{j_2j_2s_1} - a_{j_2j_2s_2})\Delta x_{j_2},\\
\Delta\dot{x}_{j_1} &= a_{j_1j_1 0}(a_{j_2j_2s_1} - a_{j_1j_2 1})^2\Delta x_{j_1}.
\end{aligned}$$

The flows $(x^*_{j_1}, x^*_{j_2}) = (a_{j_1j_1 1}, a_{j_2j_2s_1})$ in the x_{j_1}-direction are source and sink for $a_{j_1j_1 0}(a_{j_2j_2s_1} - a_{j_1j_2 1})^2 > 0$ and $a_{j_1j_1 0}(a_{j_2j_2s_1} - a_{j_1j_2 1})^2 < 0$, respectively. The flows $(x^*_{j_1}, x^*_{j_2}) = (a_{j_1j_1 1}, a_{j_2j_2s_1})$ in the x_{j_2}-direction are source and sink for $a_{j_2j_2 0}\prod_{s_2=1,s_2\neq s_1}^{3}(a_{j_2j_2s_1} - a_{j_2j_2s_2}) > 0$ and $a_{j_2j_2 0}\prod_{s_2=1,s_2\neq s_1}^{3}(a_{j_2j_2s_1} - a_{j_2j_2s_2}) < 0$, respectively.

Therefore, the equilibrium of $(x^*_{j_1}, x^*_{j_2}) = (a_{j_1j_1 1}, a_{j_2j_2s_1})$ ($s_1, s_2 \in \{1,2,3\}$, $s_1 \neq s_2$) has the following properties as in Eqs. (1.106)–(1.109).

- For $a_{j_1j_1 0}(a_{j_2j_2s_1} - a_{j_1j_2 1})^2 > 0$ and $a_{j_2j_2 0}\prod_{s_2=1,s_2\neq s_1}^{3}(a_{j_2j_2s_1} - a_{j_2j_2s_2}) > 0$,

$$(a_{j_1j_1 1}, a_{j_2j_2s_1}) = \underbrace{(\mathrm{SO}, \mathrm{SO})}_{\text{source}}.$$

The equilibrium of $(x^*_{j_1}, x^*_{j_2}) = (a_{j_1j_1 1}, a_{j_2j_2s_1})$ is an (SO,SO)-source.

- For $a_{j_1j_10}(a_{j_2j_2s_1}-a_{j_1j_21})^2<0$ and $a_{j_2j_20}\prod_{s_2=1,s_2\neq s_1}^{3}(a_{j_2j_2s_1}-a_{j_2j_2s_2})>0$,

$$(a_{j_1j_11},a_{j_2j_2s_1})=\underbrace{(\text{SI},\text{SO})}_{\text{saddle}}.$$

The equilibrium of $(x_{j_1}^*,x_{j_2}^*)=(a_{j_1j_11},a_{j_2j_2s_1})$ is an (SI,SO)-saddle.

- For $a_{j_1j_10}(a_{j_2j_2s_1}-a_{j_1j_21})^2>0$ and $a_{j_2j_20}\prod_{s_2=1,s_2\neq s_1}^{3}(a_{j_2j_2s_1}-a_{j_2j_2s_2})<0$,

$$(a_{j_1j_11},a_{j_2j_2s_1})=\underbrace{(\text{SO},\text{SI})}_{\text{saddle}}.$$

The equilibrium of $(x_{j_1}^*,x_{j_2}^*)=(a_{j_1j_11},a_{j_2j_2s_1})$ is an (SO,SI)-saddle.

- For $a_{j_1j_10}(a_{j_2j_2s_1}-a_{j_1j_21})^2<0$ and $a_{j_2j_20}\prod_{s_2=1,s_2\neq s_1}^{3}(a_{j_2j_2s_1}-a_{j_2j_2s_2})<0$,

$$(a_{j_1j_11},a_{j_2j_2s_1})=\underbrace{(\text{SI},\text{SI})}_{\text{sink}}.$$

The equilibrium of $(x_{j_1}^*,x_{j_2}^*)=(a_{j_1j_11},a_{j_2j_2s_1})$ is an (SI,SI)-sink.

(vi$_{1b}$) In phase space, at $x_{j_2}^*=a_{j_1j_21}$ with $\bar{x}_{j_1}\neq a_{j_1j_11}$,

$$\left.\frac{dx_{j_1}}{dx_{j_2}}\right|_{x_{j_2}^*=a_{j_1j_21}}=\frac{a_{j_1j_10}}{a_{j_2j_20}}\left.\frac{(\bar{x}_{j_1}-a_{j_1j_11})(x_{j_2}-a_{j_1j_21})^2}{\prod_{s_1=1}^{3}(x_{j_2}-a_{j_2j_2s_1})}\right|_{x_{j_2}^*=a_{j_1j_21}}=0,$$

$$\left.\frac{d^2x_{j_1}}{dx_{j_2}^2}\right|_{x_{j_2}^*=a_{j_1j_21}}=0.$$

If

$$\left.\frac{d^3x_{j_1}}{dx_{j_2}^3}\right|_{x_{j_2}^*=a_{j_1j_21}}=\frac{a_{j_1j_10}}{a_{j_2j_20}}\frac{2(\bar{x}_{j_1}-a_{j_1j_11})}{\prod_{s_1=1}^{3}(a_{j_1j_21}-a_{j_2j_2s_1})}>0,$$

there is a third-order increasing-inflection flow at $x_{j_2}^*=a_{j_1j_21}$ with $\bar{x}_{j_1}\neq a_{j_1j_11}$. If

$$\left.\frac{d^3x_{j_1}}{dx_{j_2}^3}\right|_{x_{j_2}^*=a_{j_1j_21}}=\frac{a_{j_1j_10}}{a_{j_2j_20}}\frac{2(\bar{x}_{j_1}-a_{j_1j_11})}{\prod_{s_1=1}^{3}(a_{j_1j_21}-a_{j_2j_2s_1})}<0,$$

there is a third-order decreasing-inflection flow at $x_{j_2}^*=a_{j_1j_21}$ with $\bar{x}_{j_1}\neq a_{j_1j_11}$. Let

$$\dot{x}_{j_2} = a_{j_2j_20}\prod\nolimits_{s_1=1}^{3}(a_{j_1j_21} - a_{j_2j_2s_1}) \neq 0,$$

The third-order inflection flows at $x_{j_2}^* = a_{j_1j_21}$ with $\bar{x}_{j_1} \neq a_{j_1j_11}$ are positive and negative for $\dot{x}_{j_2} > 0$ and $\dot{x}_{j_2} < 0$, respectively.

Thus, the equilibrium of $x_{j_2}^* = a_{j_1j_21}$ with $\bar{x}_{j_1} \neq a_{j_1j_11}$ has the following properties.

- For $a_{j_1j_10}(\bar{x}_{j_1} - a_{j_1j_11}) > 0$ and $a_{j_2j_20}\prod_{s_1=1}^{3}(a_{j_1j_21} - a_{j_2j_2s_1}) > 0$,

$$(a_{j_1j_21}, \dot{x}_{j_2}) = \underbrace{(\mathrm{II, pF})}_{\text{increasing-inflection flow }(+)}.$$

- For $a_{j_1j_10}(\bar{x}_{j_1} - a_{j_1j_11}) < 0$ and $a_{j_2j_20}\prod_{s_1=1}^{3}(a_{j_1j_21} - a_{j_2j_2s_1}) > 0$,

$$(a_{j_1j_21}, \dot{x}_{j_2}) = \underbrace{(\mathrm{DI, pF})}_{\text{decreasing-inflection flow }(+)}.$$

- For $a_{j_1j_10}(\bar{x}_{j_1} - a_{j_1j_11}) > 0$ and $a_{j_2j_20}\prod_{s_1=1}^{3}(a_{j_1j_21} - a_{j_2j_2s_1}) < 0$,

$$(a_{j_1j_21}, \dot{x}_{j_2}) = \underbrace{(\mathrm{DI, nF})}_{\text{decreasing-inflection flow }(-)}.$$

- For $a_{j_1j_10}(\bar{x}_{j_1} - a_{j_1j_11}) < 0$ and $a_{j_2j_20}\prod_{s_1=1}^{3}(a_{j_1j_21} - a_{j_2j_2s_1}) < 0$,

$$(a_{j_1j_21}, \dot{x}_{j_2}) = \underbrace{(\mathrm{II, nF})}_{\text{increasing-inflection flow }(-)}.$$

The variational equation at $x_{j_1}^* = a_{j_1j_11}$ with $\bar{x}_{j_2} \neq a_{j_1j_21}$ is

$$\Delta\dot{x}_{j_1} = a_{j_1j_10}(\bar{x}_{j_2} - a_{j_1j_21})^2\Delta x_{j_1},$$

and

$$\dot{x}_{j_2} = a_{j_2j_20}\prod\nolimits_{s_1=1}^{3}(\bar{x}_{j_2} - a_{j_2j_2s_1}).$$

Thus, the equilibrium of $x_{j_1}^* = a_{j_1j_11}$ with $\bar{x}_{j_2} \neq a_{j_1j_21}, a_{j_1j_2s_1}$ $(s_1 = 1, 2, 3)$ has the following properties.

- For $a_{j_1j_1 0}(\overline{x}_{j_2} - a_{j_1j_2 1})^2 > 0$ and $a_{j_2j_2 0}\prod_{s_1=1}^{3}(\overline{x}_{j_2} - a_{j_2j_2 s_1}) > 0$,

$$(a_{j_1j_1 1}, \dot{x}_{j_2}) = \underbrace{(\text{SO}, \text{pF})}_{\text{source flow }(+)}.$$

- For $a_{j_1j_1 0}(\overline{x}_{j_2} - a_{j_1j_2 1})^2 < 0$ and $a_{j_2j_2 0}\prod_{s_1=1}^{3}(\overline{x}_{j_2} - a_{j_2j_2 s_1}) > 0$,

$$(a_{j_1j_1 1}, \dot{x}_{j_2}) = \underbrace{(\text{SI}, \text{pF})}_{\text{sink flow }(+)}.$$

- For $a_{j_1j_1 0}(\overline{x}_{j_2} - a_{j_1j_2 1})^2 > 0$ and $a_{j_2j_2 0}\prod_{s_1=1}^{3}(\overline{x}_{j_2} - a_{j_2j_2 s_1}) < 0$,

$$(a_{j_1j_1 1}, \dot{x}_{j_2}) = \underbrace{(\text{SO}, \text{nF})}_{\text{source flow }(-)}.$$

- For $a_{j_1j_1 0}(\overline{x}_{j_2} - a_{j_1j_2 1})^2 < 0$ and $a_{j_2j_2 0}\prod_{s_1=1}^{3}(\overline{x}_{j_2} - a_{j_2j_2 s_1}) < 0$,

$$(a_{j_1j_1 1}, \dot{x}_{j_2}) = \underbrace{(\text{SI}, \text{nF})}_{\text{sink flow }(-)}.$$

From the above analysis, the equilibrium of $(x_{j_1}^*, x_{j_2}^*) = (a_{j_1j_1 1}, a_{j_1j_2 1})$ with $a_{j_1j_2 1} \neq a_{j_2j_2 1}, a_{j_2j_2 2}$ has the following properties as in Eqs. (1.110)–(1.113).

- For $a_{j_1j_1 0} > 0$ and $a_{j_2j_2 0}\prod_{s_1=1}^{3}(a_{j_1j_2 1} - a_{j_2j_2 s_1}) > 0$,

$$(a_{j_1j_1 1}, a_{j_1j_2 1}) = \underbrace{({}_{\text{DI:II}}\text{SO}, \text{pF})}_{\text{inflection-source flow }(+)}.$$

  The equilibrium of $(x_{j_1}^*, x_{j_2}^*) = (a_{j_1j_1 1}, a_{j_1j_2 1})$ is a $({}_{\text{DI:II}}\text{SO}, \text{pF})$-positive inflection-source flow.
- For $a_{j_1j_1 0} < 0$ and $a_{j_2j_2 0}\prod_{s_1=1}^{3}(a_{j_1j_2 1} - a_{j_2j_2 s_1}) > 0$,

$$(a_{j_1j_1 1}, a_{j_1j_2 1}) = \underbrace{({}_{\text{II:DI}}\text{SI}, \text{pF})}_{\text{inflection-sink flow }(+)}.$$

The equilibrium of $(x_{j_1}^*, x_{j_2}^*) = (a_{j_1j_11}, a_{j_1j_21})$ is an $({}_{\mathrm{II:DI}}\mathrm{SI}, \mathrm{pF})$-positive inflection-sink flow.

- For $a_{j_1j_10} > 0$ and $a_{j_2j_20}\prod_{s_1=1}^{3}(a_{j_1j_21} - a_{j_2j_2s_1}) < 0$,

$$(a_{j_1j_11}, a_{j_1j_21}) = \underbrace{({}_{\mathrm{II:DI}}\mathrm{SO}, \mathrm{nF})}_{\text{inflection-source flow }(-)}.$$

The equilibrium of $(x_{j_1}^*, x_{j_2}^*) = (a_{j_1j_11}, a_{j_1j_21})$ is an $({}_{\mathrm{II:DI}}\mathrm{SI}, \mathrm{nF})$-negative inflection-source flow.

- For $a_{j_1j_10} < 0$ and $a_{j_2j_20}\prod_{s_1=1}^{3}(a_{j_1j_21} - a_{j_2j_2s_1}) < 0$,

$$(a_{j_1j_11}, a_{j_1j_21}) = \underbrace{({}_{\mathrm{DI:II}}\mathrm{SI}, \mathrm{nF})}_{\text{inflection-sink flow }(-)}.$$

The equilibrium of $(x_{j_1}^*, x_{j_2}^*) = (a_{j_1j_11}, a_{j_1j_21})$ is a $({}_{\mathrm{DI:II}}\mathrm{SO}, \mathrm{nF})$-negative inflection-sink flow.

(vi$_2$) At $a_{j_2j_2s_1} = a_{j_1j_21}$ (s_1, $s_2 \in \{1,2,3\}$, $s_1 \neq s_2$), we have

$$\frac{dx_{j_1}}{dx_{j_2}} = \frac{a_{j_1j_10}}{a_{j_2j_20}} \frac{(x_{j_1} - a_{j_1j_11})(x_{j_2} - a_{j_1j_21})}{(x_{j_2} - a_{j_2j_2s_2})(x_{j_2} - a_{j_2j_2s_3})},$$

and the deformation of the foregoing equation is

$$\frac{1}{(x_{j_1} - a_{j_1j_11})}dx_{j_1} = \frac{a_{j_1j_10}}{a_{j_2j_20}} \frac{(x_{j_2} - a_{j_1j_21})}{(x_{j_2} - a_{j_2j_2s_2})(x_{j_2} - a_{j_2j_2s_3})}dx_{j_2}.$$

With an initial condition of (x_{j_10}, x_{j_20}) at $t = t_0$, the integration of the above equation gives the first integral manifold $(a_{j_2j_2s_1} = a_{j_1j_21})$ as

$$\ln\frac{|x_{j_10} - a_{j_1j_11}|}{|x_{j_1} - a_{j_1j_11}|} = \frac{a_{j_1j_10}}{a_{j_2j_20}}\sum_{s_2=1, s_2\neq s_1, s_3}^{3}\frac{(a_{j_2j_2s_2} - a_{j_1j_21})}{(a_{j_2j_2s_2} - a_{j_2j_2s_3})}\ln\frac{|x_{j_2} - a_{j_2j_2s_2}|}{|x_{j_20} - a_{j_2j_2s_2}|}.$$

(vi$_{2a}$) At $x_{j_2}^* = a_{j_1j_21} = a_{j_2j_2s_1}$ with $\bar{x}_{j_1} \neq a_{j_1j_11}$,

$$\left.\frac{dx_{j_1}}{dx_{j_2}}\right|_{x_{j_2}^* = a_{j_1j_21}} = \left.\frac{a_{j_1j_10}}{a_{j_2j_20}} \frac{(\bar{x}_{j_1} - a_{j_1j_11})(x_{j_2} - a_{j_1j_21})}{\prod_{s_2=1, s_2\neq s_1}^{3}(x_{j_2} - a_{j_2j_2s_2})}\right|_{x_{j_2}^* = a_{j_1j_21}} = 0.$$

If

$$\left.\frac{d^2x_{j_1}}{dx_{j_2}^2}\right|_{x_{j_2}^*=a_{j_1j_2 1}} = \frac{a_{j_1j_1 0}}{a_{j_2j_2 0}} \frac{(\overline{x}_{j_1} - a_{j_1j_1 1})}{\prod_{s_2=1, s_2\neq s_1}^{3}(a_{j_1j_2 1} - a_{j_2j_2s_2})} > 0,$$

there is an up-parabola flow at $x_{j_2}^* = a_{j_1j_2 1} = a_{j_2j_2s_1}$ with $\overline{x}_{j_1} \neq a_{j_1j_1 1}$. If

$$\left.\frac{d^2x_{j_1}}{dx_{j_2}^2}\right|_{x_{j_2}^*=a_{j_1j_2 1}} = \frac{a_{j_1j_1 0}}{a_{j_2j_2 0}} \frac{(\overline{x}_{j_1} - a_{j_1j_1 1})}{\prod_{s_2=1, s_2\neq s_1}^{3}(a_{j_1j_2s_1} - a_{j_2j_2s_2})} < 0,$$

there is a down-parabola flow at $x_{j_2}^* = a_{j_1j_2 1} = a_{j_2j_2s_1}$ with $\overline{x}_{j_1} \neq a_{j_1j_1 1}$. The corresponding variational equation is

$$\Delta\dot{x}_{j_2} = a_{j_2j_2 0}\prod_{s_2=1, s_2\neq s_1}^{3}(a_{j_1j_2s_1} - a_{j_2j_2s_2})\Delta x_{j_2}.$$

The parabola flow at $x_{j_2}^* = a_{j_1j_2 1} = a_{j_2j_2s_1}$ with $\overline{x}_{j_1} \neq a_{j_1j_1 1}$ are source and sink flows for $a_{j_2j_2 0}\prod_{s_2=1, s_2\neq s_1}^{3}(a_{j_1j_2s_1} - a_{j_2j_2s_2}) > 0$ and $a_{j_2j_2 0}\prod_{s_2=1, s_2\neq s_1}^{3}(a_{j_1j_2s_1} - a_{j_2j_2s_2}) < 0$, respectively.

Therefore, the infinite-equilibrium of $x_{j_2}^* = a_{j_1j_2 1} = a_{j_2j_2s_1}$ with $\overline{x}_{j_1} \neq a_{j_1j_1 1}$ has the following properties as in Eqs. (1.115)–(1.118).

- For $a_{j_1j_1 0}(\overline{x}_{j_1} - a_{j_1j_1 1}) > 0$ and $a_{j_2j_2 0}\prod_{s_2=1, s_2\neq s_1}^{3}(a_{j_1j_2s_1} - a_{j_2j_2s_2}) > 0$,

$$(\overline{x}_{j_1}, a_{j_2j_2 1}) = \underbrace{(\text{UP}, \text{SO})}_{\text{up-parabola source}}.$$

The infinite-equilibrium of $x_{j_2}^* = a_{j_1j_2 1} = a_{j_2j_2s_1}$ is a (UP,SO)-up-parabola source.

- For $a_{j_1j_1 0}(\overline{x}_{j_1} - a_{j_1j_1 1}) < 0$ and $a_{j_2j_2 0}\prod_{s_2=1, s_2\neq s_1}^{3}(a_{j_1j_2s_1} - a_{j_2j_2s_2}) > 0$,

$$(\overline{x}_{j_1}, a_{j_2j_2s_1}) = \underbrace{(\text{DP}, \text{SO})}_{\text{down-parabola source}}.$$

The infinite-equilibrium of $x_{j_2}^* = a_{j_1j_2 1} = a_{j_2j_2s_1}$ is a (DP,SO)-down-parabola source.

- For $a_{j_1j_1 0}(\overline{x}_{j_1} - a_{j_1j_1 1}) > 0$ and $a_{j_2j_2 0}\prod_{s_2=1, s_2\neq s_1}^{3}(a_{j_1j_2s_1} - a_{j_2j_2s_2}) < 0$,

$$(\overline{x}_{j_1}, a_{j_2j_2s_1}) = \underbrace{(\text{DP}, \text{SI})}_{\text{down-parabola sink}}.$$

The infinite-equilibrium of $x_{j_2}^* = a_{j_1j_2 1} = a_{j_2j_2s_1}$ is a (DP,SI)-down-parabola sink.

- For $a_{j_1j_10}(\overline{x}_{j_1} - a_{j_1j_11}) < 0$ and $a_{j_2j_20}\prod_{s_2=1,s_2\neq s_1}^{3}(a_{j_1j_2s_1} - a_{j_2j_2s_2}) < 0$,

$$(\overline{x}_{j_1}, a_{j_2j_2s_1}) = \underbrace{(\text{UP, SI})}_{\text{up-parabola sink}}.$$

The infinite-equilibrium of $x_{j_2}^* = a_{j_1j_21} = a_{j_2j_2s_1}$ is a (UP,SI)-up-parabola sink.

($\text{vi}_{2\text{b}}$) From case ($\text{vi}_{2\text{a}}$), for $\overline{x}_{j_1} = a_{j_1j_11}$, thus, the equilibrium of $(x_{j_1}^*, x_{j_2}^*) = (a_{j_1j_11}, a_{j_1j_21})$ with $a_{j_2j_2s_1} = a_{j_1j_21}$ has the following properties as in Eqs. (1.119)–(1.122).

- For $a_{j_1j_10} > 0$ and $a_{j_2j_20}\prod_{s_2=1,s_2\neq s_1}^{3}(a_{j_1j_2s_1} - a_{j_2j_2s_2}) > 0$,

$$(a_{j_1j_11}, a_{j_1j_21}) = \underbrace{(\text{DP : UP, SO})}_{\text{hyperbolic source}}.$$

The equilibrium of $(x_{j_1}^*, x_{j_2}^*) = (a_{j_1j_11}, a_{j_1j_21})$ is a (DP:UP,SO)-hyperbolic source.

- For $a_{j_1j_10} < 0$ and $a_{j_2j_20}\prod_{s_2=1,s_2\neq s_1}^{3}(a_{j_1j_2s_1} - a_{j_2j_2s_2}) > 0$,

$$(a_{j_1j_11}, a_{j_1j_21}) = \underbrace{(\text{UP : DP, SO})}_{\text{hyperbolic-secant source}}.$$

The equilibrium of $(x_{j_1}^*, x_{j_2}^*) = (a_{j_1j_11}, a_{j_1j_21})$ is a (UP:DP,SO)-hyperbolic-secant source.

- For $a_{j_1j_10} > 0$ and $a_{j_2j_20}\prod_{s_2=1,s_2\neq s_1}^{3}(a_{j_1j_2s_1} - a_{j_2j_2s_2}) < 0$,

$$(a_{j_1j_11}, a_{j_1j_21}) = \underbrace{(\text{UP : DP, SI})}_{\text{hyperbolic-secant sink}}.$$

The equilibrium of $(x_{j_1}^*, x_{j_2}^*) = (a_{j_1j_11}, a_{j_1j_21})$ is a (UP:DP,SI)-hyperbolic-secant sink.

- For $a_{j_1j_10} < 0$ and $a_{j_2j_20}\prod_{s_2=1,s_2\neq s_1}^{3}(a_{j_1j_2s_1} - a_{j_2j_2s_2}) < 0$,

$$(a_{j_1j_11}, a_{j_1j_21}) = \underbrace{(\text{DP : UP, SI})}_{\text{hyperbolic sink}}.$$

The equilibrium of $(x_{j_1}^*, x_{j_2}^*) = (a_{j_1j_11}, a_{j_1j_21})$ is a (DP:UP,SI)-hyperbolic sink.

(vii) For $\Delta_{j_1j_2} > 0$ and $\Delta_{j_2j_2} < 0$, the standard form is

$$\begin{aligned}
\dot{x}_{j_2} &= a_{j_2j_20}(x_{j_2} - a_{j_2j_21})\left[(x_{j_2} - a_{j_2j_2})^2 + b_{j_2j_2}\right],\\
\dot{x}_{j_1} &= a_{j_1j_10}(x_{j_1} - a_{j_1j_11})(x_{j_2} - a_{j_1j_21})(x_{j_2} - a_{j_1j_22})
\end{aligned}$$

where

$$
\begin{aligned}
&b_{j_1j_21}, b_{j_1j_22} = -\frac{1}{2}(B_{j_1j_2} \pm \sqrt{\Delta_{j_1j_2}}),\\
&\{a_{j_1j_21}, a_{j_1j_22}\} = \text{sort}\{b_{j_1j_21}, b_{j_1j_22}\}, a_{j_1j_21} < a_{j_1j_22};\\
&a_{j_2j_2} = -\frac{1}{2}B_{j_2j_2}, b_{j_2j_2} = \frac{1}{4}(-\Delta_{j_2j_2}), a_{j_2j_21} = b_{j_2j_21}.
\end{aligned}
$$

(vii$_1$) For $a_{j_2j_21} \neq a_{j_1j_2l_1}(l_1 = 1, 2)$, in phase space,

$$
\frac{dx_{j_1}}{dx_{j_2}} = \frac{a_{j_1j_10}}{a_{j_2j_20}} \frac{(x_{j_1} - a_{j_1j_11})\prod_{l_1=1}^{2}(x_{j_2} - a_{j_1j_2l_1})}{(x_{j_2} - a_{j_2j_21})\left[(x_{j_2} - a_{j_2j_2})^2 + b_{j_2j_2}\right]},
$$

and the deformation of the foregoing equation is

$$
\begin{aligned}
\frac{dx_{j_1}}{x_{j_1} - a_{j_1j_11}} = \frac{a_{j_1j_10}}{a_{j_2j_20}} \Bigg\{ &\frac{x_{j_2} - a_{j_2j_21}}{(x_{j_2} - a_{j_2j_2})^2 + b_{j_2j_2}} + \frac{(2a_{j_2j_21} - a_{j_1j_21} - a_{j_1j_22})}{(x_{j_2} - a_{j_2j_2})^2 + b_{j_2j_2}}\\
&+ \frac{\prod_{l_1=1}^{2}(a_{j_2j_21} - a_{j_1j_2l_1})}{(x_{j_2} - a_{j_2j_21})\left[(x_{j_2} - a_{j_2j_2})^2 + b_{j_2j_2}\right]} \Bigg\} dx_{j_2}.
\end{aligned}
$$

With

$$
\begin{aligned}
&\frac{\prod_{l_1=1}^{2}(a_{j_2j_21} - a_{j_1j_2l_1})}{(x_{j_2} - a_{j_2j_21})\left[(x_{j_2} - a_{j_2j_2})^2 + b_{j_2j_2}\right]}\\
&= \frac{\prod_{l_1=1}^{2}(a_{j_2j_21} - a_{j_1j_2l_1})}{(a_{j_2j_21} - a_{j_2j_2})^2 + b_{j_2j_2}} \left[\frac{1}{x_{j_2} - a_{j_2j_21}} - \frac{x_2 - a_{j_2j_21} + 2(a_{j_2j_21} - a_{j_2j_2})}{(x_{j_2} - a_{j_2j_2})^2 + b_{j_2j_2}} \right],
\end{aligned}
$$

we have

$$
\begin{aligned}
\frac{dx_{j_1}}{x_{j_1} - a_{j_1j_11}} = \frac{a_{j_1j_10}}{a_{j_2j_20}} \Bigg\{ &\frac{\prod_{l_1=1}^{2}(a_{j_2j_21} - a_{j_1j_2l_1})}{(a_{j_2j_21} - a_{j_2j_2})^2 + b_{j_2j_2}} \frac{1}{(x_{j_2} - a_{j_2j_21})}\\
&+ \left[1 - \frac{\prod_{l_1=1}^{2}(a_{j_2j_21} - a_{j_1j_2l_1})}{(a_{j_2j_21} - a_{j_2j_2})^2 + b_{j_2j_2}}\right] \frac{x_{j_2} - a_{j_2j_2}}{(x_{j_2} - a_{j_2j_2})^2 + b_{j_2j_2}}\\
&+ \Bigg[a_{j_2j_2} + a_{j_2j_21} - a_{j_1j_21} - a_{j_1j_22}\\
&- (a_{j_2j_21} - a_{j_2j_2}) \frac{\prod_{l_1=1}^{2}(a_{j_2j_21} - a_{j_1j_2l_1})}{(a_{j_2j_21} - a_{j_2j_2})^2 + b_{j_2j_2}}\Bigg] \frac{1}{(x_{j_2} - a_{j_2j_2})^2 + b_{j_2j_2}} \Bigg\} dx_{j_2}.
\end{aligned}
$$

With an initial condition of (x_{j_10}, x_{j_20}) at $t = t_0$, the integration of the above equation gives the first integral manifold as

$$\ln\frac{|x_{j_1}-a_{j_1j_1 1}|}{|x_{j_1 0}-a_{j_1j_1 1}|}=\frac{a_{j_1j_1 0}}{a_{j_2j_2 0}}\left\{A\ln\frac{|x_{j_2}-a_{j_2j_2 1}|}{|x_{j_2 0}-a_{j_2j_2 1}|}\right.$$
$$+\frac{1}{2}(1-A)\ln\frac{|(x_{j_2}-a_{j_2j_2})^2+b_{j_2j_2}|}{|(x_{j_2 0}-a_{j_2j_2})^2+b_{j_2j_2}|}$$
$$\left.+\frac{B}{\sqrt{b_{j_2j_2}}}\left[\arctan\frac{x_{j_2}-a_{j_2j_2}}{\sqrt{b_{j_2j_2}}}-\arctan\frac{x_{j_2 0}-a_{j_2j_2}}{\sqrt{b_{j_2j_2}}}\right]\right\}$$

where

$$A=\frac{(a_{j_2j_2 1}-a_{j_1j_2 1})(a_{j_2j_2 1}-a_{j_1j_2 2})}{(a_{j_2j_2 1}-a_{j_2j_2})^2+b_{j_2j_2}},$$
$$B=a_{j_2j_2}+a_{j_2j_2 1}-a_{j_1j_2 1}-a_{j_1j_2 2}-(a_{j_2j_2 1}-a_{j_2j_2})A.$$

(vii_{1a}) For $(x_{j_1}^*,x_{j_2}^*)=(a_{j_1j_1 1},a_{j_2j_2 1})$, the variational equation is

$$\Delta\dot{x}_{j_2}=a_{j_2j_2 0}\Delta x_{j_2}\left[(a_{j_2j_2 1}-a_{j_2j_2})^2+b_{j_2j_2}\right],$$
$$\Delta\dot{x}_{j_1}=a_{j_1j_1 0}\Delta x_{j_1}\prod\nolimits_{l_1=1}^{2}(a_{j_2j_2 1}-a_{j_1j_2 l_1}).$$

The flows at $(x_{j_1}^*,x_{j_2}^*)=(a_{j_1j_1 1},a_{j_2j_2 1})$ in the x_{j_1}-direction are source and sink for $a_{j_1j_1 0}\prod_{l_1=1}^{2}(a_{j_2j_2 1}-a_{j_1j_2 l_1})>0$ and $a_{j_1j_1 0}\prod_{l_1=1}^{2}(a_{j_2j_2 1}-a_{j_1j_2 l_1})<0$, respectively. The flows at $(x_{j_1}^*,x_{j_2}^*)=(a_{j_1j_1 1},a_{j_2j_2 s_1})$ in the x_{j_2}-direction are source and sink for $a_{j_2j_2 0}>0$ and $a_{j_2j_2 0}<0$, respectively.

Therefore, the equilibrium of $(x_{j_1}^*,x_{j_2}^*)=(a_{j_1j_1 1},a_{j_2j_2 1})$ has the following properties as in Eqs. (1.127)–(1.130).

- For $a_{j_1j_1 0}\prod_{l_1=1}^{2}(a_{j_2j_2 1}-a_{j_1j_2 l_1})>0$ and $a_{j_2j_2 0}>0$,

$$(a_{j_1j_1 1},a_{j_2j_2 1})=\underbrace{(\text{SO},\text{SO})}_{\text{source}}.$$

The equilibrium of $(x_{j_1}^*,x_{j_2}^*)=(a_{j_1j_1 1},a_{j_2j_2 1})$ is an (SO,SO)-source.

- For $a_{j_1j_1 0}\prod_{l_1=1}^{2}(a_{j_2j_2 1}-a_{j_1j_2 l_1})<0$ and $a_{j_2j_2 0}>0$,

$$(a_{j_1j_1 1},a_{j_2j_2 1})=\underbrace{(\text{SI},\text{SO})}_{\text{saddle}}.$$

The equilibrium of $(x_{j_1}^*,x_{j_2}^*)=(a_{j_1j_1 1},a_{j_2j_2 1})$ is an (SI,SO)-saddle.

- For $a_{j_1j_10}\prod_{l_1=1}^{2}(a_{j_2j_21}-a_{j_1j_2l_1})>0$ and $a_{j_2j_20}<0$,

$$(a_{j_1j_11},a_{j_2j_21})=\underbrace{(\mathrm{SO},\mathrm{SI})}_{\text{saddle}}.$$

The equilibrium of $(x^*_{j_1},x^*_{j_2})=(a_{j_1j_11},a_{j_2j_21})$ is an (SO,SI)-saddle.

- For $a_{j_1j_10}\prod_{l_1=1}^{2}(a_{j_2j_21}-a_{j_1j_2l_1})<0$ and $a_{j_2j_20}<0$,

$$(a_{j_1j_11},a_{j_2j_21})=\underbrace{(\mathrm{SI},\mathrm{SI})}_{\text{sink}}.$$

The equilibrium of $(x^*_{j_1},x^*_{j_2})=(a_{j_1j_11},a_{j_2j_21})$ is an (SI,SI)-sink.

(vii$_{1b}$) For $x^*_{j_2}=a_{j_1j_2l_1}\neq a$ and $\bar{x}_{j_1}\neq a_{j_1j_11}$ ($l_1,\,l_2\in\{1,2\}$, $l_1\neq l_2$), in phase space,

$$\left.\frac{dx_{j_1}}{dx_{j_2}}\right|_{x^*_{j_2}=a_{j_1j_2l_1}}=\left.\frac{a_{j_1j_10}}{a_{j_2j_20}}\frac{(\bar{x}_{j_1}-a_{j_1j_11})(x_{j_2}-a_{j_1j_2l_1})(x_{j_2}-a_{j_1j_2l_2})}{(x_{j_2}-a_{j_2j_21})\left[(x_{j_2}-a_{j_2j_2})^2+b_{j_2j_2}\right]}\right|_{x^*_{j_2}=a_{j_1j_2l_1}}=0.$$

If

$$\left.\frac{d^2x_{j_1}}{dx^2_{j_2}}\right|_{x^*_{j_2}=a_{j_1j_2l_1}}=\frac{a_{j_1j_10}}{a_{j_2j_20}}\frac{(\bar{x}_{j_1}-a_{j_1j_11})(a_{j_1j_2l_1}-a_{j_1j_22})}{(a_{j_1j_2l_1}-a_{j_2j_21})\left[(a_{j_1j_2l_1}-a_{j_2j_2})^2+b_{j_2j_2}\right]}>0,$$

there is an up-parabola flow at $x^*_{j_2}=a_{j_1j_2s_1}$ with $\bar{x}_{j_1}\neq a_{j_1j_11}$. If

$$\left.\frac{d^2x_{j_1}}{dx^2_{j_2}}\right|_{x^*_{j_2}=a_{j_1j_2l_1}}=\frac{a_{j_1j_10}}{a_{j_2j_20}}\frac{(\bar{x}_{j_1}-a_{j_1j_11})(a_{j_1j_2l_1}-a_{j_1j_22})}{(a_{j_1j_2l_1}-a_{j_2j_21})\left[(a_{j_1j_2l_1}-a_{j_2j_2})^2+b_{j_2j_2}\right]}<0,$$

there is a down-parabola flow at $x^*_{j_2}=a_{j_1j_2s_1}$ with $\bar{x}_{j_1}\neq a_{j_1j_11}$. Let

$$\dot{x}_{j_2}=a_{j_2j_20}(a_{j_1j_2l_1}-a_{j_2j_21})\left[(a_{j_1j_2l_1}-a_{j_2j_2})^2+b_{j_2j_2}\right],$$

with

$$(a_{j_1j_2l_1}-a_{j_2j_2})^2+b_{j_2j_2}>0.$$

The parabola flows at $x^*_{j_2}=a_{j_1j_2l_1}$ with $\bar{x}_{j_1}\neq a_{j_1j_11}$ are positive and negative for $a_{j_2j_20}>0$ and $a_{j_2j_20}<0$, respectively.

- For $a_{j_1j_10}(\overline{x}_{j_1} - a_{j_1j_11})(a_{j_1j_2l_1} - a_{j_1j_2l_2}) > 0$ and $a_{j_2j_20}(a_{j_1j_2l_1} - a_{j_2j_21}) > 0$,

$$(a_{j_1j_2l_1}, \dot{x}_{j_2}) = \underbrace{(\text{UP}, \text{pF})}_{\text{up-parabola flow } (+)}.$$

- For $a_{j_1j_10}(\overline{x}_{j_1} - a_{j_1j_11})(a_{j_1j_2l_1} - a_{j_1j_2l_2}) < 0$ and $a_{j_2j_20}(a_{j_1j_2l_1} - a_{j_2j_21}) > 0$,

$$(a_{j_1j_2l_1}, \dot{x}_{j_2}) = \underbrace{(\text{DP}, \text{pF})}_{\text{down-parabola flow } (+)}.$$

- For $a_{j_1j_10}(\overline{x}_{j_1} - a_{j_1j_11})(a_{j_1j_2l_1} - a_{j_1j_2l_2}) > 0$ and $a_{j_2j_20}(a_{j_1j_2l_1} - a_{j_2j_21}) < 0$,

$$(a_{j_1j_2l_1}, \dot{x}_{j_2}) = \underbrace{(\text{DP}, \text{nF})}_{\text{down-parabola flow } (-)}.$$

- For $a_{j_1j_10}(\overline{x}_{j_1} - a_{j_1j_11})(a_{j_1j_2l_1} - a_{j_1j_2l_2}) < 0$ and $a_{j_2j_20}(a_{j_1j_2l_1} - a_{j_2j_21}) < 0$,

$$(a_{j_1j_2l_1}, \dot{x}_{j_2}) = \underbrace{(\text{UP}, \text{nF})}_{\text{up-parabola flow } (-)}.$$

From the above analysis, thus, the equilibrium of $(x^*_{j_1}, x^*_{j_2}) = (a_{j_1j_11}, a_{j_1j_2l_1})$ (l_1, l_2 $\in \{1,2\}$, $l_1 \neq l_2$) has the following properties as in Eqs. (1.131)–(1.134).

- For $a_{j_1j_10}(a_{j_1j_2l_1} - a_{j_1j_2l_2}) > 0$ and $a_{j_1j_10}(a_{j_1j_2l_1} - a_{j_2j_21}) > 0$,

$$(a_{j_1j_11}, a_{j_1j_2l_1}) = \underbrace{(\text{DP}:\text{UP}, \text{pF})}_{\text{hyperbolic flow } (+)}.$$

  The equilibrium of $(x^*_{j_1}, x^*_{j_2}) = (a_{j_1j_11}, a_{j_1j_2l_1})$ is a (DP:UP,pF)-positive hyperbolic flow.
- For $a_{j_1j_10}(a_{j_1j_2l_1} - a_{j_1j_2l_2}) < 0$ and $a_{j_1j_10}(a_{j_1j_2l_1} - a_{j_2j_21}) > 0$,

$$(a_{j_1j_11}, a_{j_1j_2l_1}) = \underbrace{(\text{UP}:\text{DP}, \text{pF})}_{\text{hyperbolic-secant flow } (+)}.$$

  The equilibrium of $(x^*_{j_1}, x^*_{j_2}) = (a_{j_1j_11}, a_{j_1j_2l_1})$ is a (UP:DP,pF)-positive hyperbolic-secant flow.

- For $a_{j_1j_10}(a_{j_1j_2l_1} - a_{j_1j_2l_2}) > 0$ and $a_{j_1j_10}(a_{j_1j_2l_1} - a_{j_2j_21}) < 0$,

$$(a_{j_1j_11}, a_{j_1j_2l_1}) = \underbrace{(\text{UP}:\text{DP}, \text{nF})}_{\text{hyperbolic-secant flow }(-)}.$$

The equilibrium of $(x_{j_1}^*, x_{j_2}^*) = (a_{j_1j_11}, a_{j_1j_2l_1})$ is a (UP:DP,nF)-negative hyperbolic-secant flow.

- For $a_{j_1j_10}(a_{j_1j_2l_1} - a_{j_1j_2l_2}) < 0$ and $a_{j_1j_10}(a_{j_1j_2l_1} - a_{j_2j_21}) < 0$,

$$(a_{j_1j_11}, a_{j_1j_2l_1}) = \underbrace{(\text{DP}:\text{UP}, \text{nF})}_{\text{hyperbolic flow }(-)}.$$

The equilibrium of $(x_{j_1}^*, x_{j_2}^*) = (a_{j_1j_11}, a_{j_1j_2l_1})$ is a (DP:UP,nF)-negative hyperbolic flow.

(vii$_2$) For $a_{j_2j_21} = a_{j_1j_2l_1} (l_1 = 1, 2)$, in phase space,

$$\frac{dx_{j_1}}{dx_{j_2}} = \frac{a_{j_1j_10}}{a_{j_2j_20}} \frac{(x_{j_1} - a_{j_1j_11})(x_{j_2} - a_{j_1j_2l_2})}{(x_{j_2} - a_{j_2j_2})^2 + b_{j_2j_2}},$$

and the deformation of the foregoing equation is

$$\frac{dx_{j_1}}{x_{j_1} - a_{j_1j_11}} = \frac{a_{j_1j_10}}{a_{j_2j_20}} \frac{x_{j_2} - a_{j_2j_2} + a_{j_2j_2} - a_{j_1j_2l_2}}{(x_{j_2} - a_{j_2j_2})^2 + b_{j_2j_2}} dx_{j_2}.$$

With an initial condition of (x_{j_10}, x_{j_20}) at $t = t_0$, the integration of the above equation gives the first integral manifold for $a_{j_2j_21} = a_{j_1j_2l_1} (l_1, l_2 \in \{1, 2\}, l_1 \neq l_2)$ as

$$\begin{aligned}\ln \frac{|x_{j_1} - a_{j_1j_11}|}{|x_{j_10} - a_{j_1j_11}|} &= \frac{a_{j_1j_10}}{a_{j_2j_20}} \left[\frac{1}{2} \ln \frac{|(x_{j_2} - a_{j_2j_2})^2 + b_{j_2j_2}|}{|(x_{j_20} - a_{j_2j_2})^2 + b_{j_2j_2}|} \right. \\ &\quad \left. + \frac{a_{j_2j_2} - a_{j_1j_2l_2}}{\sqrt{b_{j_2j_2}}} \left(\arctan \frac{x_{j_2} - a_{j_2j_2}}{\sqrt{b_{j_2j_2}}} - \arctan \frac{x_{j_20} - a_{j_2j_2}}{\sqrt{b_{j_2j_2}}} \right) \right].\end{aligned}$$

(vii$_{2a}$) In phase space, at $x_{j_2}^* = a_{j_2j_21} = a_{j_1j_2l_1}$ with $\bar{x}_{j_1} \neq a_{j_1j_11}$, if

$$\left.\frac{dx_{j_1}}{dx_{j_2}}\right|_{x_{j_2}^* = a_{j_1j_2l_1}} = \frac{a_{j_1j_10}}{a_{j_2j_20}} \frac{(\bar{x}_{j_1} - a_{j_1j_11})(a_{j_1j_21} - a_{j_1j_2l_2})}{[(x_{j_2} - a_{j_2j_2})^2 + b_{j_2j_2}]} > 0,$$

there is an increasing-inflection flow at $x_{j_2}^* = a_{j_2j_21} = a_{j_1j_2l_1}$ with $\bar{x}_{j_1} \neq a_{j_1j_11}$, and if

$$\left.\frac{dx_{j_1}}{dx_{j_2}}\right|_{x^*_{j_2}=a_{j_1j_2l_1}} = \frac{a_{j_1j_10}}{a_{j_2j_20}}\frac{(\overline{x}_{j_1}-a_{j_1j_11})(a_{j_1j_21}-a_{j_1j_2l_2})}{\left[(x_{j_2}-a_{j_2j_2})^2+b_{j_2j_2}\right]}<0,$$

there is a decreasing-inflection flow at $x^*_{j_2}=a_{j_2j_21}=a_{j_1j_2l_1}$ with $\overline{x}_{j_1}\neq a_{j_1j_11}$.

The variational equation is

$$\Delta\dot{x}_{j_2}=a_{j_2j_20}\left[(a_{j_2j_21}-a_{j_2j_2})^2+b_{j_2j_2}\right]\Delta x_{j_2},$$

and

$$(a_{j_2j_21}-a_{j_2j_2})^2+b_{j_2j_2}>0.$$

The inflection flows for $x^*_{j_2}=a_{j_1j_21}$ with $\overline{x}_{j_1}\neq a_{j_1j_11}$ are source and sink, respectively.

Therefore, the infinite-equilibrium of $x^*_{j_2}=a_{j_1j_2l_1}=a_{j_2j_21}$ with $\overline{x}_{j_1}\neq a_{j_1j_11}$ has the following properties as in Eqs. (1.136)–(1.139).

- For $a_{j_1j_10}(\overline{x}_{j_1}-a_{j_1j_11})(a_{j_1j_2l_1}-a_{j_1j_2l_2})>0$ and $a_{j_2j_20}>0$,

$$(\overline{x}_{j_1},a_{j_2j_21})=\underbrace{(\text{II},\text{SO})}_{\text{increasing-inflection source}}.$$

  The infinite-equilibrium of $x^*_{j_2}=a_{j_1j_2l_1}=a_{j_2j_21}$ is an (II,SO)-increasing-inflection source.

- For $a_{j_1j_10}(\overline{x}_{j_1}-a_{j_1j_11})(a_{j_1j_2l_1}-a_{j_1j_2l_2})<0$ and $a_{j_2j_20}>0$,

$$(\overline{x}_{j_1},a_{j_2j_21})=\underbrace{(\text{DI},\text{SO})}_{\text{decreasing-inflection source}}.$$

  The infinite-equilibrium of $x^*_{j_2}=a_{j_1j_2l_1}=a_{j_2j_21}$ is a (DI,SO)-decreasing-inflection source.

- For $a_{j_1j_10}(\overline{x}_{j_1}-a_{j_1j_11})(a_{j_1j_2l_1}-a_{j_1j_2l_2})>0$ and $a_{j_2j_20}<0$,

$$(\overline{x}_{j_1},a_{j_2j_21})=\underbrace{(\text{DP},\text{SI})}_{\text{down-parabola sink}}.$$

  The infinite-equilibrium of $x^*_{j_2}=a_{j_1j_2l_1}=a_{j_2j_21}$ is a (DI,SI)-decreasing-inflection sink.

- For $a_{j_1j_10}(\overline{x}_{j_1}-a_{j_1j_11})(a_{j_1j_2l_1}-a_{j_1j_2l_2})<0$ and $a_{j_2j_20}<0$,

$$(\overline{x}_{j_1},a_{j_2j_21})=\underbrace{(\text{II},\text{SI})}_{\text{increasing-inflection sink}}.$$

The infinite-equilibrium of $x_{j_2}^* = a_{j_1j_2l_1} = a_{j_2j_21}$ is an (II,SI)-increasing-inflection sink.

(vii$_{2b}$) From (vii$_{2a}$), for $\bar{x}_{j_1} = a_{j_1j_11}$, the equilibrium of $(x_{j_1}^*, x_{j_2}^*) = (a_{j_1j_11}, a_{j_2j_2l_1})$ with $a_{j_1j_2l_1} = a_{j_2j_21}$ has the following properties as in Eqs. (1.140)–(1.143).

- For $a_{j_1j_10}(a_{j_1j_2l_1} - a_{j_1j_2l_2}) > 0$ and $a_{j_2j_20} > 0$,

$$(a_{j_1j_11}, a_{j_2j_21}) = \underbrace{(\text{DI : II, SO})}_{\text{inflection source}}.$$

The equilibrium of $(x_{j_1}^*, x_{j_2}^*) = (a_{j_1j_11}, a_{j_2j_21})$ is a (DI:II,SO)-inflection source.

- For $a_{j_1j_10}(a_{j_1j_2l_1} - a_{j_1j_2l_2}) < 0$ and $a_{j_2j_20} > 0$,

$$(a_{j_1j_11}, a_{j_2j_21}) = \underbrace{(\text{II : DI, SO})}_{\text{inflection source}}.$$

The equilibrium of $(x_{j_1}^*, x_{j_2}^*) = (a_{j_1j_11}, a_{j_2j_21})$ is an (II:DI,SO)-inflection source.

- For $a_{j_1j_10}(a_{j_1j_2l_1} - a_{j_1j_2l_2}) > 0$ and $a_{j_2j_20} < 0$,

$$(a_{j_1j_11}, a_{j_2j_21}) = \underbrace{(\text{II : DI, SI})}_{\text{inflection sink}}.$$

The equilibrium of $(x_{j_1}^*, x_{j_2}^*) = (a_{j_1j_11}, a_{j_2j_21})$ is an (II:,DI,SI)-inflection sink.

- For $a_{j_1j_10}(a_{j_1j_2l_1} - a_{j_1j_2l_2}) < 0$ and $a_{j_2j_20} < 0$,

$$(a_{j_1j_11}, a_{j_2j_21}) = \underbrace{(\text{DI : II, SI})}_{\text{inflection sink}}.$$

The equilibrium of $(x_{j_1}^*, x_{j_2}^*) = (a_{j_1j_11}, a_{j_2j_21})$ is a (DI:II,SI)-inflection sink.

(viii) For $\Delta_{j_1j_2} > 0$ and $\Delta_{j_2j_2} = 0$, the standard form is

$$\begin{aligned}
\dot{x}_{j_2} &= a_{j_2j_20}(x_{j_2} - a_{j_2j_2s_1})(x_{j_2} - a_{j_2j_2s_2})^2, \\
\dot{x}_{j_1} &= a_{j_1j_10}(x_{j_1} - a_{j_1j_11})(x_{j_2} - a_{j_1j_21})(x_{j_2} - a_{j_1j_22})
\end{aligned}$$

where

$$\begin{aligned}
&b_{j_1j_21}, b_{j_1j_22} = -\frac{1}{2}(B_{j_1j_2} \pm \sqrt{\Delta_{j_1j_2}}), \\
&\{a_{j_1j_21}, a_{j_1j_22}\} = \text{sort}\{b_{j_1j_21}, b_{j_1j_22}\}, a_{j_1j_21} < a_{j_1j_22}; \\
&a_{j_2j_2s_2} = -\frac{1}{2}B_{j_2j_2}, a_{j_2j_2s_1} = b_{j_2j_21}.
\end{aligned}$$

(viii$_1$) For $a_{j_2j_2s} \neq a_{j_1j_21}, a_{j_1j_22}$ $(s = s_1, s_2)$,

$$\frac{dx_{j_1}}{dx_{j_2}} = \frac{a_{j_1j_10}}{a_{j_2j_20}} \frac{(x_{j_1} - a_{j_1j_11})(x_{j_2} - a_{j_1j_21})(x_{j_2} - a_{j_1j_22})}{(x_{j_2} - a_{j_2j_2s_1})(x_{j_2} - a_{j_2j_2s_2})^2},$$

and the deformation of the foregoing equation is

$$\begin{aligned}
&\frac{1}{(x_{j_1} - a_{j_1j_11})} dx_{j_1} \\
&= \frac{a_{j_1j_10}}{a_{j_2j_20}} \Bigg\{ \frac{\prod_{l_1=1}^{2}(a_{j_2j_2s_1} - a_{j_1j_2l_1})}{(a_{j_2j_2s_1} - a_{j_2j_2s_2})^2} \frac{1}{(x_{j_2} - a_{j_2j_2s_1})} \\
&+ \Big[\frac{\sum_{l_1=1}^{2}(a_{j_2j_2s_2} - a_{j_1j_2l_1})}{(a_{j_2j_2s_2} - a_{j_2j_2s_1})} - \frac{\prod_{l_1=1}^{2}(a_{j_2j_2s_2} - a_{j_1j_2l_1})}{(a_{j_2j_2s_2} - a_{j_2j_2s_1})^2}\Big] \frac{1}{(x_{j_2} - a_{j_2j_2s_2})} \\
&+ \frac{\prod_{l_1=1}^{2}(a_{j_2j_2s_2} - a_{j_1j_2l_1})}{(a_{j_2j_2s_2} - a_{j_2j_2s_1})} \frac{1}{(x_{j_2} - a_{j_2j_2s_2})^2} \Bigg\} dx_{j_2}.
\end{aligned}$$

With an initial condition of (x_{j_10}, x_{j_20}) at $t = t_0$, the integration of the above equation gives the first integral manifold for $a_{j_2j_2s} \neq a_{j_1j_21} (s = s_1, s_2)$ as

$$\begin{aligned}
\ln \frac{|x_{j_1} - a_{j_1j_11}|}{|x_{j_10} - a_{j_1j_11}|} &= \frac{a_{j_1j_10}}{a_{j_2j_20}} \Bigg\{ \frac{\prod_{l_1=1}^{2}(a_{j_2j_2s_1} - a_{j_1j_2l_1})}{(a_{j_2j_2s_1} - a_{j_2j_2s_2})^2} \ln \frac{|x_{j_2} - a_{j_2j_2s_1}|}{|x_{j_20} - a_{j_2j_2s_1}|} \\
&+ \Big[\frac{\sum_{l_1=1}^{2}(a_{j_2j_2s_2} - a_{j_1j_2l_1})}{(a_{j_2j_2s_2} - a_{j_2j_2s_1})} - \frac{\prod_{l_1=1}^{2}(a_{j_2j_2s_2} - a_{j_1j_2l_1})}{(a_{j_2j_2s_2} - a_{j_2j_2s_1})^2}\Big] \\
&\times \ln \frac{|x_{j_2} - a_{j_2j_2s_2}|}{|x_{j_20} - a_{j_2j_2s_2}|} - \frac{\prod_{l_1=1}^{2}(a_{j_2j_2s_2} - a_{j_1j_2l_1})}{(a_{j_2j_2s_2} - a_{j_2j_2s_1})} \\
&\times \Big(\frac{1}{x_{j_2} - a_{j_2j_2s_2}} - \frac{1}{x_{j_20} - a_{j_2j_2s_2}}\Big) \Bigg\}.
\end{aligned}$$

(viii$_{1a}$) At $(x_{j_1}^*, x_{j_2}^*) = (a_{j_1j_11}, a_{j_2j_2s_1})$, the variational equation is

$$\begin{aligned}
\Delta\dot{x}_{j_2} &= a_{j_2j_20} \Delta x_{j_2} (a_{j_2j_2s_1} - a_{j_2j_2s_2})^2, \\
\Delta\dot{x}_{j_1} &= a_{j_1j_10} \prod_{l_1=1}^{2} (a_{j_2j_2s_1} - a_{j_1j_2l_1}) \Delta x_{j_1}.
\end{aligned}$$

The flows at $(x_{j_1}^*, x_{j_2}^*) = (a_{j_1j_11}, a_{j_2j_21})$ in the x_{j_1}-direction are source and sink for $a_{j_1j_10}\prod_{l_1=1}^{2}(a_{j_2j_2s_1} - a_{j_1j_2l_1}) > 0$ and $a_{j_1j_10}\prod_{l_1=1}^{2}(a_{j_2j_2s_1} - a_{j_1j_2l_1}) < 0$, respectively. The flows at $(x_{j_1}^*, x_{j_2}^*) = (a_{j_1j_11}, a_{j_2j_2s_1})$ are source and sink for $a_{j_2j_20}(a_{j_2j_2s_1} - a_{j_2j_2s_2})^2 > 0$ and $a_{j_2j_20}(a_{j_2j_2s_1} - a_{j_2j_2s_2})^2 < 0$ in the x_{j_2}-direction, respectively.

Therefore, the equilibrium of $(x_{j_1}^*, x_{j_2}^*) = (a_{j_1j_11}, a_{j_2j_2s_1})$ has the following properties as in Eqs. (1.147)–(1.150).

- For $a_{j_1j_10}\prod_{l_1=1}^{2}(a_{j_2j_2s_1}-a_{j_1j_2l_1})>0$ and $a_{j_2j_20}(a_{j_2j_2s_1}-a_{j_2j_2s_2})^2>0$,

$$(a_{j_1j_1 1},a_{j_2j_2s_1})=\underbrace{(\mathrm{SO},\mathrm{SO})}_{\text{source}}.$$

The equilibrium of $(x_{j_1}^*,x_{j_2}^*)=(a_{j_1j_1 1},a_{j_2j_2s_1})$ is an (SO,SO)-source.

- For $a_{j_1j_10}\prod_{l_1=1}^{2}(a_{j_2j_2s_1}-a_{j_1j_2l_1})<0$ and $a_{j_2j_20}(a_{j_2j_2s_1}-a_{j_2j_2s_2})^2>0$,

$$(a_{j_1j_1 1},a_{j_2j_2s_1})=\underbrace{(\mathrm{SI},\mathrm{SO})}_{\text{saddle}}.$$

The equilibrium of $(x_{j_1}^*,x_{j_2}^*)=(a_{j_1j_1 1},a_{j_2j_2s_1})$ is an (SI,SO)-saddle.

- For $a_{j_1j_10}\prod_{l_1=1}^{2}(a_{j_2j_2s_1}-a_{j_1j_2l_1})>0$ and $a_{j_2j_20}(a_{j_2j_2s_1}-a_{j_2j_2s_2})^2<0$,

$$(a_{j_1j_1 1},a_{j_2j_2s_1})=\underbrace{(\mathrm{SO},\mathrm{SI})}_{\text{saddle}}.$$

The equilibrium of $(x_{j_1}^*,x_{j_2}^*)=(a_{j_1j_1 1},a_{j_2j_2 1})$ is an (SO,SI)-saddle.

- For $a_{j_1j_10}\prod_{l_1=1}^{2}(a_{j_2j_2s_1}-a_{j_1j_2l_1})<0$ and $a_{j_2j_20}(a_{j_2j_2s_1}-a_{j_2j_2s_2})^2<0$,

$$(a_{j_1j_1 1},a_{j_2j_2s_1})=\underbrace{(\mathrm{SI},\mathrm{SI})}_{\text{sink}}.$$

The equilibrium of $(x_{j_1}^*,x_{j_2}^*)=(a_{j_1j_1 1},a_{j_2j_2s_1})$ is an (SI,SI)-sink.

(viii$_{1b}$) At $(x_{j_1}^*,x_{j_2}^*)=(a_{j_1j_1 1},a_{j_2j_2s_2})$, the variational equations are

$$\begin{aligned}\Delta\dot{x}_{j_2}&=a_{j_2j_20}(a_{j_2j_2s_2}-a_{j_2j_2s_1})(\Delta x_{j_2})^2,\\ \Delta\dot{x}_{j_1}&=a_{j_1j_10}\prod\nolimits_{l_1=1}^{2}(a_{j_2j_2s_2}-a_{j_1j_2l_1})\Delta x_{j_1}.\end{aligned}$$

The flows at $(x_{j_1}^*,x_{j_2}^*)=(a_{j_1j_1 1},a_{j_2j_2s_2})$ in the x_{j_1}-direction are source and sink for $a_{j_1j_10}\prod_{l_1=1}^{2}(a_{j_2j_2s_2}-a_{j_1j_2l_1})>0$ and $a_{j_1j_10}\prod_{l_1=1}^{2}(a_{j_2j_2s_2}-a_{j_1j_2l_1})<0$, respectively. The flows at $(x_{j_1}^*,x_{j_2}^*)=(a_{j_1j_1 1},a_{j_2j_2s_2})$ in the x_{j_2}-direction are upper-saddle and lower-saddle for $a_{j_2j_20}(a_{j_2j_2s_2}-a_{j_2j_2s_1})>0$ and $a_{j_2j_20}(a_{j_2j_2s_2}-a_{j_2j_2s_1})<0$, respectively.

Therefore, the equilibrium of $(x_{j_1}^*,x_{j_2}^*)=(a_{j_1j_1 1},a_{j_2j_2s_2})$ has the following properties as in Eqs. (1.151)–(1.154).

- For $a_{j_1j_10}\prod_{l_1=1}^{2}(a_{j_2j_2s_2}-a_{j_1j_2l_1})>0$ and $a_{j_2j_20}(a_{j_2j_2s_2}-a_{j_2j_2s_1})>0$,

$$(a_{j_1j_1 1},a_{j_2j_2s_2})=\underbrace{(\text{SO},\text{US})}_{\text{upper-saddle source}}.$$

The equilibrium of $(x_{j_1}^*,x_{j_2}^*)=(a_{j_1j_1 1},a_{j_2j_2s_2})$ is an (SO,US)-upper-saddle source.

- For $a_{j_1j_10}\prod_{l_1=1}^{2}(a_{j_2j_2s_2}-a_{j_1j_2l_1})<0$ and $a_{j_2j_20}(a_{j_2j_2s_2}-a_{j_2j_2s_1})>0$,

$$(a_{j_1j_1 1},a_{j_2j_2s_2})=\underbrace{(\text{SI},\text{US})}_{\text{upper-saddle-sink}}.$$

The equilibrium of $(x_{j_1}^*,x_{j_2}^*)=(a_{j_1j_1 1},a_{j_2j_2s_2})$ is an (SI,US)-upper-saddle sink.

- For $a_{j_1j_10}\prod_{l_1=1}^{2}(a_{j_2j_2s_2}-a_{j_1j_2l_1})>0$ and $a_{j_2j_20}(a_{j_2j_2s_2}-a_{j_2j_2s_1})<0$,

$$(a_{j_1j_1 1},a_{j_2j_2s_2})=\underbrace{(\text{SO},\text{LS})}_{\text{lower-saddle source}}.$$

The equilibrium of $(x_{j_1}^*,x_{j_2}^*)=(a_{j_1j_1 1},a_{j_2j_2s_2})$ is an (SO,LS)-lower-saddle source.

- For $a_{j_1j_10}\prod_{l_1=1}^{2}(a_{j_2j_2s_2}-a_{j_1j_2l_1})<0$ and $a_{j_2j_20}(a_{j_2j_2s_2}-a_{j_2j_2s_1})<0$,

$$(a_{j_1j_1 1},a_{j_2j_2s_2})=\underbrace{(\text{SI},\text{LS})}_{\text{lower-saddle source}}.$$

The equilibrium of $(x_{j_1}^*,x_{j_2}^*)=(a_{j_1j_1 1},a_{j_2j_2s_2})$ is an (SI,LS)-lower-saddle sink.

(viii$_{1c}$) For $x_{j_2}^*=a_{j_1j_2l_1}\neq a_{j_2j_2s_1},a_{j_2j_2s_2}$ and $\bar{x}_{j_1}\neq a_{j_1j_1 1}$ (l_1, $l_2\in\{1,2\}$, $l_1\neq l_2$), in phase space,

$$\left.\frac{dx_{j_1}}{dx_{j_2}}\right|_{x_{j_2}^*=a_{j_1j_2s_1}}=\frac{a_{j_1j_10}}{a_{j_2j_20}}\left.\frac{(\bar{x}_{j_1}-a_{j_1j_1 1})(x_{j_2}-a_{j_1j_2l_1})(x_{j_2}-a_{j_1j_2l_2})}{(x_{j_2}-a_{j_2j_2s_1})(x_{j_2}-a_{j_2j_2s_2})^2}\right|_{x_{j_2}^*=a_{j_1j_2s_1}}=0.$$

If

$$\left.\frac{d^2x_{j_1}}{dx_{j_2}^2}\right|_{x_{j_2}^*=a_{j_1j_2l_1}}=\frac{a_{j_1j_10}}{a_{j_2j_20}}\frac{(\bar{x}_{j_1}-a_{j_1j_1 1})(a_{j_1j_2l_1}-a_{j_1j_2l_2})}{(a_{j_1j_2l_1}-a_{j_2j_2s_1})(a_{j_1j_2l_1}-a_{j_2j_2s_2})^2}>0,$$

there is an up-parabola flow at $x_{j_2}^*=a_{j_1j_2l_1}\neq a_{j_2j_2s_1},a_{j_2j_2s_2}$ with $\bar{x}_{j_1}\neq a_{j_1j_1 1}$. If

$$\left.\frac{d^2x_{j_1}}{dx_{j_2}^2}\right|_{x_{j_2}^*=a_{j_1j_2l_1}}=\frac{a_{j_1j_10}}{a_{j_2j_20}}\frac{(\bar{x}_{j_1}-a_{j_1j_1 1})(a_{j_1j_2l_1}-a_{j_1j_2l_2})}{(a_{j_1j_2l_1}-a_{j_2j_2s_1})(a_{j_1j_2l_1}-a_{j_2j_2s_2})^2}<0,$$

there is a down-parabola flow at $x_{j_2}^*=a_{j_1j_2l_1}\neq a_{j_2j_2s_1},a_{j_2j_2s_2}$ with $\bar{x}_{j_1}\neq a_{j_1j_1 1}$. Let

$$\dot{x}_{j_2} = (a_{j_1j_2l_1} - a_{j_2j_2s_1})(a_{j_1j_2l_1} - a_{j_2j_2s_2})^2 \neq 0.$$

The parabola flows at $x^*_{j_2} = a_{j_1j_2l_1} \neq a_{j_2j_2s_1}, a_{j_2j_2s_2}$ with $\overline{x}_{j_1} \neq a_{j_1j_11}$ are positive and negative for $a_{j_2j_20} > 0$ and $a_{j_2j_20} < 0$, respectively.

- For $a_{j_1j_10}(\overline{x}_{j_1} - a_{j_1j_11})(a_{j_1j_2l_1} - a_{j_1j_2l_2}) > 0$ and $a_{j_2j_20}(a_{j_1j_2l_1} - a_{j_2j_2s_1})(a_{j_1j_2l_1} - a_{j_2j_2s_2})^2 > 0$,

$$(a_{j_1j_2s_1}, \dot{x}_{j_2}) = \underbrace{(\text{UP}, \text{pF})}_{\text{up-parabola flow } (+)}.$$

- For $a_{j_1j_10}(\overline{x}_{j_1} - a_{j_1j_11})(a_{j_1j_2l_1} - a_{j_1j_2l_2}) < 0$ and $a_{j_2j_20}(a_{j_1j_2l_1} - a_{j_2j_2s_1})(a_{j_1j_2l_1} - a_{j_2j_2s_2})^2 > 0$,

$$(a_{j_1j_2s_1}, \dot{x}_{j_2}) = \underbrace{(\text{DP}, \text{pF})}_{\text{down-parabola flow } (+)}.$$

- For $a_{j_1j_10}(\overline{x}_{j_1} - a_{j_1j_11})(a_{j_1j_2l_1} - a_{j_1j_2l_2}) > 0$ and $a_{j_2j_20}(a_{j_1j_2l_1} - a_{j_2j_2s_1})(a_{j_1j_2l_1} - a_{j_2j_2s_2})^2 < 0$,

$$(a_{j_1j_2s_1}, \dot{x}_{j_2}) = \underbrace{(\text{DP}, \text{nF})}_{\text{down-parabola flow } (-)}.$$

- For $a_{j_1j_10}(\overline{x}_{j_1} - a_{j_1j_11})(a_{j_1j_2l_1} - a_{j_1j_2l_2}) < 0$ and $a_{j_2j_20}(a_{j_1j_2l_1} - a_{j_2j_2s_1})(a_{j_1j_2l_1} - a_{j_2j_2s_2})^2 < 0$,

$$(a_{j_1j_2s_1}, \dot{x}_{j_2}) = \underbrace{(\text{UP}, \text{nF})}_{\text{up-parabola flow } (-)}.$$

Therefore, from the above analysis, for $\overline{x}_{j_1} = a_{j_1j_11}$, the equilibrium of $(x^*_{j_1}, x^*_{j_2}) = (a_{j_1j_11}, a_{j_1j_2l_1})$ with $a_{j_1j_2l_1} \neq a_{j_2j_2s_1}, a_{j_2j_2s_2}$ (l_1, $l_2 \in \{1,2\}$, $l_1 \neq l_2$) has the following properties as in Eqs. (1.155)–(1.158).

- For $a_{j_1j_10}(a_{j_1j_2l_1} - a_{j_1j_2l_2}) > 0$ and $a_{j_2j_20}(a_{j_1j_2l_1} - a_{j_2j_2s_1})(a_{j_1j_2l_1} - a_{j_2j_2s_2})^2 > 0$,

$$(a_{j_1j_11}, a_{j_1j_2l_1}) = \underbrace{(\text{DP} : \text{UP}, \text{pF})}_{\text{hyperbolic flow } (+)}.$$

The equilibrium of $(x^*_{j_1}, x^*_{j_2}) = (a_{j_1j_11}, a_{j_1j_2l_1})$ is a (DP:UP, pF)-positive hyperbolic flow.

- For $a_{j_1j_10}(a_{j_1j_2l_1}-a_{j_1j_2l_2})<0$ and $a_{j_2j_20}(a_{j_1j_2l_1}-a_{j_2j_2s_1})(a_{j_1j_2l_1}-a_{j_2j_2s_2})^2>0$,

$$(a_{j_1j_11},a_{j_1j_2l_1})=\underbrace{(\mathrm{UP}:\mathrm{DP},\mathrm{pF})}_{\text{hyperbolic-secant flow }(+)}.$$

The equilibrium of $(x^*_{j_1},x^*_{j_2})=(a_{j_1j_11},a_{j_1j_2l_1})$ is a (UP:DP, pF)-positive hyperbolic-secant flow.

- For $a_{j_1j_10}(a_{j_1j_2l_1}-a_{j_1j_2l_2})>0$ and $a_{j_2j_20}(a_{j_1j_2l_1}-a_{j_2j_2s_1})(a_{j_1j_2l_1}-a_{j_2j_2s_2})^2<0$,

$$(a_{j_1j_11},a_{j_1j_2l_1})=\underbrace{(\mathrm{UP}:\mathrm{DP},\mathrm{nF})}_{\text{hyperbolic-secant flow }(-)}.$$

The equilibrium of $(x^*_{j_1},x^*_{j_2})=(a_{j_1j_11},a_{j_1j_2l_1})$ is a (UP:DP,nF)-negative hyperbolic-secant flow.

- For $a_{j_1j_10}(a_{j_1j_2l_1}-a_{j_1j_2l_2})<0$ and $a_{j_2j_20}(a_{j_1j_2l_1}-a_{j_2j_2s_1})(a_{j_1j_2l_1}-a_{j_2j_2s_2})^2<0$,

$$(a_{j_1j_11},a_{j_1j_2l_1})=\underbrace{(\mathrm{DP}:\mathrm{UP},\mathrm{nF})}_{\text{hyperbolic flow }(-)}.$$

The equilibrium of $(x^*_{j_1},x^*_{j_2})=(a_{j_1j_11},a_{j_1j_2l_1})$ is a (DP:UP,nF)-negative hyperbolic flow.

(viii$_2$) For $a_{j_2j_2s_1}=a_{j_2j_2s_2}=a_{j_2j_21}$, the standard form is

$$\begin{aligned}\dot{x}_{j_2}&=a_{j_2j_20}(x_{j_2}-a_{j_2j_21})^3,\\ \dot{x}_{j_1}&=a_{j_1j_10}(x_{j_1}-a_{j_1j_11})(x_{j_2}-a_{j_1j_21})(x_{j_2}-a_{j_1j_22}).\end{aligned}$$

In phase space,

$$\frac{dx_{j_1}}{dx_{j_2}}=\frac{a_{j_1j_10}}{a_{j_2j_20}}\frac{(x_{j_1}-a_{j_1j_11})(x_{j_2}-a_{j_1j_21})(x_{j_2}-a_{j_1j_22})}{(x_{j_2}-a_{j_2j_21})^3},$$

and the deformation of the foregoing equation is

$$\frac{dx_{j_1}}{x_{j_1}-a_{j_1j_1}}=\frac{a_{j_1j_10}}{a_{j_2j_20}}\left[\frac{1}{x_{j_2}-a_{j_2j_21}}+\frac{\sum_{l_1=1}^{2}(a_{j_2j_21}-a_{j_1j_2l_1})}{(x_{j_2}-a_{j_2j_21})^2}+\frac{\prod_{l_1=1}^{2}(a_{j_2j_21}-a_{j_1j_2l_1})}{(x_{j_2}-a_{j_2j_21})^3}\right]dx_{j_2}.$$

With an initial condition of (x_{j_10},x_{j_20}) at $t=t_0$, the integration of the above equation gives the first integral manifold as

$$\ln\frac{|x_{j_1}-a_{j_1j_11}|}{|x_{j_10}-a_{j_1j_11}|}=\frac{a_{j_1j_10}}{a_{j_2j_20}}\Big\{\ln\frac{|x_{j_2}-a_{j_2j_21}|}{|x_{j_20}-a_{j_2j_21}|}-\sum\nolimits_{l_1=1}^{2}(a_{j_2j_21}-a_{j_1j_2l_1})\Big[\frac{1}{x_{j_2}-a_{j_2j_21}}-\frac{1}{x_{j_20}-a_{j_2j_21}}\Big]$$
$$-\frac{1}{2}\prod\nolimits_{l_1=1}^{2}(a_{j_2j_21}-a_{j_1j_2l_1})\Big[\frac{1}{(x_{j_2}-a_{j_2j_21})^2}-\frac{1}{(x_{j_20}-a_{j_2j_21})^2}\Big]\Big\}.$$

For $a_{j_2j_2s_1}=a_{j_2j_2s_2}=a_{j_1j_21}$, at $(x_{j_1}^*,x_{j_2}^*)=(a_{j_1j_11},a_{j_2j_21})$, the variational equations are

$$\Delta\dot{x}_{j_2}=a_{j_2j_20}(\Delta x_{j_2})^3,$$
$$\Delta\dot{x}_{j_1}=a_{j_1j_10}\Delta x_{j_1}\prod\nolimits_{l_1=1}^{2}(a_{j_2j_21}-a_{j_1j_2l_1}).$$

Because of

$$\prod\nolimits_{l_1=1}^{2}(a_{j_2j_21}-a_{j_1j_2l_1})\neq 0,$$

the flows at $(x_{j_1}^*,x_{j_2}^*)=(a_{j_1j_11},a_{j_2j_2s_1})$ in the x_{j_1}-direction are source and sink for $a_{j_1j_10}\prod_{l_1=1}^{2}(a_{j_2j_21}-a_{j_1j_2l_1})>0$ and $a_{j_1j_10}\prod_{l_1=1}^{2}(a_{j_2j_2l_1}-a_{j_1j_2l_1})<0$, respectively; and the flows at $(x_{j_1}^*,x_{j_2}^*)=(a_{j_1j_11},a_{j_2j_2s_1})$ in the x_{j_2}-direction are third-order source and sink for $a_{j_2j_20}>0$ and $a_{j_2j_20}<0$, respectively.

Therefore, the equilibrium of $(x_{j_1}^*,x_{j_2}^*)=(a_{j_1j_11},a_{j_2j_21})$ has the following properties as in Eqs. (1.161)–(1.164).

- For $a_{j_1j_10}\prod_{l_1=1}^{2}(a_{j_2j_21}-a_{j_1j_2l_1})>0$ and $a_{j_2j_20}>0$,

$$(a_{j_1j_11},a_{j_2j_21})=\underbrace{(\text{SO},3^{\text{rd}}\text{SO})}_{\text{third-order source}}.$$

  The equilibrium of $(x_{j_1}^*,x_{j_2}^*)=(a_{j_1j_11},a_{j_2j_21})$ is an (SO, 3rdSO)-third-order source, which is for an appearing bifurcation from source to source, saddle, and source.

- For $a_{j_1j_10}\prod_{l_1=1}^{2}(a_{j_2j_21}-a_{j_1j_2l_1})<0$ and $a_{j_2j_20}>0$,

$$(a_{j_1j_11},a_{j_2j_21})=\underbrace{(\text{SI},3^{\text{rd}}\text{SO})}_{\text{third-order saddle}}.$$

  The equilibrium of $(x_{j_1}^*,x_{j_2}^*)=(a_{j_1j_11},a_{j_2j_21})$ is an (SI,3rdSO)-third-order saddle, which is for an appearing bifurcation from saddle to saddle, sink, and saddle.

- For $a_{j_1j_10}\prod_{l_1=1}^{2}(a_{j_2j_21}-a_{j_1j_2l_1})>0$ and $a_{j_2j_20}<0$,

$$(a_{j_1j_11},a_{j_2j_21})=\underbrace{(\text{SO},3^{\text{rd}}\text{SI})}_{\text{saddle}}.$$

The equilibrium of $(x_{j_1}^{*},x_{j_2}^{*})=(a_{j_1j_11},a_{j_2j_21})$ is an (SO,3rdSI)-third-order saddle, which is for an appearing bifurcation from saddle to saddle, source, and saddle.

- For $a_{j_1j_10}\prod_{l_1=1}^{2}(a_{j_2j_21}-a_{j_1j_2l_1})<0$ and $a_{j_2j_20}<0$,

$$(a_{j_1j_11},a_{j_2j_21})=\underbrace{(\text{SI},3^{\text{rd}}\text{SI})}_{\text{sink}}.$$

The equilibrium of $(x_{j_1}^{*},x_{j_2}^{*})=(a_{j_1j_11},a_{j_2j_21})$ is an (SI,3rdSI)-third-order sink, which is for an appearing bifurcation from sink to sink, saddle, and sink.

(viii$_3$) For $a_{j_2j_2s_1}=a_{j_2j_2l_1}\cdot(l_1,l_2\in\{1,2\},l_1\neq l_2)$,

$$\frac{dx_{j_1}}{dx_{j_2}}=\frac{a_{j_1j_10}}{a_{j_2j_20}}\frac{(x_{j_1}-a_{j_1j_11})(x_{j_2}-a_{j_1j_2l_2})}{(x_{j_2}-a_{j_2j_2s_2})^2},$$

and the deformation of the foregoing equation is

$$\frac{1}{(x_{j_1}-a_{j_1j_11})}dx_{j_1}=\frac{a_{j_1j_10}}{a_{j_2j_20}}\Big[\frac{1}{(x_{j_2}-a_{j_2j_2s_2})}+\frac{(a_{j_2j_2s_2}-a_{j_1j_2l_2})}{(x_{j_2}-a_{j_2j_2s_2})^2}\Big]dx_{j_2}.$$

With an initial condition of (x_{j_10},x_{j_20}) at $t=t_0$, the integration of the above equation gives the first integral manifold for $a_{j_2j_2s_1}\neq a_{j_1j_2l_1}(s_1,s_2,l_2,l_2\in\{1,2\},s_1\neq s_2,l_2\neq l_2)$ as

$$\begin{aligned}\ln\frac{|x_{j_1}-a_{j_1j_11}|}{|x_{j_10}-a_{j_1j_11}|}=&\frac{a_{j_1j_10}}{a_{j_2j_20}}\Big[\ln\frac{|x_{j_2}-a_{j_2j_2s_2}|}{|x_{j_20}-a_{j_2j_2s_2}|}\\&-(a_{j_2j_2s_2}-a_{j_1j_2l_2})\Big(\frac{1}{x_{j_2}-a_{j_2j_2s_2}}-\frac{1}{x_{j_20}-a_{j_2j_2s_2}}\Big)\Big].\end{aligned}$$

(viii$_{3a}$) For $x_{j_2}^{*}=a_{j_1j_2l_1}=a_{j_2j_2s_1}$ with $\bar{x}_{j_1}\neq a_{j_1j_11}$, if

$$\frac{dx_{j_1}}{dx_{j_2}}\Big|_{x_{j_2}^{*}=a_{j_1j_2l_1}}=\frac{a_{j_1j_10}}{a_{j_2j_20}}\frac{(\bar{x}_{j_1}-a_{j_1j_11})(a_{j_1j_2l_1}-a_{j_1j_2l_2})}{(a_{j_2j_2s_1}-a_{j_2j_2s_2})^2}>0,$$

there is an increasing-inflection flow, if

$$\left.\frac{dx_{j_1}}{dx_{j_2}}\right|_{x_{j_2}^*=a_{j_2j_2l_1}} = \frac{a_{j_1j_10}}{a_{j_2j_20}} \frac{(\overline{x}_{j_1} - a_{j_1j_11})(a_{j_1j_2l_1} - a_{j_1j_2l_2})}{(a_{j_2j_2s_1} - a_{j_2j_2s_2})^2} < 0,$$

there is a decreasing-inflection flow. At $x_{j_2}^* = a_{j_1j_2l_1} = a_{j_2j_2s_1}$, the variational equation is

$$\Delta\dot{x}_{j_2} = a_{j_2j_20}(a_{j_2j_2s_1} - a_{j_2j_2s_2})^2 \Delta x_{j_2}.$$

The inflection flows are source and sink flows for $a_{j_2j_20}(a_{j_2j_2s_1} - a_{j_2j_2s_2})^2 > 0$ and $a_{j_2j_20}(a_{j_2j_2s_1} - a_{j_2j_2s_2})^2 < 0$, respectively.

Therefore, the infinite-equilibrium of $x_{j_2}^* = a_{j_1j_2l_1} = a_{j_2j_2s_1}$ with $\overline{x}_{j_1} \neq a_{j_1j_11}$ has the following properties as in Eqs. (1.166)–(1.169).

- For $a_{j_1j_10}(\overline{x}_{j_1} - a_{j_1j_11})(a_{j_1j_2l_1} - a_{j_jj_2l_2}) > 0$ and $a_{j_2j_20}(a_{j_2j_2s_1} - a_{j_2j_2s_2})^2 > 0$,

$$(\overline{x}_{j_1}, a_{j_2j_2s_1}) = \underbrace{(\text{II}, \text{SO})}_{\text{increasing-inflection source}}.$$

  The infinite-equilibrium of $x_{j_2}^* = a_{j_1j_2l_1} = a_{j_2j_2s_1}$ is an (II,SO)-increasing-inflection source.
- For $a_{j_1j_10}(\overline{x}_{j_1} - a_{j_1j_11})(a_{j_1j_2l_1} - a_{j_1j_2j_2}) < 0$ and $a_{j_2j_20}(a_{j_2j_2s_1} - a_{j_2j_2s_2})^2 > 0$,

$$(\overline{x}_{j_1}, a_{j_2j_2s_1}) = \underbrace{(\text{DI}, \text{SO})}_{\text{decreasing-inflection source}}.$$

  The infinite-equilibrium of $x_{j_2}^* = a_{j_1j_2l_1} = a_{j_2j_2s_1}$ is a (DI,SO)-decreasing-inflection source.
- For $a_{j_1j_10}(\overline{x}_{j_1} - a_{j_1j_11})(a_{j_1j_2l_1} - a_{j_1j_2l_2}) > 0$ and $a_{j_2j_20}(a_{j_2j_2s_1} - a_{j_2j_2s_2})^2 < 0$,

$$(\overline{x}_{j_1}, a_{j_2j_2s_1}) = \underbrace{(\text{DI}, \text{SI})}_{\text{decreasing-inflection sink}}.$$

  The infinite-equilibrium of $x_{j_2}^* = a_{j_1j_2l_1} = a_{j_2j_2s_1}$ is a (DI,SI)-decreasing-inflection sink.
- For $a_{j_1j_10}(\overline{x}_{j_1} - a_{j_1j_11})(a_{j_1j_2l_1} - a_{j_1j_2l_2}) < 0$ and $a_{j_2j_20}(a_{j_2j_2s_1} - a_{j_2j_2s_2})^2 < 0$,

$$(\overline{x}_{j_1}, a_{j_2j_2s_1}) = \underbrace{(\text{II}, \text{SI})}_{\text{increasing-inflection sink}}.$$

  The infinite-equilibrium of $x_{j_2}^* = a_{j_1j_2l_1} = a_{j_2j_2s_1}$ is an (II,SI)-increasing-inflection sink.

(viii$_{3b}$) From (viii$_{3a}$), for $\overline{x}_{j_1} = a_{j_1j_11}$, the equilibrium of $(x^*_{j_1}, x^*_{j_2}) = (a_{j_1j_11}, a_{j_2j_2s_1})$ with $a_{j_1j_2l_1} = a_{j_2j_2s_1}$ has the following properties as in Eqs. (1.170)–(1.173).

- For $a_{j_1j_10}(a_{j_1j_2l_1} - a_{j_1j_2l_2}) > 0$ and $a_{j_2j_20}(a_{j_2j_2s_1} - a_{j_2j_2s_2})^2 > 0$,

$$(a_{j_1j_11}, a_{j_2j_2s_1}) = \underbrace{(\text{DI} : \text{II}, \text{SO})}_{\text{inflection source}}.$$

The equilibrium of $(x^*_{j_1}, x^*_{j_2}) = (a_{j_1j_11}, a_{j_2j_2s_1})$ is a (DI:II,SO)-inflection source.

- For $a_{j_1j_10}(a_{j_1j_2l_1} - a_{j_1j_2l_2}) < 0$ and $a_{j_2j_20}(a_{j_2j_2s_1} - a_{j_2j_2s_2})^2 > 0$,

$$(a_{j_1j_1}, a_{j_2j_2s_1}) = \underbrace{(\text{II} : \text{DI}, \text{SO})}_{\text{infection source}}.$$

The equilibrium of $(x^*_{j_1}, x^*_{j_2}) = (a_{j_1j_11}, a_{j_2j_2s_1})$ is an (II:DI,SO)-inflection source.

- For $a_{j_1j_10}(a_{j_1j_2l_1} - a_{j_1j_2l_2}) > 0$ and $a_{j_2j_20}(a_{j_2j_2s_1} - a_{j_2j_2s_2})^2 < 0$,

$$(a_{j_1j_11}, a_{j_2j_2s_1}) = \underbrace{(\text{II} : \text{DI}, \text{SI})}_{\text{inffection sink}}.$$

The equilibrium of $(x^*_{j_1}, x^*_{j_2}) = (a_{j_1j_11}, a_{j_2j_2s_1})$ is an (II:DI,SI)-inflection sink.

- For $a_{j_1j_10}(a_{j_1j_2l_1} - a_{j_1j_2l_2}) < 0$ and $a_{j_2j_20}(a_{j_2j_2s_1} - a_{j_2j_2s_2})^2 < 0$,

$$(a_{j_1j_11}, a_{j_2j_2s_1}) = \underbrace{(\text{DI} : \text{II}, \text{SI})}_{\text{inflection sink}}.$$

The equilibrium of $(x^*_{j_1}, x^*_{j_2}) = (a_{j_1j_11}, a_{j_2j_2s_1})$ is a (DI:II,SI)-inflection sink.

(viii$_4$) For $a_{j_2j_2s_2} = a_{j_1j_2l_1} (l_1, l_2 \in \{1, 2\}, l_1 \neq l_2)$, in phase space,

$$\frac{dx_{j_1}}{dx_{j_2}} = \frac{a_{j_1j_10}}{a_{j_2j_20}} \frac{(x_{j_1} - a_{j_1j_11})(x_{j_2} - a_{j_1j_2l_2})}{(x_{j_2} - a_{j_2j_2s_1})(x_{j_2} - a_{j_2j_2s_2})},$$

and the deformation of the foregoing equation is

$$\begin{aligned}\frac{1}{(x_{j_1} - a_{j_1j_11})} dx_{j_1} = & \frac{a_{j_1j_10}}{a_{j_2j_20}} \Big[\frac{a_{j_2j_2s_1} - a_{j_1j_2l_2}}{a_{j_2j_2s_1} - a_{j_2j_2s_2}} \frac{1}{x_{j_2} - a_{j_2j_2s_1}} \\ & + \frac{a_{j_2j_2s_2} - a_{j_1j_2l_2}}{a_{j_2j_2s_2} - a_{j_2j_2s_2}} \frac{1}{x_{j_2} - a_{j_2j_2s_2}} \Big] dx_{j_2}.\end{aligned}$$

With an initial condition of (x_{j_10}, x_{j_20}) at $t = t_0$, the integration of the above equation gives the first integral manifold for $a_{j_2j_2s_2} = a_{j_1j_2l_1} (l_1, l_2 \in \{1, 2\}, l_1 \neq l_2)$ as

$$\ln \frac{|x_{j_1} - a_{j_1j_1 1}|}{|x_{j_1 0} - a_{j_1j_1 1}|} = \frac{a_{j_1j_1 0}}{a_{j_2j_2 0}} \Big[\frac{(a_{j_2j_2s_1} - a_{j_1j_2l_2})}{(a_{j_2j_2s_1} - a_{j_2j_2s_2})} \ln \frac{|x_{j_2} - a_{j_2j_2s_1}|}{|x_{j_2 0} - a_{j_2j_2s_1}|}$$
$$+ \frac{(a_{j_2j_2s_2} - a_{j_1j_2l_2})}{(a_{j_2j_2s_2} - a_{j_2j_2s_1})} \ln \frac{|x_{j_2} - a_{j_2j_2s_2}|}{|x_{j_2 0} - a_{j_2j_2s_2}|}\Big].$$

(viii$_{4a}$) For $x^*_{j_2} = a_{j_1j_2l_1} = a_{j_2j_2s_2}$ with $\overline{x}_{j_1} \neq a_{j_1j_1 1}$ and $\varepsilon \to 0$,

$$\left.\frac{dx_{j_1}}{dx_{j_2}}\right|_{x^*{}_{j_2} = a_{j_2j_2s_2} \pm \varepsilon} = \frac{a_{j_1j_1 0}}{a_{j_2j_2 0}} \frac{(\overline{x}_{j_1} - a_{j_1j_1 1})(a_{j_2j_2s_2} - a_{j_1j_2l_2})}{(a_{j_2j_2s_2} - a_{j_2j_2s_1})\varepsilon}.$$

If

$$\frac{a_{j_1j_1 0}}{a_{j_2j_2 0}} \frac{(\overline{x}_{j_1} - a_{j_1j_1 1})(a_{j_2j_2s_2} - a_{j_1j_2l_2})}{(a_{j_2j_2s_2} - a_{j_2j_2s_1})} > 0,$$

there is a down-up flow at $x^*_{j_2} = a_{j_1j_2l_1} = a_{j_2j_2s_2}$ with $\overline{x}_{j_1} \neq a_{j_1j_1 1}$. If

$$\frac{a_{j_1j_1 0}}{a_{j_2j_2 0}} \frac{(\overline{x}_{j_1} - a_{j_1j_1 1})(a_{j_2j_2s_2} - a_{j_1j_2l_2})}{(a_{j_2j_2s_2} - a_{j_2j_2s_1})} < 0,$$

there is an up-down flow at $x^*_{j_2} = a_{j_1j_2l_1} = a_{j_2j_2s_2}$ with $\overline{x}_{j_1} \neq a_{j_1j_1 1}$.

The corresponding variational equation is

$$\Delta\dot{x}_{j_2} = a_{j_2j_2 0}(a_{j_2j_2s_2} - a_{j_2j_2s_1})(\Delta x_{j_2})^2.$$

The up-down and down-up flows in the x_{j_2}-direction are upper-saddle and lower-saddle for $a_{j_2j_2 0}(a_{j_2j_2s_2} - a_{j_2j_2s_1}) > 0$ and $a_{j_2j_0 0}(a_{j_2j_2s_2} - a_{j_2j_2s_1}) < 0$.

Therefore, the infinite-equilibrium of $x^*_{j_2} = a_{j_1j_2l_1} = a_{j_2j_2s_2}$ with $\overline{x}_{j_1} \neq a_{j_1j_1 1}$ has the following properties as in Eqs. (1.175)–(1.178).

- For $a_{j_1j_1 1}(\overline{x}_{j_1} - a_{j_1j_1 1})(a_{j_1j_2l_1} - a_{j_1j_2l_2}) > 0$ and $a_{j_2j_2 0}(a_{j_2j_2s_2} - a_{j_2j_2s_1}) > 0$,

$$(\overline{x}_{j_1}, a_{j_1j_2l_1}) = \underbrace{(\text{DU}, \text{US})}_{\text{down-up upper-saddle}}.$$

  The infinite-equilibrium of $x^*_{j_2} = a_{j_1j_2l_1} = a_{j_2j_2s_2}$ is a (DU,US) down-up upper-saddle.
- For $a_{j_1j_1 0}(\overline{x}_{j_1} - a_{j_1j_1 1})(a_{j_1j_2l_1} - a_{j_jj_2l_2}) < 0$ and $a_{j_2j_2 0}(a_{j_2j_2s_2} - a_{j_2j_2s_1}) > 0$,

$$(\overline{x}_{j_1}, a_{j_1j_2l_1}) = \underbrace{(\text{UD}, \text{US})}_{\text{up-down upper-saddle}}.$$

The infinite-equilibrium of $x_{j_2}^* = a_{j_1 j_2 l_1} = a_{j_2 j_2 s_2}$ is a (UD,US)-up-down upper-saddle.

- For $a_{j_1 j_1 0}(\overline{x}_{j_1} - a_{j_1 j_1 1})(a_{j_1 j_2 l_1} - a_{j_1 j_2 l_2}) > 0$ and $a_{j_2 j_2 0}(a_{j_2 j_2 s_2} - a_{j_2 j_2 s_1}) < 0$,

$$(\overline{x}_{j_1}, a_{j_1 j_2 l_1}) = \underbrace{(\mathrm{UD}, \mathrm{LS})}_{\text{up-down lower-saddle}}.$$

The infinite-equilibrium of $x_{j_2}^* = a_{j_1 j_2 l_1} = a_{j_2 j_2 s_2}$ is a (UD,LS)-up-down lower-saddle.

- For $a_{j_1 j_1 0}(\overline{x}_{j_1} - a_{j_1 j_1 1})(a_{j_1 j_2 l_1} - a_{j_1 j_2 l_2}) < 0$ and $a_{j_2 j_2 0}(a_{j_2 j_2 s_2} - a_{j_2 j_2 s_1}) < 0$,

$$(\overline{x}_{j_1}, a_{j_1 j_2 l_1}) = \underbrace{(\mathrm{DU}, \mathrm{LS})}_{\text{down-up lower-saddle}}.$$

The infinite-equilibrium of $x_{j_2}^* = a_{j_1 j_2 l_1} = a_{j_2 j_2 s_2}$ is a (DU,LS)-down-up lower-saddle.

(viii$_{4b}$) From (viii$_{4a}$), for $\overline{x}_{j_1} = a_{j_1 j_1 1}$, the equilibrium of $(x_{j_1}^*, x_{j_2}^*) = (a_{j_1 j_1 1}, a_{j_2 j_2 s_2})$ with $a_{j_2 j_2 s_2} = a_{j_1 j_2 l_1}$ has the following properties as in Eqs. (1.189)–(1.192).

- For $a_{j_1 j_1 0}(a_{j_1 j_2 l_1} - a_{j_1 j_2 l_2}) > 0$ and $a_{j_2 j_2 0}(a_{j_2 j_2 s_2} - a_{j_2 j_2 s_1}) > 0$,

$$(a_{j_1 j_1 1}, a_{j_2 j_2 s_2}) = \underbrace{(\mathrm{DU} : \mathrm{UD}, \mathrm{US})}_{\text{down-up-to-up-down upper-saddle}}.$$

The equilibrium of $(x_{j_1}^*, x_{j_2}^*) = (a_{j_1 j_1 1}, a_{j_2 j_2 1})$ is a (DU:UD,US)-down-up-to-up-down upper-saddle.

- For $a_{j_1 j_1 0}(a_{j_1 j_2 1} - a_{j_1 j_2 l_2}) < 0$ and $a_{j_2 j_2 0}(a_{j_2 j_2 s_2} - a_{j_2 j_2 s_1}) > 0$,

$$(a_{j_1 j_1 1}, a_{j_2 j_2 s_2}) = \underbrace{(\mathrm{UD} : \mathrm{DU}, \mathrm{US})}_{\text{up-down-to-down-up upper-ssaddle}}.$$

The equilibrium of $(x_{j_1}^*, x_{j_2}^*) = (a_{j_1 j_1 1}, a_{j_2 j_2 1})$ is a (UD:DU,US)-up-down-to-down-up upper-saddle.

- For $a_{j_1 j_1 0}(a_{j_1 j_2 l_1} - a_{j_1 j_2 l_2}) > 0$ and $a_{j_2 j_2 0}(a_{j_2 j_2 s_2} - a_{j_2 j_2 s_1}) < 0$,

$$(a_{j_1 j_1 1}, a_{j_2 j_2 s_2}) = \underbrace{(\mathrm{UD} : \mathrm{DU}, \mathrm{LS})}_{\text{up-down-to-down-up lower-saddle}}.$$

The equilibrium of $(x_{j_1}^*, x_{j_2}^*) = (a_{j_1 j_1 1}, a_{j_2 j_2 1})$ is a (UD:DU,LS)-up-down-to-down-up lower-saddle.

- For $a_{j_1j_10}(a_{j_1j_21}-a_{j_1j_2l_2})<0$ and $a_{j_2j_20}(a_{j_2j_2s_2}-a_{j_2j_2s_1})<0$,

$$(a_{j_1j_11},a_{j_2j_2s_2})=\underbrace{(\mathrm{DU}:\mathrm{UD},\mathrm{LS})}_{\text{down-up-to-up-down lower-saddle}}.$$

The equilibrium of $(x^*_{j_1},x^*_{j_2})=(a_{j_1j_11},a_{j_2j_21})$ is a (DU:UD,LS)-down-up-to-up-down lower-saddle.

(viii$_5$) For $a_{j_2j_2s_1}=a_{j_2j_2s_2}=a_{j_2j_21}=a_{j_1j_2l_1}$, in phase space,

$$\frac{dx_{j_1}}{dx_{j_2}}=\frac{a_{j_1j_10}}{a_{j_2j_20}}\frac{(x_{j_1}-a_{j_1j_11})(x_{j_2}-a_{j_1j_2l_2})}{(x_{j_2}-a_{j_2j_21})^2},$$

and the deformation of the foregoing equation is

$$\frac{dx_{j_1}}{x_{j_1}-a_{j_1j_11}}=\frac{a_{j_1j_10}}{a_{j_2j_20}}\Big[\frac{1}{(x_{j_2}-a_{j_2j_21})}+\frac{(a_{j_2j_21}-a_{j_1j_2l_2})}{(x_{j_2}-a_{j_2j_21})^2}\Big]dx_{j_2}.$$

With an initial condition of (x_{j_10},x_{j_20}) at $t=t_0$, the integration of the above equation gives the first integral manifold as

$$\ln\frac{|x_{j_1}-a_{j_1j_11}|}{|x_{j_10}-a_{j_1j_11}|}=\frac{a_{j_1j_10}}{a_{j_2j_20}}\Big[\ln\frac{|x_{j_2}-a_{j_2j_21}|}{|x_{j_20}-a_{j_2j_21}|}$$
$$-(a_{j_2j_21}-a_{j_1j_2l_2})\Big(\frac{1}{x_{j_2}-a_{j_2j_21}}-\frac{1}{x_{j_20}-a_{j_2j_21}}\Big)\Big].$$

(viii$_{5a}$) At $x^*_{j_2}=a_{j_1j_2l_1}=a_{j_2j_21}$ with $\bar{x}_{j_1}\neq a_{j_1j_11}$ and $\varepsilon\to 0$,

$$\frac{dx_{j_1}}{dx_{j_2}}\Big|_{x^*{j_2}=a_{j_2j_21}\pm\varepsilon}=\frac{a_{j_1j_10}}{a_{j_2j_00}}\frac{(\bar{x}_{j_1}-a_{j_1j_1})(a_{j_2j_21}-a_{j_1j_2l_2})}{\varepsilon^2}.$$

If

$$\frac{a_{j_1j_10}(\bar{x}_{j_1}-a_{j_1j_11})(a_{j_2j_21}-a_{j_1j_2l_2})}{a_{j_2j_20}}>0,$$

there is an increasing-concave flow at $x^*_{j_2}=a_{j_1j_2l_1}=a_{j_2j_11}$ with $\bar{x}_{j_1}\neq a_{j_1j_1}$. If

$$\frac{a_{j_1j_10}(\overline{x}_{j_1}-a_{j_1j_11})(a_{j_2j_21}-a_{j_1j_2j_2})}{a_{j_2j_20}}<0,$$

there is a decreasing-concave flow at $x^*_{j_2}=a_{j_1j_2l_1}=a_{j_2j_21}$ with $\overline{x}_{j_1}\neq a_{j_1j_11}$.

The corresponding variational equation is

$$\Delta\dot{x}_{j_2}=a_{j_2j_20}(\Delta x_{j_2})^3,$$

The increasing and decreasing concave flows are third-order source and sink for $a_{j_2j_20}>0$ and $a_{j_2j_20}<0$, respectively.

Therefore, the infinite-equilibrium of $x^*_{j_2}=a_{j_1j_2l_1}=a_{j_2j_21}$ with $\overline{x}_{j_1}\neq a_{j_1j_11}$. has the following properties as in Eqs. (1.184)–(1.187).

- For $a_{j_1j_10}(\overline{x}_{j_1}-a_{j_1j_11})(a_{j_2j_21}-a_{j_1j_2l_2})>0$ and $a_{j_2j_20}>0$,

$$(\overline{x}_{j_1},a_{j_2j_21})=\underbrace{(\text{IC},3^{\text{rd}}\text{SO})}_{\text{third-order increasing-concave source}}.$$

  The infinite-equilibrium of $x^*_{j_2}=a_{j_1j_21}=a_{j_2j_21}$ is an (IC,3rdSO)-third-order increasing-concave source.
- For $a_{j_1j_10}(\overline{x}_{j_1}-a_{j_1j_11})(a_{j_2j_21}-a_{j_1j_2l_2})<0$ and $a_{j_2j_20}>0$,

$$(\overline{x}_{j_1},a_{j_2j_21})=\underbrace{(\text{DC},3^{\text{rd}}\text{SO})}_{\text{third-order decreasing-concave source}}.$$

  The infinite-equilibrium of $x^*_{j_2}=a_{j_1j_2l_1}=a_{j_2j_21}$ is a (DC,3rdSO)-third-order decreasing-concave source.
- For $a_{j_1j_10}(\overline{x}_{j_1}-a_{j_1j_11})(a_{j_2j_21}-a_{j_1j_2l_2})>0$ and $a_{j_2j_20}<0$,

$$(\overline{x}_{j_1},a_{j_2j_21})=\underbrace{(\text{DC},3^{\text{rd}}\text{SI})}_{\text{third-order decreasing-concave sink}}.$$

  The infinite-equilibrium of $x^*_{j_2}=a_{j_1j_2l_1}=a_{j_2j_21}$ is a (DC,3rdSI)-third-order decreasing-concave sink.
- For $a_{j_1j_10}(\overline{x}_{j_1}-a_{j_1j_11})(a_{j_2j_21}-a_{j_1j_2j_2})<0$ and $a_{j_2j_20}<0$,

$$(\overline{x}_{j_1},a_{j_2j_21})=\underbrace{(\text{IC},3^{\text{rd}}\text{SI})}_{\text{third-order increasing-concave sink}}.$$

  The infinite-equilibrium of $x^*_{j_2}=a_{j_1j_2l_1}=a_{j_2j_21}$ is an (IC,3rdSO)-third-order increasing-concave sink.

(viii$_{5b}$) From (viii$_{5b}$), for $\overline{x}_{j_1} = a_{j_1 j_1 1}$, the equilibrium of $(x^*_{j_1}, x^*_{j_2}) = (a_{j_1 j_1 1}, a_{j_2 j_2 1})$ with $a_{j_1 j_2 1} = a_{j_2 j_2 1}$ has the following properties as in Eqs. (1.188)–(1.191).

- For $a_{j_1 j_1 0}(a_{j_2 j_1 1} - a_{j_1 j_2 l_2}) > 0$ and $a_{j_2 j_2 0} > 0$,

$$(a_{j_1 j_1 1}, a_{j_2 j_2 1}) = \underbrace{(\text{DC} : \text{IC}, 3^{\text{rd}}\text{SO})}_{\text{third-order decreasing-to-increasing concave source}}.$$

  The equilibrium of $x^*_{j_2} = a_{j_1 j_2 l_1} = a_{j_2 j_2 1}$ is a (DC:IC,3rdSO)-third-order decreasing-to-increasing concave source.
- For $a_{j_1 j_1 0}(a_{j_2 j_2 1} - a_{j_1 j_2 l_2}) < 0$ and $a_{j_2 j_2 0} > 0$,

$$(a_{j_1 j_1 1}, a_{j_2 j_2 1}) = \underbrace{(\text{IC} : \text{DC}, 3^{\text{rd}}\text{SO})}_{\text{third-order increasing-to-decreasing concave source}}.$$

  The equilibrium of $(x^*_{j_1}, x^*_{j_2}) = (a_{j_1 j_1 1}, a_{j_2 j_2 1})$ is an (IC:DC,3rdSO)-third-order increasing-to-decreasing concave source.
- For $a_{j_1 j_1 0}(a_{j_2 j_2 1} - a_{j_1 j_2 l_2}) > 0$ and $a_{j_2 j_2 0} < 0$,

$$(a_{j_1 j_1 1}, a_{j_2 j_2 1}) = \underbrace{(\text{IC} : \text{DC}, 3^{\text{rd}}\text{SI})}_{\text{third-order increasing-to-decreasing concave sink}}.$$

  The equilibrium of $(x^*_{j_1}, x^*_{j_2}) = (a_{j_1 j_1 1}, a_{j_2 j_2 1})$ is an (IC:DC,3rdSI)-third-order increasing-to-deceasing concave sink.
- For $a_{j_1 j_1 0}(a_{j_2 j_2 1} - a_{j_1 j_2 l_2}) < 0$ and $a_{j_2 j_2 0} < 0$,

$$(a_{j_1 j_1 1}, a_{j_2 j_2 1}) = \underbrace{(\text{DC} : \text{IC}, 3^{\text{rd}}\text{SI})}_{\text{third-order deceasing-to-incerasing concave sink}}.$$

  The equilibrium of $(x^*_{j_1}, x^*_{j_2}) = (a_{j_1 j_1 1}, a_{j_2 j_2 1})$ is a (DC:IC,3rdSI)-third-order decreasing-to-increasing concave sink.

(ix) For $\Delta_{j_1 j_2} > 0$ and $\Delta_{j_2 j_2} > 0$, the standard form is

$$\begin{aligned}
\dot{x}_{j_2} &= a_{j_2 j_2 0}(x_{j_2} - a_{j_2 j_2 1})(x_{j_2} - a_{j_2 j_2 2})(x_{j_2} - a_{j_2 j_2 3}),\\
\dot{x}_{j_1} &= a_{j_1 j_1 0}(x_{j_1} - a_{j_1 j_1 1})(x_{j_2} - a_{j_1 j_2 1})(x_{j_2} - a_{j_1 j_2 2})
\end{aligned}$$

where

$$
\begin{aligned}
& b_{j_1j_2 1}, b_{j_1j_2 2} = -\frac{1}{2}(B_{j_1j_2} \pm \sqrt{\Delta_{j_1j_2}}), \\
& \{a_{j_1j_2 1}, a_{j_1j_2 2}\} = \operatorname{sort}\{b_{j_1j_2 1}, b_{j_1j_2 2}\}, a_{j_1j_2 1} < a_{j_1j_2 2}, \\
& b_{j_2j_2 2}, b_{j_2j_2 3} = -\frac{1}{2}(B_{j_2j_2} \pm \sqrt{\Delta^2_{j_2j_2}}), \\
& \{a_{j_2j_2 1}, a_{j_2j_2 2}, a_{j_2j_2 3}\} = \operatorname{sort}\{b_{j_2j_2 1}, b_{j_2j_2 2}, b_{j_2j_2 3}\}, \\
& a_{j_2j_2 s_1} < a_{j_2j_2 s_2}; s_1, s_2 \in \{1, 2, 3\}, s_1 < s_2.
\end{aligned}
$$

(ix$_1$) For $a_{j_2j_2s_1} \neq a_{j_1j_2l_1} (l_1, l_2 \in \{1,2\}, l_1 \neq l_2; s_1, s_2 \in \{1,2,3\}, s_1 \neq s_2)$, in phase space,

$$
\frac{dx_{j_1}}{dx_{j_2}} = \frac{a_{j_1j_1 0}}{a_{j_2j_2 0}} \frac{(x_{j_1} - a_{j_1j_1 1}) \prod_{l_1=1}^{2} (x_{j_2} - a_{j_1j_2 l_1})}{\prod_{s_1=1}^{3} (x_{j_2} - a_{j_2j_2 s_1})},
$$

and the deformation of the above equation yields

$$
\frac{dx_{j_1}}{x_{j_1} - a_{j_1j_1 1}} = \frac{a_{j_jj_1 0}}{a_{j_2j_2 0}} \sum_{s_1=1}^{3} \frac{\prod_{l_1=1}^{2} (a_{j_2j_2 s_1} - a_{j_1j_2 l_1})}{\prod_{s_2=1, s_2 \neq s_1}^{3} (a_{j_2j_2 s_1} - a_{j_2j_2 s_2})} \frac{1}{x_{j_1} - a_{j_2j_2 s_1}} dx_{j_2}.
$$

With an initial condition of $(x_{j_1 0}, x_{j_2 0})$ at $t = t_0$, the integration of the above equation gives the first integral manifold $(a_{j_2j_2s_1} \neq a_{j_1j_2l_1})$ as

$$
\ln \frac{|x_{j_1} - a_{j_1j_1 1}|}{|x_{j_1 0} - a_{j_1j_1 1}|} = \frac{a_{j_1j_1 0}}{a_{j_2j_2 0}} \sum_{s_2=1, s_2 \neq s_1}^{3} \frac{\prod_{l_1=1}^{2} (a_{j_2j_2 s_1} - a_{j_1j_2 l_1})}{\prod_{s_2=1, s_2 \neq s_1}^{3} (a_{j_2j_2 s_1} - a_{j_2j_2 s_2})} \ln \frac{|x_{j_2} - a_{j_2j_2 s_2}|}{|x_{j_2 0} - a_{j_2j_2 s_2}|}.
$$

(ix$_{1a}$) For $(x^*_{j_1}, x^*_{j_2}) = (a_{j_1j_1 1}, a_{j_2j_2 s_1})$, the variational equation is

$$
\begin{aligned}
\Delta \dot{x}_{j_2} &= a_{j_2j_2 0} \prod_{s_2=1, s_2 \neq s_1}^{3} (a_{j_2j_2 s_1} - a_{j_2j_2 s_2}) \Delta x_{j_2}, \\
\Delta \dot{x}_{j_1} &= a_{j_1j_1 0} \prod_{l_1=1}^{2} (a_{j_2j_2 s_1} - a_{j_1j_2 l_1}) \Delta x_{j_1}.
\end{aligned}
$$

The flows at $(x^*_{j_1}, x^*_{j_2}) = (a_{j_1j_1 1}, a_{j_2j_2 1})$ in the x_{j_1}-direction are source and sink for $a_{j_1j_1 0} \prod_{l_1=1}^{2} (a_{j_2j_2 s_1} - a_{j_1j_2 l_1}) > 0$ and $a_{j_1j_1 0} \prod_{l_1=1}^{2} (a_{j_2j_2 s_1} - a_{j_1j_1 l_1}) < 0$, respectively. The flows at $(x^*_{j_1}, x^*_{j_2}) = (a_{j_1j_1 1}, a_{j_2j_2 s_1})$ in the x_{j_2}-direction are source and sink for $a_{j_2j_2 0} \prod_{s_2=1, s_2 \neq s_1}^{3} (a_{j_2j_2 s_1} - a_{j_2j_2 s_2}) > 0$ and $a_{j_2j_2 0} \prod_{s_2=1, s_2 \neq s_1}^{3} (a_{j_2j_2 s_1} - a_{j_2j_2 s_2}) < 0$, respectively.

Therefore, the equilibrium of $(x^*_{j_1}, x^*_{j_2}) = (a_{j_1j_1 1}, a_{j_2j_2 s_1})$ $(s_1, s_2 \in \{1, 2\}, s_1 \neq s_2)$ has the following properties as in Eqs. (1.195)–(1.198).

- For $a_{j_1j_10}\prod_{l_1=1}^{2}(a_{j_2j_2s_1}-a_{j_1j_2l_1})>0$ and $a_{j_2j_20}\prod_{s_2=1,s_2\neq s_1}^{3}(a_{j_2j_2s_1}-a_{j_2j_2s_2})>0$,

$$(a_{j_1j_11},a_{j_2j_2s_1})=\underbrace{(\text{SO},\text{SO})}_{\text{source}}.$$

The equilibrium of $(x_{j_1}^*,x_{j_2}^*)=(a_{j_1j_11},a_{j_2j_2s_1})$ is an (SO,SO)-source.

- For $a_{j_1j_10}\prod_{l_1=1}^{2}(a_{j_2j_2s_1}-a_{j_1j_2l_1})<0$ and $a_{j_2j_20}\prod_{s_2=1,s_2\neq s_1}^{3}(a_{j_2j_2s_1}-a_{j_2j_2s_2})>0$,

$$(a_{j_1j_11},a_{j_2j_2s_1})=\underbrace{(\text{SI},\text{SO})}_{\text{saddle}}.$$

The equilibrium of $(x_{j_1}^*,x_{j_2}^*)=(a_{j_1j_11},a_{j_2j_2s_1})$ is an (SI,SO)-saddle.

- For $a_{j_1j_10}\prod_{l_1=1}^{2}(a_{j_2j_2s_1}-a_{j_1j_2l_1})>0$ and $a_{j_2j_20}\prod_{s_2=1,s_2\neq s_1}^{3}(a_{j_2j_2s_1}-a_{j_2j_2s_2})<0$,

$$(a_{j_1j_11},a_{j_2j_2s_1})=\underbrace{(\text{SO},\text{SI})}_{\text{saddle}}.$$

The equilibrium of $(x_{j_1}^*,x_{j_2}^*)=(a_{j_1j_11},a_{j_2j_2s_1})$ is an (SO,SI)-saddle.

- For $a_{j_1j_10}\prod_{l_1=1}^{2}(a_{j_2j_2s_1}-a_{j_1j_2l_1})<0$ and $a_{j_2j_20}\prod_{s_2=1,s_2\neq s_1}^{3}(a_{j_2j_2s_1}-a_{j_2j_2s_2})<0$,

$$(a_{j_1j_11},a_{j_2j_2s_1})=\underbrace{(\text{SI},\text{SI})}_{\text{sink}}.$$

The equilibrium of $(x_{j_1}^*,x_{j_2}^*)=(a_{j_1j_11},a_{j_2j_2s_1})$ is an (SI,SI)-sink.

(ix$_{1b}$) For $x_{j_2}^*=a_{j_1j_2l_1}\neq a_{j_2j_21},a_{j_2j_22}$ and $\overline{x}_{j_1}\neq a_{j_1j_11}(s_1,s_2\in\{1,2\},s_1\neq s_2)$, in phase space,

$$\left.\frac{dx_{j_1}}{dx_{j_2}}\right|_{x_{j_2}^*=a_{j_1j_2l_1}}=\left.\frac{a_{j_1j_10}}{a_{j_2j_20}}\,\frac{(\overline{x}_{j_1}-a_{j_1j_11})(x_{j_2}-a_{j_1j_2l_1})(x_{j_2}-a_{j_1j_2l_2})}{\prod_{s_1=1}^{3}(x_{j_2}-a_{j_2j_2s_1})}\right|_{x_2^*=a_{j_1j_2l_1}}=0.$$

If

$$\left.\frac{d^2x_{j_1}}{dx_{j_2}^2}\right|_{x_{j_2}^*=a_{j_1j_2l_1}}=\frac{a_{j_1j_10}}{a_{j_2j_20}}\,\frac{(\overline{x}_{j_1}-a_{j_1j_11})(a_{j_1j_2l_1}-a_{j_1j_2l_2})}{\prod_{s_1=1}^{3}(a_{j_1j_2l_1}-a_{j_2j_2s_1})}>0,$$

there is an up-parabola flow at $x_{j_2}^*=a_{j_1j_2l_1}\neq a_{j_2j_21},a_{j_2j_22}$ with $\overline{x}_{j_1}\neq a_{j_1j_11}$. If

$$\left.\frac{d^2x_{j_1}}{dx_{j_2}^2}\right|_{x_{j_2}^*=a_{j_1j_2l_1}} = \frac{a_{j_1j_10}}{a_{j_2j_20}} \frac{(\bar{x}_{j_1} - a_{j_1j_11})(a_{j_1j_2l_1} - a_{j_1j_2l_2})}{\prod_{s_1=1}^{3}(a_{j_1j_2l_1} - a_{j_2j_2s_1})} < 0,$$

there is a down-parabola flow at $x_{j_2}^* = a_{j_1j_2l_1}$ with $\bar{x}_{j_1} \neq a_{j_1j_11}$. Let

$$\dot{x}_{j_2} = a_{j_2j_20}\prod\nolimits_{s_1=1}^{3}(a_{j_1j_2l_1} - a_{j_2j_2s_1}) \neq 0.$$

The parabola flows at $x_{j_2}^* = a_{j_1j_2s_1} \neq a_{j_2j_21}$ with $\bar{x}_{j_1} \neq a_{j_1j_11}$ are positive and negative for $a_{j_2j_20} > 0$ and $a_{j_2j_20} < 0$, respectively.

- For $a_{j_1j_1j_0}(\bar{x}_{j_1} - a_{j_1j_11})(a_{j_1j_2l_1} - a_{j_1j_2l_2}) > 0$ and $a_{j_2j_20}\prod_{s_1=1}^{3}(a_{j_1j_2l_1} - a_{j_2j_2s_1}) > 0$,

$$(a_{j_1j_2s_1}, \dot{x}_{j_2}) = \underbrace{(\text{UP}, \text{pF})}_{\text{up-parabola flow } (+)}.$$

- For $a_{j_1j_10}(\bar{x}_{j_1} - a_{j_1j_11})(a_{j_1j_2l_1} - a_{j_1j_2l_2}) < 0$ and $a_{j_2j_20}\prod_{s_1=1}^{3}(a_{j_1j_2l_1} - a_{j_2j_2s_1}) > 0$,

$$(a_{j_1j_2s_1}, \dot{x}_{j_2}) = \underbrace{(\text{DP}, \text{pF})}_{\text{down-parabola flow } (+)}.$$

- For $a_{j_1j_10}(\bar{x}_{j_1} - a_{j_1j_11})(a_{j_1j_2l_1} - a_{j_1j_2l_2}) > 0$ and $a_{j_2j_20}\prod_{s_1=1}^{3}(a_{j_1j_2l_1} - a_{j_2j_2s_1}) < 0$,

$$(a_{j_1j_2s_1}, \dot{x}_{j_2}) = \underbrace{(\text{DP}, \text{nF})}_{\text{down-parabola flow } (-)}.$$

- For $a_{j_1j_10}(\bar{x}_{j_1} - a_{j_1j_11})(a_{j_1j_2l_1} - a_{j_1j_2l_2}) < 0$ and $a_{j_2j_20}\prod_{s_1=1}^{3}(a_{j_1j_2l_1} - a_{j_2j_2s_1}) < 0$,

$$(a_{j_1j_2s_1}, \dot{x}_{j_2}) = \underbrace{(\text{UP}, \text{nF})}_{\text{up-parabola flow } (-)}.$$

From the above analysis, for $\bar{x}_{j_1} = a_{j_1j_11}$, therefore, the equilibrium of $(x_{j_1}^*, x_{j_2}^*) = (a_{j_1j_11}, a_{j_1j_2l_1})(l_1, l_2 \in \{1,2\}, l_1 \neq l_2)$ with $a_{j_1j_2l_1} \neq a_{j_2j_21}, a_{j_2j_22}$ has the following properties as in Eqs. (1.199)–(1.202).

- For $a_{j_1j_10}(a_{j_1j_2l_1} - a_{j_1j_2l_2}) > 0$ and $a_{j_2j_20}\prod_{s_1=1}^{3}(a_{j_1j_2l_1} - a_{j_1j_2s_1}) > 0$,

$$(a_{j_1j_11}, a_{j_1j_2l_1}) = \underbrace{(\text{DP}:\text{UP}, \text{pF})}_{\text{hyperbolic flow } (+)}.$$

  The equilibrium of $(x_{j_1}^*, x_{j_2}^*) = (a_{j_1j_11}, a_{j_1j_2l_1})$ is a (DP:UP,pF)-positive hyperbolic flow.

- For $a_{j_1j_1 0}(a_{j_1j_2l_1}-a_{j_1j_2l_2})<0$ and $a_{j_2j_2 0}\prod_{s_1=1}^{3}(a_{j_1j_2l_1}-a_{j_1j_2s_1})>0$,

$$(a_{j_1j_1 1},a_{j_1j_2l_1})=\underbrace{(\text{UP}:\text{DP},\text{pF})}_{\text{hyperbolic-seceant flow }(+)}.$$

The equilibrium of $(x_{j_1}^*,x_{j_2}^*)=(a_{j_1j_1 1},a_{j_1j_2l_1})$ is a (UP:DP,pF)-positive hyperbolic-secant flow.

- For $a_{j_1j_1 0}(a_{j_1j_2l_1}-a_{j_1j_2l_2})>0$ and $a_{j_2j_2 0}\prod_{s_1=1}^{3}(a_{j_1j_2l_1}-a_{j_2j_2s_1})<0$,

$$(a_{j_1j_1 1},a_{j_1j_2l_1})=\underbrace{(\text{UP}:\text{DP},\text{nF})}_{\text{hyperbolic-secant flow }(-)}.$$

The equilibrium of $(x_{j_1}^*,x_{j_2}^*)=(a_{j_1j_1 1},a_{j_1j_2l_1})$ is a (UP:DP,nF)-negative hyperbolic-secant flow.

- For $a_{j_1j_1 0}(a_{j_1j_2l_1}-a_{j_1j_2l_2})<0$ and $a_{j_2j_2 0}\prod_{s_1=1}^{3}(a_{j_1j_2l_1}-a_{j_1j_2s_1})<0$,

$$(a_{j_1j_1 1},a_{j_1j_2l_1})=\underbrace{(\text{DP}:\text{UP},\text{nF})}_{\text{hyperbolic flow }(-)}.$$

The equilibrium of $(x_{j_1}^*,x_{j_2}^*)=(a_{j_1j_1 1},a_{j_1j_2l_1})$ is a (DP:UP,nF)-negative hyperbolic flow.

(ix$_2$) For $a_{j_2j_2s_1}=a_{j_1j_2l_1}$ and $a_{j_2j_2s_2}\neq a_{j_1j_2l_2}$, $l_1,l_2\in\{1,2\}$; $l_1\neq l_2$; $s_1,s_2,s_3\in\{1,2,3\}$; $s_1\neq s_2\neq s_3$, in phase space,

$$\begin{aligned}\frac{dx_{j_1}}{dx_{j_2}}&=\frac{a_{j_1j_1 0}}{a_{j_2j_2 0}}\frac{(x_{j_1}-a_{j_1j_1 1})(x_{j_2}-a_{j_1j_2l_1})(x_{j_2}-a_{j_1j_2l_2})}{(x_{j_2}-a_{j_2j_2s_1})(x_{j_2}-a_{j_2j_2s_2})(x_{j_2}-a_{j_2j_2s_3})}\\&=\frac{a_{j_1j_1 0}}{a_{j_2j_2 0}}\frac{(x_{j_1}-a_{j_1j_1 1})(x_{j_2}-a_{j_1j_2l_2})}{(x_{j_2}-a_{j_2j_2s_2})(x_{j_2}-a_{j_2j_2s_3})},\end{aligned}$$

and the deformation of the foregoing equation is

$$\frac{dx_{j_1}}{(x_{j_1}-a_{j_1j_1 1})}=\frac{a_{j_1j_1 0}}{a_{j_2j_2 0}}\sum_{s_2=1,s_2\neq s_3}^{3}\frac{(a_{j_2j_2s_2}-a_{j_1j_2l_2})}{(a_{j_2j_2s_2}-a_{j_2j_2s_3})}\frac{1}{x_{j_2}-a_{j_2j_2s_2}}dx_{j_2}.$$

With an initial condition of $(x_{j_1 0},x_{j_2 0})$ at $t=t_0$, the integration of the above equation gives the first integral manifold ($a_{j_2j_2s_1}=a_{j_1j_2l_1}$ and $a_{j_2j_2s_1}\neq a_{j_1j_2l_2},l_1,l_2\in\{1,2\};l_1\neq l_2;s_1,s_2,s_3\in\{1,2,3\};s_1\neq s_2\neq s_3$) as

$$\ln\frac{|x_{j_1}-a_{j_1j_11}|}{|x_{j_10}-a_{j_1j_11}|}=\frac{a_{j_1j_10}}{a_{j_2j_20}}\sum\nolimits_{s_2=1,s_2\neq s_1,s_3}^{3}\frac{(a_{j_2j_2s_2}-a_{j_1j_2l_2})}{(a_{j_2j_2s_2}-a_{j_2j_2s_3})}\ln\frac{|x_{j_2}-a_{j_2j_2s_2}|}{|x_{j_20}-a_{j_2j_2s_2}|}.$$

(ix$_{2a}$) At $x^*_{j_2}=a_{j_1j_2l_1}=a_{j_2j_2s_1}$ with $\bar{x}_{j_1}\neq a_{j_1j_11}$, if

$$\left.\frac{dx_{j_1}}{dx_{j_2}}\right|_{x^*j_2=a_{j_1j_2l_1}}=\frac{a_{j_1j_10}}{a_{j_2j_20}}\frac{(x_{j_1}-a_{j_1j_11})(a_{j_1j_2l_1}-a_{j_1j_2l_2})}{\prod_{s_2=1,s_2\neq s_1}^{3}(a_{j_2j_2s_1}-a_{j_2j_2s_2})}>0,$$

there is an increasing-inflection flow, and if

$$\left.\frac{dx_{j_1}}{dx_{j_2}}\right|_{x^*j_2=a_{j_1j_2l_1}}=\frac{a_{j_1j_10}}{a_{j_2j_20}}\frac{(x_{j_1}-a_{j_1j_11})(a_{j_1j_2l_1}-a_{j_1j_2l_2})}{\prod_{s_2=1,s_2\neq s_1}^{3}(a_{j_2j_2s_1}-a_{j_2j_2s_2})}<0,$$

there is a decreasing-inflection flow.

The corresponding variational equation is

$$\Delta\dot{x}_{j_2}=a_{j_2j_20}\prod\nolimits_{s_2=1,s_2\neq s_1}^{3}(a_{j_2j_2s_1}-a_{j_2j_2s_2})\Delta x_{j_2}.$$

The inflection flows at $x^*_{j_2}=a_{j_1j_2l_1}=a_{j_2j_2s_1}$ with $\bar{x}_{j_1}\neq a_{j_1j_11}$ are source and sink $a_{j_2j_20}\prod_{s_2=1,s_2\neq s_1}^{3}(a_{j_2j_2s_1}-a_{j_2j_2s_2})>0$ and $a_{j_2j_20}\prod_{s_2=1,s_2\neq s_1}^{3}(a_{j_2j_2s_1}-a_{j_2j_2s_2})<0$, respectively.

Therefore, the infinite-equilibrium of $x^*_{j_2}=a_{j_1j_2l_1}=a_{j_2j_2s_1}$ with $\bar{x}_{j_1}\neq a_{j_1j_11}$ has the following properties as in Eqs. (1.204)–(1.207).

- For $a_{j_1j_10}(\bar{x}_{j_1}-a_{j_1j_11})(a_{j_1j_2l_1}-a_{j_1j_2l_2})>0$ and $a_{j_2j_20}\prod_{s_2=1,s_2\neq s_1}^{3}(a_{j_2j_2s_1}-a_{j_1j_2s_2})>0$,

$$(\bar{x}_{j_1},a_{j_2j_2s_1})=\underbrace{(\text{II},\text{SO})}_{\text{increasing-inffection source}}.$$

  The infinite-equilibrium of $x^*_{j_2}=a_{j_1j_2l_1}=a_{j_2j_2s_1}$ is an (II,SO)-increasing-inflection source.
- For $a_{j_1j_10}(\bar{x}_{j_1}-a_{j_1j_11})(a_{j_1j_2l_1}-a_{j_1j_2l_2})<0$ and $a_{j_2j_20}\prod_{s_2=1,s_2\neq s_1}^{3}(a_{j_2j_2s_1}-a_{j_2j_2s_2})>0$,

$$(\bar{x}_{j_1},a_{j_2j_2s_1})=\underbrace{(\text{DI},\text{SO})}_{\text{decreasing-inflection source}}.$$

  The infinite-equilibrium of $x^*_{j_2}=a_{j_1j_2l_1}=a_{j_2j_2s_1}$ is a (DI,SO)-decreasing-inflection source.

- For $a_{j_1j_10}(\overline{x}_{j_1}-a_{j_1j_11})(a_{j_1j_2l_1}-a_{j_1j_2l_2})>0$ and $a_{j_2j_20}\prod_{s_2=1,s_2\neq s_1}^{3}(a_{j_2j_2s_1}-a_{j_1j_2s_2})<0$,

$$(\overline{x}_{j_1},a_{j_2j_2s_1})=\underbrace{(\mathrm{DI},\mathrm{SI})}_{\text{decreasing-inflection sink}}.$$

The infinite-equilibrium of $x^*_{j_2}=a_{j_1j_2l_1}=a_{j_2j_2s_1}$ is a (DI,SI)-decreasing-inflection sink.

- For $a_{j_1j_10}(\overline{x}_{j_1}-a_{j_1j_11})(a_{j_1j_2l_1}-a_{j_1j_2l_2})<0$ and $a_{j_2j_20}\prod_{s_2=1,s_2\neq s_1}^{3}(a_{j_2j_2s_1}-a_{j_2j_2s_2})<0$,

$$(\overline{x}_{j_1},a_{j_2j_2s_1})=\underbrace{(\mathrm{II},\mathrm{SI})}_{\text{increasing-inflection sink}}.$$

The infinite-equilibrium of $x^*_{j_2}=a_{j_1j_2l_1}=a_{j_2j_2s_1}$ is an (II,SI)-increasing-inflection sink.

(ix_{2b}) From (ix_{2a}), for $x^*_{j_2}=a_{j_1j_11}$, the equilibrium of $x^*_{j_1}x^*_{j_2}=(a_{j_1j_11},a_{j_2j_2s_1})$ with $a_{j_2j_2s_1}=a_{j_1j_2l_1}$ has the following properties as in Eqs. (1.208)–(1.211).

- For $a_{j_1j_10}(a_{j_1j_2l_1}-a_{j_1j_2l_2})>0$ and $a_{j_2j_20}\prod_{s_2=1,s_2\neq s_1}^{3}(a_{j_2j_2s_1}-a_{j_2j_2s_2})>0$,

$$(a_{j_1j_11},a_{j_2j_2s_1})=\underbrace{(\mathrm{DI}:\mathrm{II},\mathrm{SO})}_{\text{inflection source}}.$$

The equilibrium of $(x^*_{j_1},x^*_{j_2})=(a_{j_1j_11},a_{j_2j_2s_1})$ is a (DI:II,SO)-inflection source.

- For $a_{j_1j_10}(a_{j_1j_2l_1}-a_{j_1j_2l_2})<0$ and $a_{j_2j_20}\prod_{s_2=1,s_2\neq s_1}^{3}(a_{j_2j_2s_1}-a_{j_2j_2s_2})>0$,

$$(a_{j_1j_11},a_{j_2j_2s_1})=\underbrace{(\mathrm{II}:\mathrm{DI},\mathrm{SO})}_{\text{inflection source}}.$$

The equilibrium of $(x^*_{j_1},x^*_{j_2})=(a_{j_1j_11},a_{j_2j_2s_1})$ is an (II:DI,SO)-inflection source.

- For $a_{j_1j_10}(a_{j_1j_2l_1}-a_{j_1j_2l_2})>0$ and $a_{j_2j_20}\prod_{s_2=1,s_2\neq s_1}^{3}(a_{j_2j_2s_1}-a_{j_2j_2s_2})<0$,

$$(a_{j_1j_11},a_{j_2j_2s_1})=\underbrace{(\mathrm{II}:\mathrm{DI},\mathrm{SI})}_{\text{inflection sink}}.$$

The equilibrium of $(x^*_{j_1},x^*_{j_2})-(a_{j_1j_11},a_{j_2j_2s_1})$ is an (II:DI,SI)-inflection sink.

- For $a_{j_1j_10}(a_{j_1j_2l_1}-a_{j_1j_2l_2})<0$ and $a_{j_2j_20}\prod_{s_2=1,s_2\neq s_1}^{3}(a_{j_2j_2s_1}-a_{j_2j_2s_2})<0$,

$$(a_{j_1j_1 1}, a_{j_2j_2s_1}) = \underbrace{(\text{DI} : \text{II}, \text{SI})}_{\text{inflection sink}}.$$

The equilibrium of $(x^*_{j_1}, x^*_{j_2}) = (a_{j_1j_1 1}, a_{j_2j_2s_1})$ is a (DI:II,SI)-inflection sink.

(ix$_3$) For $a_{j_2j_2s_1} = a_{j_1j_2l_1}$ and $a_{j_2j_2s_2} = a_{j_1j_2l_2}$ ($l_1, l_2 \in \{1,2\}$; $l_1 \neq l_2$; $s_1, s_2, s_3 \in \{1,2,3\}$; $s_1 \neq s_2 \neq s_3$), in phase space,

$$\begin{aligned}\frac{dx_{j_1}}{dx_{j_2}} &= \frac{a_{j_1j_1 0}}{a_{j_2j_2 0}} \frac{(x_{j_1} - a_{j_1j_1})(x_{j_2} - a_{j_1j_2l_1})(x_{j_2} - a_{j_1j_2l_2})}{(x_{j_2} - a_{j_2j_2s_1})(x_{j_2} - a_{j_2j_2s_2})(x_{j_2} - a_{j_2j_2s_3})} \\ &= \frac{a_{j_1j_1 0}}{a_{j_2j_2 0}} \frac{(x_{j_1} - a_{j_1j_1 1})}{(x_{j_2} - a_{j_2j_2s_3})},\end{aligned}$$

and the deformation of the foregoing equation is

$$\frac{dx_{j_1}}{(x_{j_1} - a_{j_1j_1 1})} = \frac{a_{j_1j_1 0}}{a_{j_2j_2 0}} \frac{1}{x_{j_2} - a_{j_2j_2s_3}} dx_{j_2}.$$

With an initial condition of $(x_{j_1 0}, x_{j_2 0})$ at $t = t_0$, the integration of the above equation gives the first integral manifold $\left(a_{j_2j_2s_1} = a_{j_1j_2l_1}\right.$ and $a_{j_2j_2s_2} = a_{j_1j_2l_2}$, $l_1, l_2 \in \{1,2\}$; $l_1 \neq l_2$; $\cdot$ $s_1, s_2, s_3 \in \{1,2,3\}$; $s_1 \neq s_2 \neq s_3$) as

$$\ln \frac{|x_{j_1} - a_{j_1j_1 1}|}{|x_{j_1 0} - a_{j_1j_1}|} = \frac{a_{j_1j_1 0}}{a_{j_2j_2 0}} \ln \frac{|x_{j_2} - a_{j_2j_2s_3}|}{|x_{j_2 0} - a_{j_2j_2s_3}|}.$$

(ix$_{3a}$) Similar to (ix$_{2a}$), for $x^*_{j_2} = a_{j_1j_2l_\alpha} = a_{j_2j_2s_\alpha} (\alpha, \beta \in \{1,2\}, \gamma \in \{1,2,3\}, \beta \neq \alpha, \gamma \neq \alpha), \bar{x}_{j_1} \neq a_{j_1j_1 1}$, in phase space, if

$$\begin{aligned}\left.\frac{dx_{j_1}}{dx_{j_2}}\right|_{x^*_{j_2} = a_{j_1j_2l_\alpha}} &= \frac{a_{j_1j_1 0}}{a_{j_2j_2 0}} \frac{(\bar{x}_{j_1} - a_{j_1j_1 1})}{(a_{j_2j_2s_\alpha} - a_{j_2j_2s_3})} \\ &= \frac{a_{j_1j_1 0}}{a_{j_2j_2 0}} \frac{(\bar{x}_{j_1} - a_{j_1j_1 1})(a_{j_1j_2l_\alpha} - a_{j_1j_2l_\beta})}{\prod_{s_\gamma = 1, s_\gamma \neq s_\alpha}^{3} (a_{j_2j_2s_\alpha} - a_{j_2j_2s_\gamma})} > 0,\end{aligned}$$

there is an increasing-inflection flow, and if

$$\left.\frac{dx_{j_1}}{dx_{j_2}}\right|_{x_{j_2}^*=a_{j_1j_2l_\alpha}} = \frac{a_{j_1j_10}}{a_{j_2j_20}}\frac{(\bar{x}_{j_1}-a_{j_1j_11})(a_{j_1j_2l_\alpha}-a_{j_1j_2l_\beta})}{\prod_{s_\gamma=1,s_\gamma\neq s_\alpha}^{3}(a_{j_2j_2s_\alpha}-a_{j_2j_2s_\gamma})}<0,$$

there is a decreasing-inflection flow.

The corresponding variational equation is

$$\Delta\dot{x}_{j_2}=a_{j_2j_20}\prod\nolimits_{s_\gamma=1,s_\gamma\neq s_\alpha}^{3}(a_{j_2j_2s_\alpha}-a_{j_2j_2s_\gamma})\Delta x_{j_2}.$$

The inflection flows at $x_{j_2}^*=a_{j_1j_2l_1}=a_{j_2j_2s_1}$ with $\bar{x}_{j_1}\neq a_{j_1j_11}$ are source and sink $a_{j_2j_20}\prod_{s_\gamma=1,s_\gamma\neq s_\alpha}^{3}(a_{j_2j_2s_\alpha}-a_{j_2j_2s_\gamma})>0$ and $a_{j_2j_20}\prod_{s_\gamma=1,s_\gamma\neq s_\alpha}^{3}(a_{j_2j_2s_\alpha}-a_{j_2j_2s_\gamma})<0$, respectively.

Therefore, the infinite-equilibrium of $x_{j_2}^*=a_{j_1j_2l_\alpha}=a_{j_2j_2s_\alpha}$ $(\alpha=1,2)$ with $\bar{x}_{j_1}\neq a_{j_1j_11}$ has the following properties as in Eqs. (1.213)–(1.216).

- For $a_{j_1j_10}(\bar{x}_{j_1}-a_{j_1j_11})(a_{j_1j_2l_\alpha}-a_{j_1j_2l_\beta})>0$ and $a_{j_2j_20}\prod_{s_\gamma=1,s_\gamma\neq s_\alpha}^{3}(a_{j_2j_2s_\alpha}-a_{j_2j_2s_\gamma})>0$,

$$(\bar{x}_{j_1},a_{j_2j_2s_\alpha})=\underbrace{(\text{II, SO})}_{\text{increasing-inflection source}}.$$

  The infinite-equilibrium of $x_{j_2}^*=a_{j_1j_2l_\alpha}=a_{j_2j_2s_\alpha}$ is an (II,SO)-increasing-inflection source.
- For $a_{j_1j_10}(\bar{x}_{j_1}-a_{j_1j_11})(a_{j_1j_2l_\alpha}-a_{j_1j_2l_\beta})<0$ and $a_{j_2j_20}\prod_{s_\gamma=1,s_\gamma\neq s_\alpha}^{3}(a_{j_2j_2s_\alpha}-a_{j_2j_2s_\gamma})>0$,

$$(\bar{x}_{j_1},a_{j_2j_2s_\alpha})=\underbrace{(\text{DI, SO})}_{\text{decreasing-inflection source}}.$$

  The infinite-equilibrium of $x_{j_2}^*=a_{j_1j_2l_\alpha}=a_{j_2j_2s_\alpha}$ is a (DI,SO)-decreasing-inflection source.
- For $a_{j_1j_10}(\bar{x}_{j_1}-a_{j_1j_11})(a_{j_1j_2l_\alpha}-a_{j_1j_2l_\beta})>0$ and $a_{j_2j_20}\prod_{s_\gamma=1,s_\gamma\neq s_\alpha}^{3}(a_{j_2j_2s_\alpha}-a_{j_2j_2s_\gamma})<0$,

$$(\bar{x}_{j_1},a_{j_2j_2s_\alpha})=\underbrace{(\text{DI, SI})}_{\text{decreasing-inflection sink}}.$$

  The infinite-equilibrium of $x_{j_2}^*=a_{j_1j_2l_\alpha}=a_{j_2j_2s_\alpha}$ is a (DI,SI)-decreasing-inflection sink.
- For $a_{j_1j_10}(\bar{x}_{j_1}-a_{j_1j_11})(a_{j_1j_2l_\alpha}-a_{j_1j_2l_\beta})<0$ and $a_{j_2j_20}\prod_{s_\gamma=1,s_\gamma\neq s_\alpha}^{3}(a_{j_2j_2s_\alpha}-a_{j_2j_2s_\gamma})<0$,

$$(\bar{x}_{j_1},a_{j_2j_2s_\alpha})=\underbrace{(\text{II, SI})}_{\text{increasing-inflection sink}}.$$

  The infinite-equilibrium of $x_{j_2}^*=a_{j_1j_2l_\alpha}=a_{j_2j_2s_\alpha}$ is an (II,SI)-increasing-inflection sink.

(ix$_{3b}$) From (ix$_{3a}$), for $\bar{x}_{j_1} \neq a_{j_1 j_1 1}$, the equilibrium of $(x^*_{j_1}, x^*_{j_2}) = (a_{j_1 j_1 1}, a_{j_2 j_2 s_\alpha})$ with $a_{j_2 j_2 s_\alpha} = a_{j_1 j_2 l_\alpha}$ has the following properties as in Eqs. (1.217)–(1.220).

- For $a_{j_1 j_1 0}(a_{j_1 j_2 l_\alpha} - a_{j_1 j_2 l_\beta}) > 0$ and $a_{j_2 j_2 0} \prod_{s_\gamma = 1, s_\gamma \neq s_\alpha}^{3} (a_{j_2 j_2 s_\alpha} - a_{j_2 j_2 s_\gamma}) > 0$,

$$(a_{j_1 j_1 1}, a_{j_2 j_2 s_\alpha}) = \underbrace{(\mathrm{DI}:\mathrm{II},\mathrm{SO})}_{\text{inflection source}}.$$

The equilibrium of $(x^*_{j_1}, x^*_{j_2}) = (a_{j_1 j_1}, a_{j_2 j_2 s_\alpha})$ is a (DI:II,SO)-inflection source.

- For $a_{j_1 j_1 0}(a_{j_1 j_2 l_\alpha} - a_{j_1 j_2 l_\beta}) < 0$ and $a_{j_2 j_2 0} \prod_{s_\gamma = 1, s_\gamma \neq s_\alpha}^{3} (a_{j_2 j_2 s_\alpha} - a_{j_2 j_2 s_\gamma}) > 0$,

$$(a_{j_1 j_1 1}, a_{j_2 j_2 s_\alpha}) = \underbrace{(\mathrm{II}:\mathrm{DI},\mathrm{SO})}_{\text{inflection source}}.$$

The equilibrium of $(x^*_{j_1}, x^*_{j_2}) = (a_{j_1 j_1}, a_{j_2 j_2 s_\alpha})$ is an (II:DI,SO)-inflection source.

- For $a_{j_1 j_1 0}(a_{j_1 j_2 l_\alpha} - a_{j_1 j_2 l_\beta}) > 0$ and $a_{j_2 j_2 0} \prod_{s_\gamma = 1, s_\gamma \neq s_\alpha}^{3} (a_{j_2 j_2 s_\alpha} - a_{j_2 j_2 s_\gamma}) < 0$,

$$(a_{j_1 j_1 1}, a_{j_2 j_2 s_\alpha}) = \underbrace{(\mathrm{II}:\mathrm{DI},\mathrm{SI})}_{\text{inflection sink}}.$$

The equilibrium of $(x^*_{j_1}, x^*_{j_2}) = (a_{j_1 j_1 1}, a_{j_2 j_2 s_\alpha})$ is an (II:DI,SI)-inflection sink.

- For $a_{j_1 j_1 0}(a_{j_1 j_2 l_\alpha} - a_{j_1 j_2 l_\beta}) < 0$ and $a_{j_2 j_2 0} \prod_{s_\gamma = 1, s_\gamma \neq s_\alpha}^{3} (a_{j_2 j_2 s_\alpha} - a_{j_2 j_2 s_\gamma}) < 0$,

$$(a_{j_1 j_1 1}, a_{j_2 j_2 s_\alpha}) = \underbrace{(\mathrm{DI}:\mathrm{II},\mathrm{SI})}_{\text{inflection sink}}.$$

The equilibrium of $(x^*_{j_1}, x^*_{j_2}) = (a_{j_1 j_1}, a_{j_2 j_2 s_\alpha})$ is a (DI:II,SI)-inflection sink.

In the end, this theorem is proved. ■

Chapter 2
Second and Third-Order Equilibriums

In this chapter, saddle-nodes (i.e., saddle-source and saddle-sink) and inflection-source and sink flows with saddle, source, and sink are discussed. The corresponding switching bifurcations are also discussed in this chapter. The saddle-source (sink) is the appearing bifurcation of saddle and source (sink). The inflection-source and sink flows are for the appearing bifurcations of the connected hyperbolic and hyperbolic-secant flows. In addition, the third-order saddles, source, and sink exist for the appearing bifurcations of saddle-nodes with source and sink, and they exist also for saddle, source, and sink. The switching bifurcations are through the inflection-saddle, parabola-source and sink, third-order trumpet source, and sink infinite-equilibriums. The inflection-saddles are for the switching bifurcations of saddle with inflection-source and sink flows. The hyperbolic-source and sink are the switching bifurcations of the inflection-source flow with source and the inflection-sink flow with sink, respectively. The hyperbolic-secant source and sink are the switching bifurcations of the inflection-source and sink flows with saddles. The trumpet sink and source infinite-equilibriums are the switching bifurcations of the inflection-source and sink flows with third-order saddle, sink, and source.

2.1 Saddle-Nodes and Third-Order Saddles, Sink, and Source

In this section, inflection singularity and bifurcation dynamics will be discussed, and the corresponding phase portraits will be illustrated.

2.1.1 Saddle-Nodes and Appearing Bifurcations

Consider a dynamical system as

A. C. J. Luo, *Two-dimensional Self and Product Cubic Systems, Vol. I*,
https://doi.org/10.1007/978-3-031-57096-4_2

$$\begin{aligned}\dot{x}_1 &= a_{110}(x_1 - a_{111})(x_2 - a_{121})^2,\\ \dot{x}_2 &= a_{220}(x_2 - a_{22s_1})(x_2 - a_{22s_2})^2,\end{aligned} \tag{2.1}$$

and the corresponding first integral manifold is

$$\begin{aligned}\ln\frac{|x_1 - a_{111}|}{|x_{10} - a_{111}|} &= \frac{a_{110}}{a_{220}}\left\{\frac{(a_{22s_1} - a_{121})^2}{(a_{22s_1} - a_{22s_2})^2}\ln\frac{|x_2 - a_{22s_1}|}{|x_{20} - a_{22s_1}|}\right.\\ &\quad + \left[\frac{2(a_{22s_2} - a_{121})}{(a_{22s_2} - a_{22s_1})} - \frac{(a_{22s_2} - a_{121})^2}{(a_{22s_2} - a_{22s_1})^2}\right]\ln\frac{|x_2 - a_{22s_2}|}{|x_{20} - a_{22s_2}|}\\ &\quad \left. - \frac{(a_{22s_2} - a_{121})^2}{(a_{22s_2} - a_{22s_1})^2}\left(\frac{1}{x_2 - a_{22s_2}} - \frac{1}{x_{20} - a_{22s_2}}\right)\right\}.\end{aligned} \tag{2.2}$$

For $a_{22s_2} < a_{22s_1}$ with $a_{221} = a_{22s_2}$ and $a_{222} = a_{22s_1}$, there are three cases: $a_{121} \in (a_{221}, a_{222})$, $a_{121} < a_{221}$, and $a_{121} > a_{222}$. The corresponding phase portraits are presented as follows.

(i) For $a_{121} \in (a_{221}, a_{222})$, the equilibriums of $x_1^* = a_{111}$ with $x_2^* = a_{121}, a_{221}, a_{222}$ have the following properties, as shown in Fig. 2.1.

$$\left\{\begin{array}{c}(a_{111}, a_{222})\\ (a_{111}, a_{121})\\ (a_{111}, a_{221})\end{array}\right\} = \left\{\begin{array}{c}\underbrace{(\text{SO, US})}_{\text{upper-saddle source}}\\ \underbrace{({}_{\text{DI:II}}\text{SO, pF})}_{\text{inflection source flow }(+)}\\ \underbrace{(\text{SO, SO})}_{\text{source}}\end{array}\right\} \text{ for } a_{110} > 0 \text{ and } a_{220} > 0; \tag{2.3}$$

$$\left\{\begin{array}{c}(a_{111}, a_{222})\\ (a_{111}, a_{121})\\ (a_{111}, a_{221})\end{array}\right\} = \left\{\begin{array}{c}\underbrace{(\text{SI, US})}_{\text{upper-saddle sink}}\\ \underbrace{({}_{\text{II:DI}}\text{SI, pF})}_{\text{inflection sink flow }(+)}\\ \underbrace{(\text{SI, SO})}_{\text{saddle}}\end{array}\right\} \text{ for } a_{110} < 0 \text{ and } a_{220} > 0; \tag{2.4}$$

$$\left\{\begin{array}{c}(a_{111}, a_{222})\\ (a_{111}, a_{121})\\ (a_{111}, a_{221})\end{array}\right\} = \left\{\begin{array}{c}\underbrace{(\text{SO, LS})}_{\text{lower-saddle source}}\\ \underbrace{({}_{\text{II:DI}}\text{SO, nF})}_{\text{inflection source flow }(-)}\\ \underbrace{(\text{SO, SI})}_{\text{saddle}}\end{array}\right\} \text{ for } a_{110} > 0 \text{ and } a_{220} < 0; \tag{2.5}$$

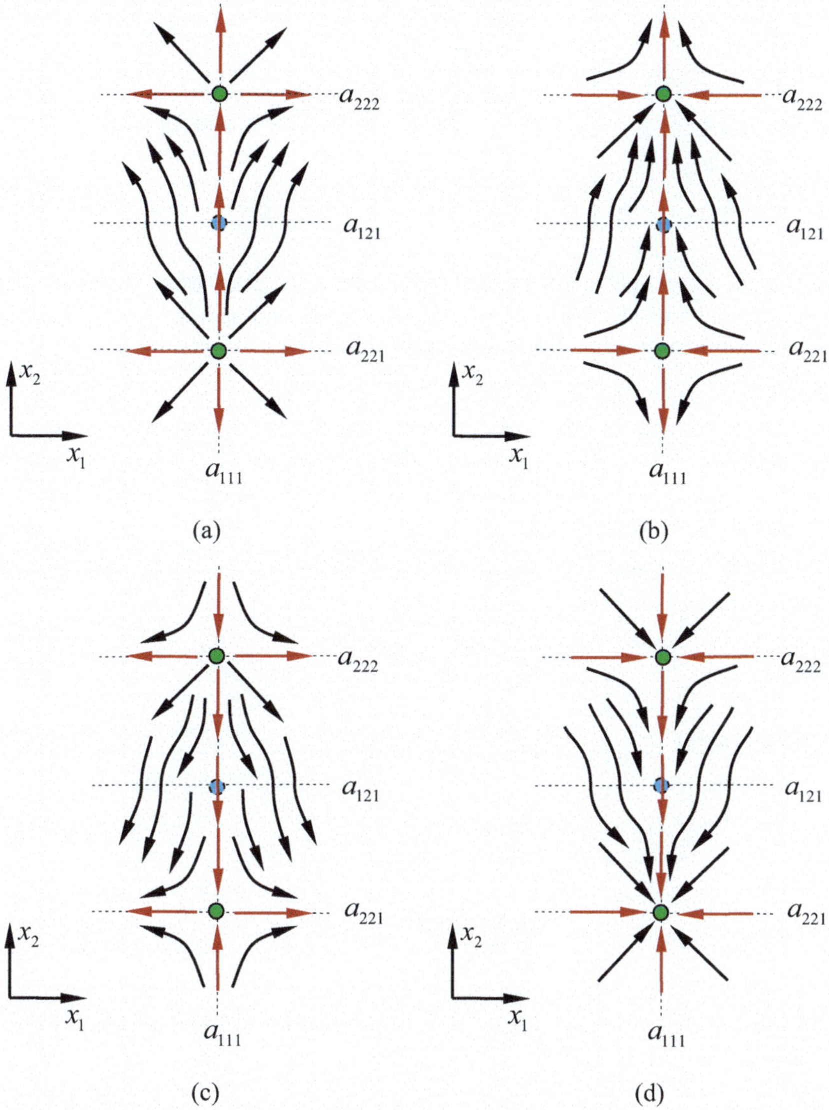

Fig. 2.1 Phase portraits ($a_{121} \in (a_{221}, a_{222})$) for two-dimensional systems on the x_1-direction with $x_1^* = a_{111}$ and on the x_2-direction with $x_2^* = a_{121}, a_{221}, a_{222}$. The equilibriums flows: (**a**) ($a_{110} > 0$, $a_{220} > 0$), (**b**) ($a_{110} < 0$, $a_{220} > 0$), (**c**) ($a_{110} > 0$, $a_{220} < 0$), (**d**) ($a_{110} < 0$, $a_{220} < 0$)

$$\left\{\begin{array}{l}(a_{111}, a_{222}) \\ (a_{111}, a_{121}) \\ (a_{111}, a_{221})\end{array}\right\} = \left\{\begin{array}{c}\underbrace{(\mathrm{SI}, \mathrm{LS})}_{\text{lower-saddle sink}} \\ \underbrace{({}_{\mathrm{DI:II}}\mathrm{SI}, \mathrm{nF})}_{\text{inflection sink flow }(-)} \\ \underbrace{(\mathrm{SI}, \mathrm{SI})}_{\text{sink}}\end{array}\right\} \text{for } a_{110} < 0 \text{ and } a_{220} < 0. \quad (2.6)$$

The saddle-sinks and saddle-sources are for the appearing bifurcations of saddle with source and sink. In addition, two saddles, sink, and sources exist. The third-order inflection-source and sink flows are also presented.

(ii) For $a_{222} < a_{121}$, the equilibriums of $x_1^* = a_{111}$ with $x_2^* = a_{121}, a_{221}$ have the properties described as in Eqs. (2.3)–(2.6), and the corresponding phase portraits are also sketched in Fig. 2.2. Compared to the case of $a_{121} \in (a_{221}, a_{222})$, just the locations of equilibriums are exchanged, and the corresponding dynamical behaviors of the equilibriums are identical.

(iii) For $a_{121} < a_{221}$, the equilibriums of $x_1^* = a_{111}$ with $x_2^* = a_{121}, a_{221}, a_{222}$ have the following properties, as shown in Fig. 2.3.

$$\left\{\begin{array}{l}(a_{111}, a_{222}) \\ (a_{111}, a_{221}) \\ (a_{111}, a_{121})\end{array}\right\} = \left\{\begin{array}{c}\underbrace{(\mathrm{SO}, \mathrm{US})}_{\text{upper-saddle source}} \\ \underbrace{(\mathrm{SO}, \mathrm{SO})}_{\text{source}} \\ \underbrace{({}_{\mathrm{II:DI}}\mathrm{SO}, \mathrm{nF})}_{\text{inflection source flow }(-)}\end{array}\right\} \text{for } a_{110} > 0 \text{ and } a_{220} > 0; \quad (2.7)$$

$$\left\{\begin{array}{l}(a_{111}, a_{222}) \\ (a_{111}, a_{221}) \\ (a_{111}, a_{121})\end{array}\right\} = \left\{\begin{array}{c}\underbrace{(\mathrm{SI}, \mathrm{US})}_{\text{upper-saddle sink}} \\ \underbrace{(\mathrm{SI}, \mathrm{SO})}_{\text{saddle}} \\ \underbrace{({}_{\mathrm{DI:II}}\mathrm{SI}, \mathrm{nF})}_{\text{inflection sink flow }(-)}\end{array}\right\} \text{for } a_{110} < 0 \text{ and } a_{220} > 0; \quad (2.8)$$

$$\left\{\begin{array}{l}(a_{111}, a_{222}) \\ (a_{111}, a_{221}) \\ (a_{111}, a_{121})\end{array}\right\} = \left\{\begin{array}{c}\underbrace{(\mathrm{SO}, \mathrm{LS})}_{\text{lower-saddle source}} \\ \underbrace{(\mathrm{SO}, \mathrm{SI})}_{\text{saddle}} \\ \underbrace{({}_{\mathrm{DI:II}}\mathrm{SO}, \mathrm{pF})}_{\text{inflection sink flow }(+)}\end{array}\right\} \text{for } a_{110} > 0 \text{ and } a_{220} < 0; \quad (2.9)$$

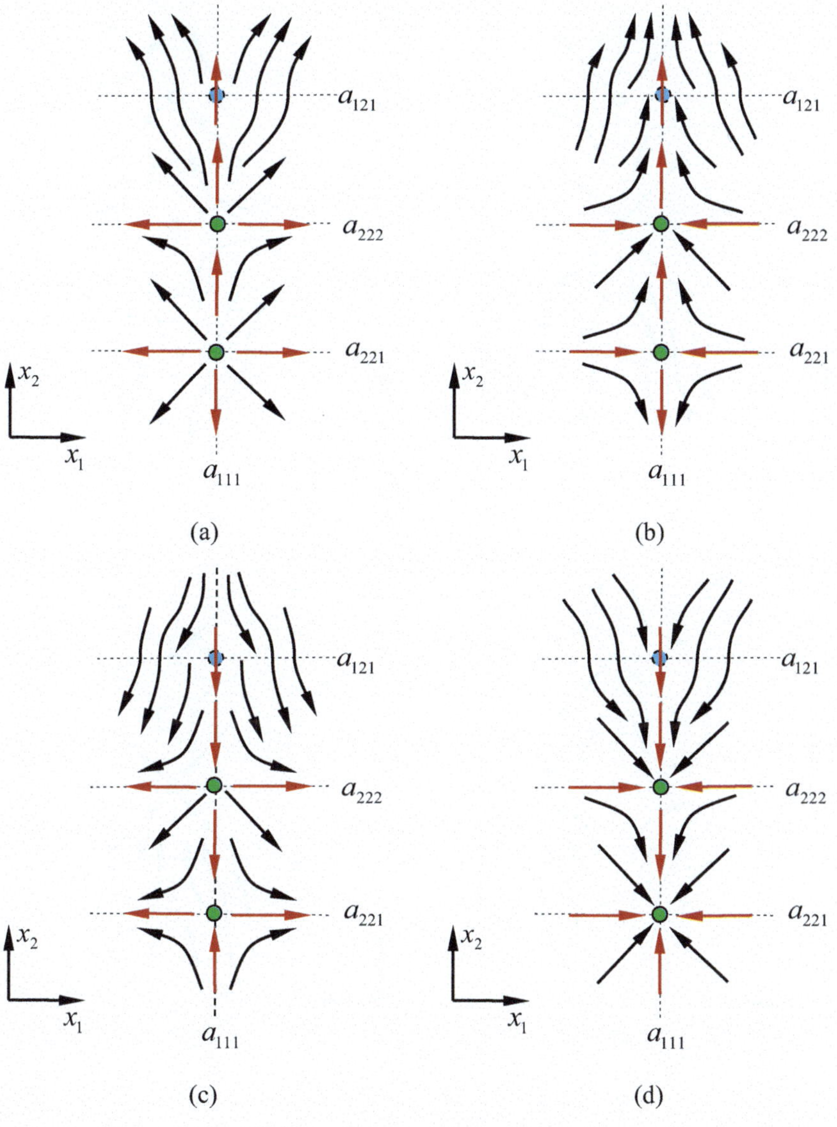

Fig. 2.2 Phase portraits ($a_{222} < a_{121}$) for two-dimensional systems on the x_1-direction with $x_1^* = a_{111}$ and on the x_2-direction with $x_2^* = a_{121}, a_{221}, a_{222}$. The equilibriums flows: (**a**) ($a_{110} > 0$, $a_{220} > 0$), (**b**) ($a_{110} < 0$, $a_{220} > 0$), (**c**) ($a_{110} > 0$, $a_{220} < 0$), (**d**) ($a_{110} < 0$, $a_{220} < 0$)

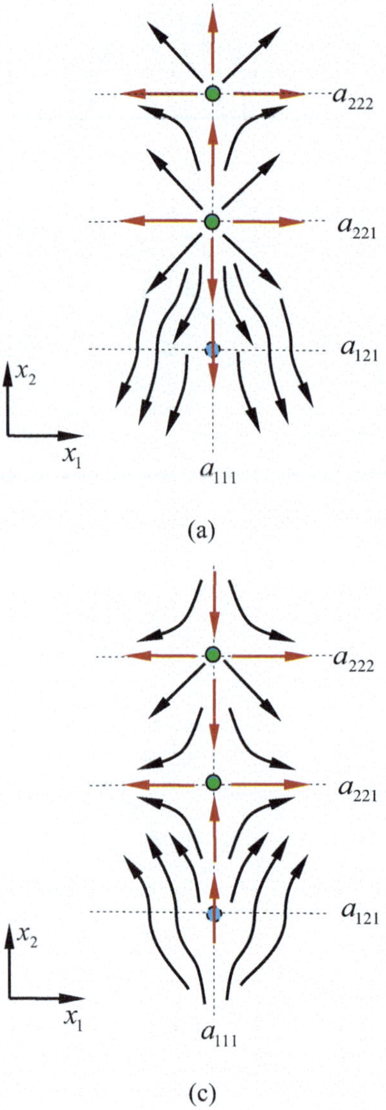

Fig. 2.3 Phase portraits ($a_{121} < a_{221}$) for two-dimensional systems on the x_1-direction with $x_1^* = a_{111}$ and on the x_2-direction with $x_2^* = a_{121}, a_{221}, a_{222}$. The equilibriums flows: (**a**) ($a_{110} > 0$, $a_{220} > 0$), (**b**) ($a_{110} < 0$, $a_{220} > 0$), (**c**) ($a_{110} > 0$, $a_{220} < 0$), (**d**) ($a_{110} < 0$, $a_{220} < 0$)

$$\left\{\begin{array}{c}(a_{111},a_{222})\\(a_{111},a_{221})\\(a_{111},a_{121})\end{array}\right\}=\left\{\begin{array}{c}\underbrace{(\mathrm{SI},\mathrm{LS})}_{\text{lower-saddle sink}}\\\underbrace{(\mathrm{SI},\mathrm{SI})}_{\text{sink}}\\\underbrace{({}_{\mathrm{II:DI}}\mathrm{SI},\mathrm{pF})}_{\text{inflection source flow }(+)}\end{array}\right\}\text{ for } a_{110}<0 \text{ and } a_{220}<0. \tag{2.10}$$

The saddle-sinks and saddle-sources are also for the appearing bifurcations of saddle with source and sink. Two saddles, sink, and sources exist. The third-order inflection source and sink flows are also presented. Compared to Figs. 2.1 and 2.2, only the third-order inflection source and sink flows are different.

Readers can compete the phase portraits for $a_{22s_2} < a_{22s_1}$ with $a_{221} = a_{22s_2}$ and $a_{222} = a_{22s_1}$ as an exercise.

2.1.2 *Third-Order Saddles, Sink, and Source*

In this section, the phase portraits for the third-order saddles, sink, and sources are presented, and the corresponding appearing bifurcations will be discussed for source, saddles, and sink.

Consider a dynamical system as

$$\begin{aligned}\dot{x}_1 &= a_{110}(x_1 - a_{111})(x_2 - a_{121})^2,\\ \dot{x}_2 &= a_{220}(x_2 - a_{221})^3,\end{aligned} \tag{2.11}$$

and the corresponding first integral manifold is

$$\begin{aligned}\ln\frac{|x_1 - a_{111}|}{|x_{10} - a_{111}|} = \frac{a_{110}}{a_{220}}\Bigg\{ &\ln\frac{|x_2 - a_{221}|}{|x_{20} - a_{221}|}\\ &- 2(a_{221} - a_{121})\Big(\frac{1}{x_2 - a_{221}} - \frac{1}{x_{20} - a_{221}}\Big)\\ &- \frac{1}{2}(a_{221} - a_{121})^2\left[\frac{1}{(x_2 - a_{221})^2} - \frac{1}{(x_{20} - a_{221})^2}\right]\Bigg\}.\end{aligned} \tag{2.12}$$

(i) For $a_{121} < a_{221}$, the equilibriums of $x_1^* = a_{111}$ with $x_2^* = a_{121}, a_{221}$ have the following properties, as shown in Fig. 2.4.

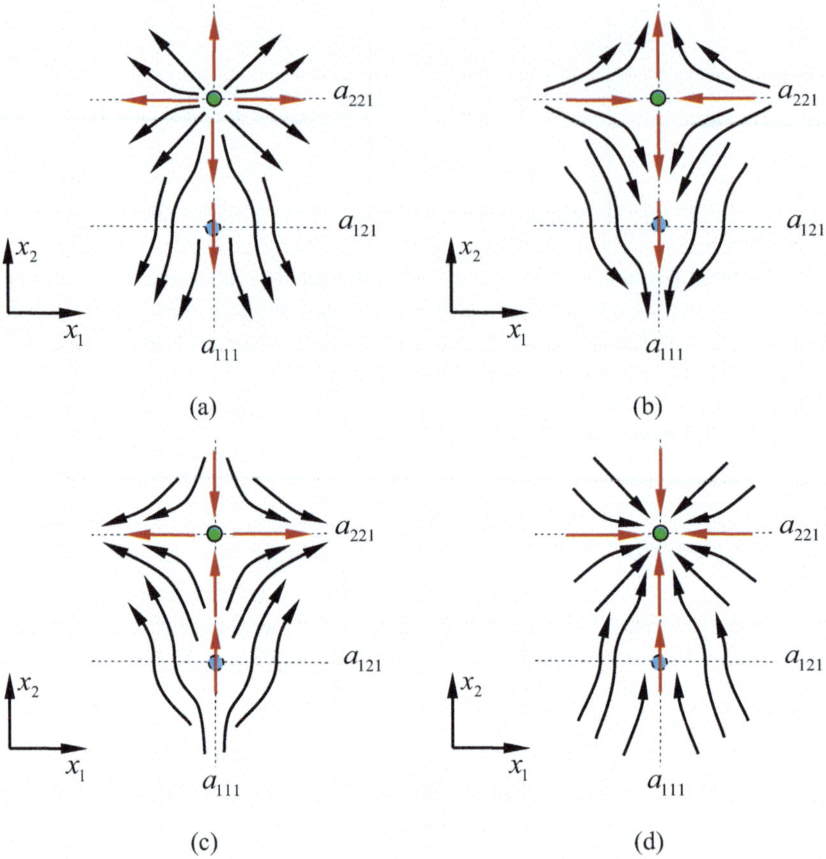

Fig. 2.4 Phase portraits ($a_{121} < a_{221}$) for two-dimensional systems on the x_1-direction with $x_1^* = a_{111}$ and on the x_2-direction with $x_2^* = a_{121}, a_{221}$. The equilibriums flows: (**a**) ($a_{110} > 0$, $a_{220} > 0$), (**b**) ($a_{110} < 0$, $a_{220} > 0$), (**c**) ($a_{110} > 0$, $a_{220} < 0$), (**d**) ($a_{110} < 0$, $a_{220} < 0$)

$$\left\{\begin{array}{l}(a_{111}, a_{221}) \\ (a_{111}, a_{121})\end{array}\right\} = \left\{\begin{array}{c}\underbrace{\underline{(\text{SO}, 3^{\text{rd}}\text{SO})}}_{\text{source}} \\ \underbrace{\underline{({}_{\text{II:DI}}\text{SO}, \text{nF})}}_{\text{inflection-source flow }(-)}\end{array}\right\} \text{for } a_{110} > 0 \text{ and } a_{220} > 0; \tag{2.13}$$

$$\left\{\begin{array}{l}(a_{111}, a_{221}) \\ (a_{111}, a_{121})\end{array}\right\} = \left\{\begin{array}{c}\underbrace{\underline{(\text{SI}, 3^{\text{rd}}\text{SO})}}_{\text{saddle}} \\ \underbrace{\underline{({}_{\text{DI:II}}\text{SI}, \text{nF})}}_{\text{inflection-sink flow }(-)}\end{array}\right\} \text{for } a_{110} < 0 \text{ and } a_{220} > 0; \tag{2.14}$$

$$\left\{\begin{array}{l}(a_{111},a_{221})\\(a_{111},a_{121})\end{array}\right\}=\left\{\begin{array}{c}\underbrace{(\mathrm{SO},3^{\mathrm{rd}}\mathrm{SI})}_{\text{saddle}}\\\underbrace{({}_{\mathrm{DI:II}}\mathrm{SO},\mathrm{pF})}_{\text{inflection-source flow }(+)}\end{array}\right\}\text{ for }a_{110}>0\text{ and }a_{220}<0; \tag{2.15}$$

$$\left\{\begin{array}{l}(a_{111},a_{221})\\(a_{111},a_{121})\end{array}\right\}=\left\{\begin{array}{c}\underbrace{(\mathrm{SI},3^{\mathrm{rd}}\mathrm{SI})}_{\text{sink}}\\\underbrace{({}_{\mathrm{II:DI}}\mathrm{SI},\mathrm{pF})}_{\text{inflection-sink flow }(+)}\end{array}\right\}\text{ for }a_{110}<0\text{ and }a_{220}<0. \tag{2.16}$$

The third-order source is an appearing bifurcation from source to source, (SO,SI)-saddle and source. The third-order sink is an appearing bifurcation from sink to sink, (SI,SO)-saddle and sink. The (SI,3rdSO)-third-order saddle is an appearing bifurcation from (SI,SO)-saddle to (SI,SO)-saddle, sink, and (SI,SO)-saddle. The (SO,3rdSI)-third-order saddle is an appearing bifurcation from (SO,SI)-saddle to (SO,SI)-saddle, source, and (SO,SI)-saddle. The third-order inflection-source and sink flows are also presented for the appearing bifurcation of sink and source flows.

(ii) For $a_{221} < a_{121}$, the equilibriums of $x_1^* = a_{111}$ with $x_2^* = a_{121}, a_{221}$ have the following properties, as shown in Fig. 2.5.

$$\left\{\begin{array}{l}(a_{111},a_{121})\\(a_{111},a_{221})\end{array}\right\}=\left\{\begin{array}{c}\underbrace{({}_{\mathrm{DI:II}}\mathrm{SO},\mathrm{pF})}_{\text{inflection-sink flow }(+)}\\\underbrace{(\mathrm{SO},3^{\mathrm{rd}}\mathrm{SO})}_{\text{source}}\end{array}\right\}\text{ for }a_{110}>0\text{ and }a_{220}>0; \tag{2.17}$$

$$\left\{\begin{array}{l}(a_{111},a_{221})\\(a_{111},a_{121})\end{array}\right\}=\left\{\begin{array}{c}\underbrace{({}_{\mathrm{II:DI}}\mathrm{SI},\mathrm{pF})}_{\text{inflection-sink flow }(+)}\\\underbrace{(\mathrm{SI},3^{\mathrm{rd}}\mathrm{SO})}_{\text{saddle}}\end{array}\right\}\text{ for }a_{110}<0\text{ and }a_{220}>0; \tag{2.18}$$

$$\left\{\begin{array}{l}(a_{111},a_{221})\\(a_{111},a_{121})\end{array}\right\}=\left\{\begin{array}{c}\underbrace{({}_{\mathrm{II:DI}}\mathrm{SO},\mathrm{nF})}_{\text{inflection-source flow }(-)}\\\underbrace{(\mathrm{SO},3^{\mathrm{rd}}\mathrm{SI})}_{\text{saddle}}\end{array}\right\}\text{ for }a_{110}>0\text{ and }a_{220}<0; \tag{2.19}$$

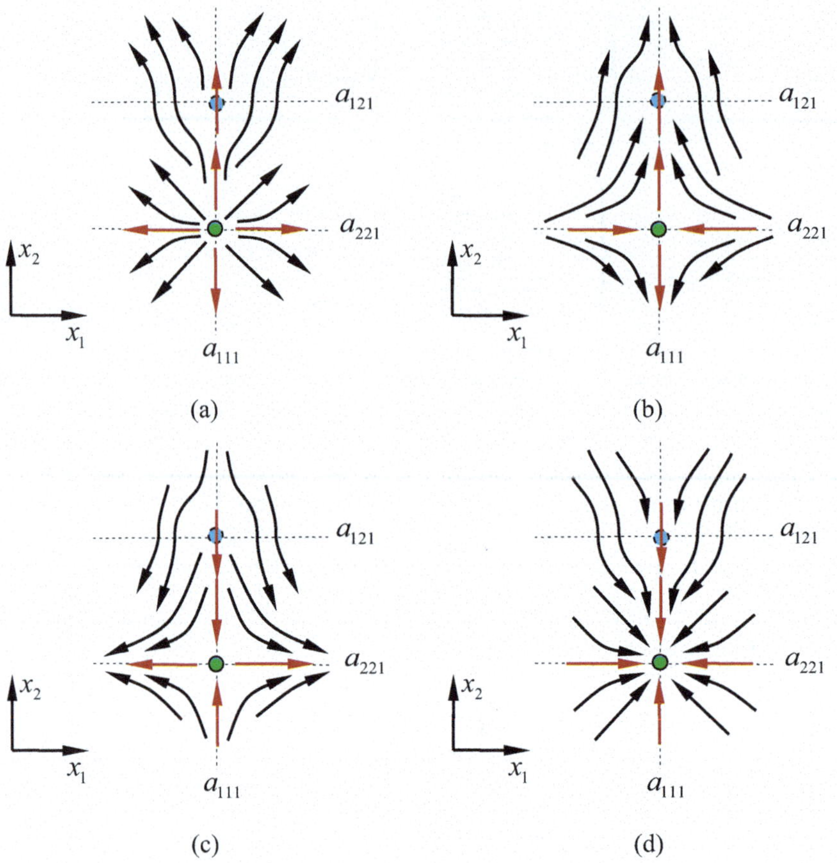

Fig. 2.5 Phase portraits ($a_{221} < a_{121}$) for two-dimensional systems on the x_1-direction with $x_1^* = a_{111}$ and on the x_2-direction with $x_2^* = a_{121}, a_{221}$. The equilibriums flows: (**a**) ($a_{110} > 0$, $a_{220} > 0$), (**b**) ($a_{110} < 0$, $a_{220} > 0$), (**c**) ($a_{110} > 0$, $a_{220} < 0$), (**d**) ($a_{110} < 0$, $a_{220} < 0$)

$$\left\{ \begin{array}{l} (a_{111}, a_{221}) \\ (a_{111}, a_{121}) \end{array} \right\} = \left\{ \begin{array}{c} \underbrace{({}_{\text{DI:II}}\text{SI, nF})}_{\text{inflection-sink flow } (-)} \\ \underbrace{(\text{SI, } 3^{\text{rd}}\text{SI})}_{\text{sink}} \end{array} \right\} \text{ for } a_{110} < 0 \text{ and } a_{220} < 0. \tag{2.20}$$

The third-order source, third-order sink, (SI,3rdSO)-third-order saddle, and (SO,3rdSI)-third-order saddle are presented with third-order inflection source and sink flows. Compared to the case of $a_{121} < a_{221}$, the third-order sink, source, and saddles for $a_{121} < a_{221}$ are unchanged, but the third-order source and sink flows are changed.

2.2 Infinite-Equilibriums and Switching Bifurcations

The infinite-equilibriums for product-cubic systems and the corresponding switching bifurcation are discussed.

2.2.1 *Inflection-Saddle Infinite-Equilibriums*

In this section, the inflection-saddle infinite-equilibrium will be discussed first, and the switching bifurcations are based on the inflection-saddle infinite-equilibriums.

Consider a dynamical system for $a_{22s_2} = a_{121}$ as

$$\begin{aligned} \dot{x}_1 &= a_{110}(x_1 - a_{111})(x_2 - a_{121})^2, \\ \dot{x}_2 &= a_{220}(x_2 - a_{22s_1})(x_2 - a_{22s_2})^2, \end{aligned} \tag{2.21}$$

and the corresponding first integral manifold for $a_{22s_2} = a_{121}$ is

$$\ln\frac{|x_1 - a_{111}|}{|x_{10} - a_{111}|} = \frac{a_{110}}{a_{220}} \ln\frac{|x_2 - a_{22s_1}|}{|x_{20} - a_{22s_1}|}. \tag{2.22}$$

(i) For $a_{22s_2} = a_{222} > a_{22s_1} = a_{221}$, the equilibriums have the following properties.

$$\left\{ \begin{matrix} (a_{111}, a_{222}) \\ (a_{111}, a_{221}) \end{matrix} \right\} = \left\{ \begin{matrix} \underbrace{(\text{DI}:\text{II},\text{US})}_{\text{inflection upper-saddle}} \\ \underbrace{(\text{SO},\text{SO})}_{\text{source}} \end{matrix} \right\} \text{ for } a_{110} > 0, a_{220} > 0, \tag{2.23}$$

$$\left\{ \begin{matrix} (a_{111}, a_{222}) \\ (a_{111}, a_{221}) \end{matrix} \right\} = \left\{ \begin{matrix} \underbrace{(\text{II}:\text{DI},\text{US})}_{\text{inflection upper-saddle}} \\ \underbrace{(\text{SI},\text{SO})}_{\text{saddle}} \end{matrix} \right\} \text{ for } a_{110} < 0, a_{220} > 0, \tag{2.24}$$

$$\left\{ \begin{matrix} (a_{111}, a_{222}) \\ (a_{111}, a_{221}) \end{matrix} \right\} = \left\{ \begin{matrix} \underbrace{(\text{II}:\text{DI},\text{LS})}_{\text{inflection lower-saddle}} \\ \underbrace{(\text{SO},\text{SI})}_{\text{saddle}} \end{matrix} \right\} \text{ for } a_{110} > 0, a_{220} < 0, \tag{2.25}$$

$$\left\{ \begin{matrix} (a_{111}, a_{222}) \\ (a_{111}, a_{221}) \end{matrix} \right\} = \left\{ \begin{matrix} \underbrace{(\text{DI}:\text{II},\text{LS})}_{\text{inflection lower-saddle}} \\ \underbrace{(\text{SI},\text{SI})}_{\text{sink}} \end{matrix} \right\} \text{ for } a_{110} < 0, a_{220} < 0. \tag{2.26}$$

The infinite-equilibrium of $x_2^* = a_{121} = a_{222}$ is summarized as follows.

$$
\begin{aligned}
(\bar{x}_1, a_{222}) &= \underbrace{(\mathrm{DI}, \mathrm{US})}_{\text{decreasing-infection upper-saddle}} && \text{if } \bar{x}_1 \in (-\infty, a_{111}),\\
(\bar{x}_1, a_{222}) &= \underbrace{(\mathrm{II}, \mathrm{US})}_{\text{increasing-inflection upper-saddle}} && \text{if } \bar{x}_1 \in (a_{111}, \infty),
\end{aligned}
\tag{2.27}
$$

for $a_{110} > 0, a_{220} > 0$;

$$
\begin{aligned}
(\bar{x}_1, a_{222}) &= \underbrace{(\mathrm{II}, \mathrm{US})}_{\text{increasing-infection upper-saddle}} && \text{if } \bar{x}_1 \in (-\infty, a_{111}),\\
(\bar{x}_1, a_{222}) &= \underbrace{(\mathrm{DI}, \mathrm{US})}_{\text{decreasing-inflection upper-saddle}} && \text{if } \bar{x}_1 \in (a_{111}, \infty),
\end{aligned}
\tag{2.28}
$$

for $a_{110} < 0, a_{220} > 0$;

$$
\begin{aligned}
(\bar{x}_1, a_{222}) &= \underbrace{(\mathrm{II}, \mathrm{LS})}_{\text{increasing-infection lower-saddle}} && \text{if } \bar{x}_1 \in (-\infty, a_{111}),\\
(\bar{x}_1, a_{222}) &= \underbrace{(\mathrm{DI}, \mathrm{LS})}_{\text{decreasing-inflection lower-saddle}} && \text{if } \bar{x}_1 \in (a_{111}, \infty),
\end{aligned}
\tag{2.29}
$$

for $a_{110} > 0, a_{220} < 0$;

$$
\begin{aligned}
(\bar{x}_1, a_{222}) &= \underbrace{(\mathrm{DI}, \mathrm{LS})}_{\text{decreasing-infection lower-saddle}} && \text{if } \bar{x}_1 \in (-\infty, a_{111}),\\
(\bar{x}_1, a_{222}) &= \underbrace{(\mathrm{II}, \mathrm{LS})}_{\text{increasing-inflection lower-saddle}} && \text{if } \bar{x}_1 \in (a_{111}, \infty),
\end{aligned}
\tag{2.30}
$$

for $a_{110} < 0, a_{220} < 0$.

The infinite-equilibrium of $x_2^* = a_{121} = a_{222}$ is increasing-inflection and decreasing-inflection upper-saddle and lower-saddle, as shown in Fig. 2.6. The infinite-equilibrium is the switching bifurcation for $a_{121} \in (a_{221}, a_{222})$ and $a_{121} > a_{222}$.

(ii) For $a_{22s_2} = a_{221} < a_{22s_1} = a_{222}$, the equilibriums have the following properties.

$$
\left\{\begin{array}{c}(a_{111}, a_{222})\\ (a_{111}, a_{221})\end{array}\right\} = \left\{\begin{array}{c}\underbrace{(\mathrm{SO}, \mathrm{SO})}_{\text{source}}\\ \underbrace{(\mathrm{II} : \mathrm{DI}, \mathrm{LS})}_{\text{inflection lower-saddle}}\end{array}\right\} \text{for } a_{110} > 0, a_{220} > 0, \tag{2.31}
$$

$$
\left\{\begin{array}{c}(a_{111}, a_{222})\\ (a_{111}, a_{221})\end{array}\right\} = \left\{\begin{array}{c}\underbrace{(\mathrm{SI}, \mathrm{SO})}_{\text{saddle}}\\ \underbrace{(\mathrm{DI} : \mathrm{II}, \mathrm{LS})}_{\text{inflection lower-saddle}}\end{array}\right\} \text{for } a_{110} < 0, a_{220} > 0, \tag{2.32}
$$

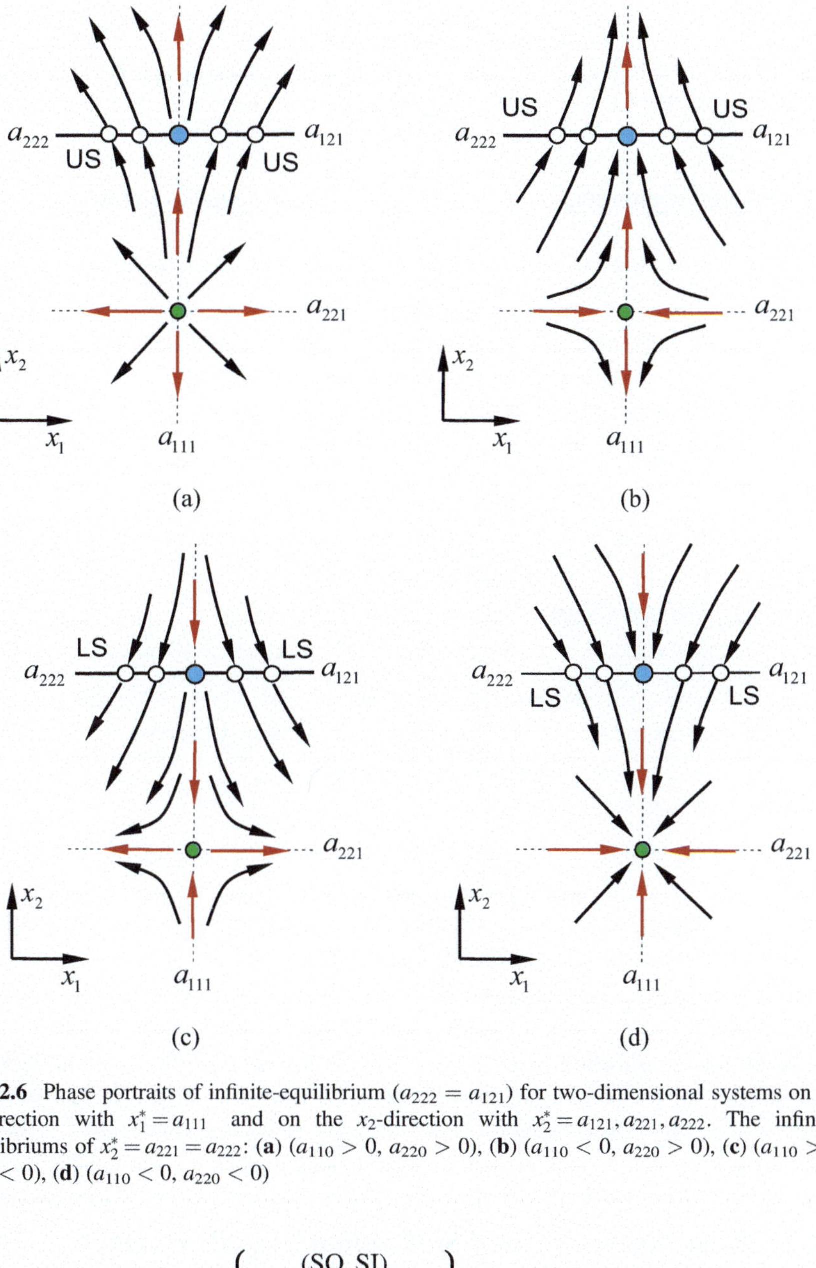

Fig. 2.6 Phase portraits of infinite-equilibrium ($a_{222} = a_{121}$) for two-dimensional systems on the x_1-direction with $x_1^* = a_{111}$ and on the x_2-direction with $x_2^* = a_{121}, a_{221}, a_{222}$. The infinite-equilibriums of $x_2^* = a_{221} = a_{222}$: **(a)** ($a_{110} > 0$, $a_{220} > 0$), **(b)** ($a_{110} < 0$, $a_{220} > 0$), **(c)** ($a_{110} > 0$, $a_{220} < 0$), **(d)** ($a_{110} < 0$, $a_{220} < 0$)

$$\left\{ \begin{array}{c} (a_{111}, a_{222}) \\ (a_{111}, a_{221}) \end{array} \right\} = \left\{ \begin{array}{c} \underbrace{(\mathrm{SO}, \mathrm{SI})}_{\text{saddle}} \\ \underbrace{(\mathrm{DI} : \mathrm{II}, \mathrm{US})}_{\text{inflection upper-saddle}} \end{array} \right\} \text{ for } a_{110} > 0, a_{220} < 0, \tag{2.33}$$

$$\left\{\begin{array}{l}(a_{111},a_{222})\\(a_{111},a_{221})\end{array}\right\}=\left\{\begin{array}{c}\underbrace{(\mathrm{SI},\mathrm{SI})}_{\text{sink}}\\\underbrace{(\mathrm{II}:\mathrm{DI},\mathrm{US})}_{\text{inflection upper-saddle}}\end{array}\right\}\text{ for }a_{110}<0,a_{220}<0. \tag{2.34}$$

The infinite-equilibrium of $x_2^* = a_{121} = a_{221}$ is summarized as follows.

$$\begin{aligned}&(\bar{x}_1,a_{221})=\underbrace{(\mathrm{II},\mathrm{LS})}_{\text{increasing-infection lower-saddle}}\quad\text{if }\bar{x}_1\in(-\infty,a_{111}),\\&(\bar{x}_1,a_{221})=\underbrace{(\mathrm{DI},\mathrm{LS})}_{\text{decreasing-inflection lower-saddle}}\quad\text{if }\bar{x}_1\in(a_{111},\infty),\\&\text{for }a_{110}>0,a_{220}>0;\end{aligned} \tag{2.35}$$

$$\begin{aligned}&(\bar{x}_1,a_{221})=\underbrace{(\mathrm{DI},\mathrm{LS})}_{\text{decreasing-infection lower-saddle}}\quad\text{if }\bar{x}_1\in(-\infty,a_{111}),\\&(\bar{x}_1,a_{221})=\underbrace{(\mathrm{II},\mathrm{LS})}_{\text{increasing-inflection lower-saddle}}\quad\text{if }\bar{x}_1\in(a_{111},\infty),\\&\text{for }a_{110}<0,a_{220}>0;\end{aligned} \tag{2.36}$$

$$\begin{aligned}&(\bar{x}_1,a_{221})=\underbrace{(\mathrm{DI},\mathrm{US})}_{\text{decreasing-infection upper-saddle}}\quad\text{if }\bar{x}_1\in(-\infty,a_{111}),\\&(\bar{x}_1,a_{221})=\underbrace{(\mathrm{II},\mathrm{US})}_{\text{increasing-inflection upper-saddle}}\quad\text{if }\bar{x}_1\in(a_{111},\infty),\\&\text{for }a_{110}>0,a_{220}<0;\end{aligned} \tag{2.37}$$

$$\begin{aligned}&(\bar{x}_1,a_{221})=\underbrace{(\mathrm{II},\mathrm{US})}_{\text{increasing-infection upper-saddle}}\quad\text{if }\bar{x}_1\in(-\infty,a_{111}),\\&(\bar{x}_1,a_{221})=\underbrace{(\mathrm{DI},\mathrm{US})}_{\text{decreasing-inflection upper-saddle}}\quad\text{if }\bar{x}_1\in(a_{111},\infty),\\&\text{for }a_{110}<0,a_{220}<0.\end{aligned} \tag{2.38}$$

The infinite-equilibrium of $x_2^* = a_{121} = a_{222}$ is increasing-inflection and decreasing-inflection upper-saddle and lower-saddle, as shown in Fig. 2.7. The infinite-equilibrium is the switching bifurcation for $a_{121} \in (a_{221}, a_{222})$ and $a_{121} < a_{221}$.

2.2.2 *Parabola-Source and Sink Infinite-Equilibriums*

In this section, the parabola-source and sink infinite-equilibrium will be discussed first, and the switching bifurcations are based the parabola-source and sink infinite-equilibriums.

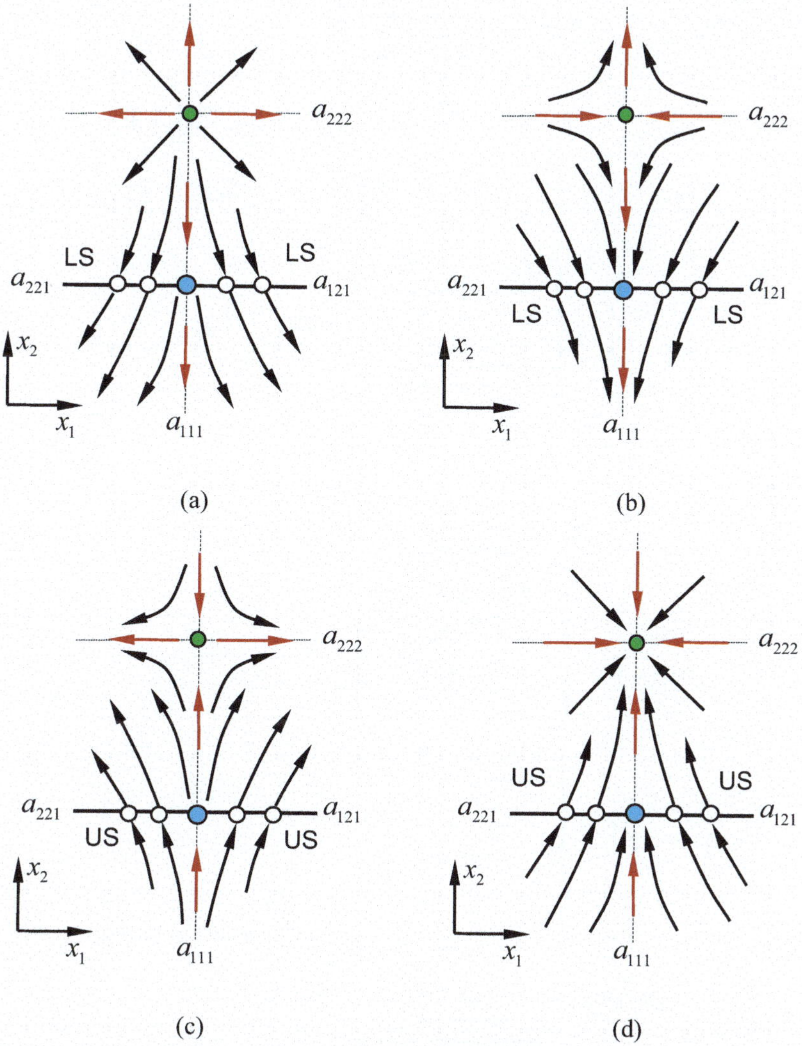

Fig. 2.7 Phase portraits of infinite-equilibrium ($a_{221} = a_{121}$) for two-dimensional systems on the x_1-direction with $x_1^* = a_{111}$ and on the x_2-direction with $x_2^* = a_{121}, a_{221}, a_{222}$. The infinite-equilibriums of $x_2^* = a_{221} = a_{221}$: (**a**) ($a_{110} > 0$, $a_{220} > 0$), (**b**) ($a_{110} < 0$, $a_{220} > 0$), (**c**) ($a_{110} > 0$, $a_{220} < 0$), (**d**) ($a_{110} < 0$, $a_{220} < 0$)

Consider a dynamical system for $a_{22s_1} = a_{121}$ as

$$\begin{aligned} \dot{x}_1 &= a_{110}(x_1 - a_{111})(x_2 - a_{121})^2, \\ \dot{x}_2 &= a_{220}(x_2 - a_{22s_1})(x_2 - a_{22s_2})^2, \end{aligned} \tag{2.39}$$

and the corresponding first integral manifold for $a_{22s_1} = a_{121}$ is

$$\ln\frac{|x_1-a_{111}|}{|x_{10}-a_{111}|}=\frac{a_{110}}{a_{220}}\Big[\ln\frac{|x_2-a_{22s_2}|}{|x_{20}-a_{22s_2}|} \\ -(a_{22s_2}-a_{121})(\frac{1}{x_2-a_{22s_2}}-\frac{1}{x_{20}-a_{22s_2}})\Big]. \tag{2.40}$$

(i) For $a_{22s_2}=a_{121}<a_{22s_1}=a_{222}$, the equilibriums have the following properties,

$$\left\{\begin{array}{l}(a_{111},a_{222})\\(a_{111},a_{221})\end{array}\right\}=\left\{\begin{array}{c}\underbrace{(\text{DP}:\text{UP},\text{SO})}_{\text{hyperbolic source}}\\\underbrace{(\text{SO},\text{LS})}_{\text{lower-saddle source}}\end{array}\right\}\text{ for } a_{110}>0, a_{220}>0, \tag{2.41}$$

$$\left\{\begin{array}{l}(a_{111},a_{222})\\(a_{111},a_{221})\end{array}\right\}=\left\{\begin{array}{c}\underbrace{(\text{UP}:\text{DP},\text{SO})}_{\text{hyperbolic-secant source}}\\\underbrace{(\text{SI},\text{LS})}_{\text{lower-saddle sink}}\end{array}\right\}\text{ for } a_{110}<0, a_{220}>0, \tag{2.42}$$

$$\left\{\begin{array}{l}(a_{111},a_{222})\\(a_{111},a_{221})\end{array}\right\}=\left\{\begin{array}{c}\underbrace{(\text{UP}:\text{DP},\text{SI})}_{\text{hyperbolic-secant sink}}\\\underbrace{(\text{SO},\text{US})}_{\text{upper-saddle source}}\end{array}\right\}\text{ for } a_{110}>0, a_{220}<0, \tag{2.43}$$

$$\left\{\begin{array}{l}(a_{111},a_{222})\\(a_{111},a_{221})\end{array}\right\}=\left\{\begin{array}{c}\underbrace{(\text{UP}:\text{DP},\text{SI})}_{\text{hyperbolic sink}}\\\underbrace{(\text{SI},\text{US})}_{\text{upper-saddle sink}}\end{array}\right\}\text{ for } a_{110}<0, a_{220}<0. \tag{2.44}$$

The infinite-equilibrium of $x_2^*=a_{121}=a_{222}$ is summarized as follows.

$$\begin{array}{l}(\bar{x}_1,a_{222})=\underbrace{(\text{DP},\text{SO})}_{\text{down-parabola source}}\quad\text{if } \bar{x}_1\in(-\infty,a_{111}),\\(\bar{x}_1,a_{222})=\underbrace{(\text{UP},\text{SO})}_{\text{up-parabola source}}\quad\text{if } \bar{x}_1\in(a_{111},\infty),\\\text{for } a_{110}>0, a_{220}>0;\end{array} \tag{2.45}$$

$$\begin{array}{l}(\bar{x}_1,a_{222})=\underbrace{(\text{UP},\text{SO})}_{\text{up-parabola source}}\quad\text{if } \bar{x}_1\in(-\infty,a_{111}),\\(\bar{x}_1,a_{222})=\underbrace{(\text{DP},\text{SO})}_{\text{down-parabola source}}\quad\text{if } \bar{x}_1\in(a_{111},\infty),\\\text{for } a_{110}<0, a_{220}>0;\end{array} \tag{2.46}$$

$$(\bar{x}_1, a_{222}) = \underbrace{(\text{UP}, \text{SI})}_{\text{up-parabola sink}} \text{ if } \bar{x}_1 \in (-\infty, a_{111}),$$
$$(\bar{x}_1, a_{222}) = \underbrace{(\text{DP}, \text{SI})}_{\text{down-parabola sink}} \text{ if } \bar{x}_1 \in (a_{111}, \infty), \tag{2.47}$$
$$\text{for } a_{110} > 0, a_{220} < 0;$$
$$(\bar{x}_1, a_{222}) = \underbrace{(\text{DP}, \text{SI})}_{\text{down-parabola sink}} \text{ if } \bar{x}_1 \in (-\infty, a_{111}),$$
$$(\bar{x}_1, a_{222}) = \underbrace{(\text{UP}, \text{SI})}_{\text{up-parabola sink}} \text{ if } \bar{x}_1 \in (a_{111}, \infty), \tag{2.48}$$
$$\text{for } a_{110} < 0, a_{220} < 0.$$

The infinite-equilibrium of $x_2^* = a_{121} = a_{222}$ are parabola-sources and sinks, as shown in Fig. 2.8. The infinite-equilibrium is the switching bifurcation for $a_{121} \in (a_{221}, a_{222})$ and $a_{121} > a_{222}$.

(ii) For $a_{22s_2} = a_{121} < a_{22s_1} = a_{222}$, the equilibriums have the following properties.

$$\left\{\begin{array}{l}(a_{111}, a_{222})\\(a_{111}, a_{221})\end{array}\right\} = \left\{\begin{array}{c}\underbrace{(\text{SO}, \text{US})}_{\text{upper-saddle source}}\\\underbrace{(\text{DP}:\text{UP}, \text{SO})}_{\text{hyperbolic source}}\end{array}\right\} \text{for } a_{110} > 0, a_{220} > 0, \tag{2.49}$$

$$\left\{\begin{array}{l}(a_{111}, a_{222})\\(a_{111}, a_{221})\end{array}\right\} = \left\{\begin{array}{c}\underbrace{(\text{SI}, \text{US})}_{\text{upper-saddle sink}}\\\underbrace{(\text{UP}:\text{DP}, \text{SO})}_{\text{hyperbolic-secant source}}\end{array}\right\} \text{for } a_{110} < 0, a_{220} > 0, \tag{2.50}$$

$$\left\{\begin{array}{l}(a_{111}, a_{222})\\(a_{111}, a_{221})\end{array}\right\} = \left\{\begin{array}{c}\underbrace{(\text{SO}, \text{LS})}_{\text{lower-saddle source}}\\\underbrace{(\text{UP}:\text{DP}, \text{SI})}_{\text{hyperbolic-secant sink}}\end{array}\right\} \text{for } a_{110} > 0, a_{220} < 0, \tag{2.51}$$

$$\left\{\begin{array}{l}(a_{111}, a_{222})\\(a_{111}, a_{221})\end{array}\right\} = \left\{\begin{array}{c}\underbrace{(\text{SI}, \text{LS})}_{\text{lower-saddle sink}}\\\underbrace{(\text{DP}:\text{UP}, \text{SI})}_{\text{hyperbolic sink}}\end{array}\right\} \text{for } a_{110} < 0, a_{220} < 0. \tag{2.52}$$

The infinite-equilibrium of $x_2^* = a_{121} = a_{222}$ is summarized as follows.

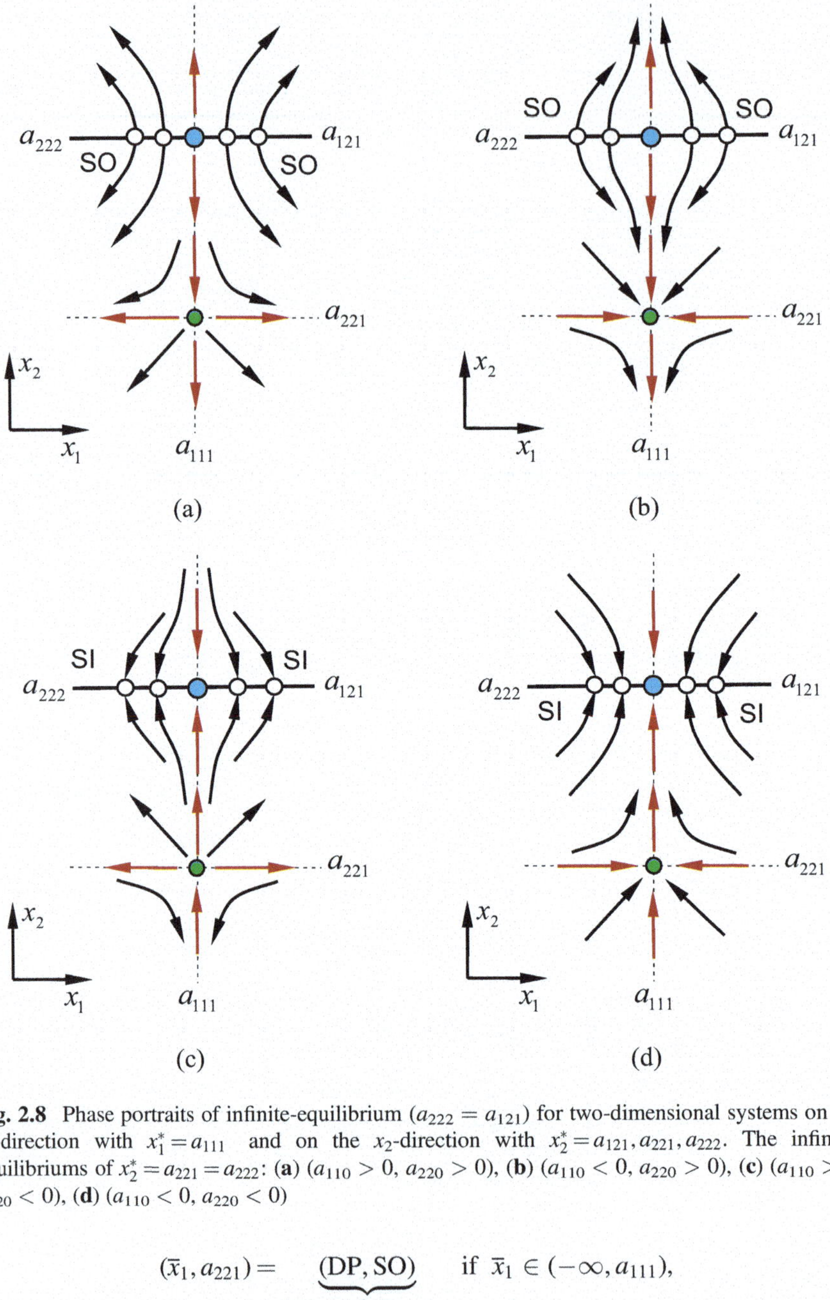

Fig. 2.8 Phase portraits of infinite-equilibrium ($a_{222} = a_{121}$) for two-dimensional systems on the x_1-direction with $x_1^* = a_{111}$ and on the x_2-direction with $x_2^* = a_{121}, a_{221}, a_{222}$. The infinite-equilibriums of $x_2^* = a_{221} = a_{222}$: (**a**) ($a_{110} > 0$, $a_{220} > 0$), (**b**) ($a_{110} < 0$, $a_{220} > 0$), (**c**) ($a_{110} > 0$, $a_{220} < 0$), (**d**) ($a_{110} < 0$, $a_{220} < 0$)

$$
\begin{aligned}
(\bar{x}_1, a_{221}) &= \underbrace{(\mathrm{DP}, \mathrm{SO})}_{\text{down-parabola source}} \quad \text{if } \bar{x}_1 \in (-\infty, a_{111}),\\
(\bar{x}_1, a_{221}) &= \underbrace{(\mathrm{UP}, \mathrm{SO})}_{\text{up-parabola source}} \quad \text{if } \bar{x}_1 \in (a_{111}, \infty),
\end{aligned}
\tag{2.53}
$$

for $a_{110} > 0, a_{220} > 0$;

$$
\begin{aligned}
(\bar{x}_1, a_{221}) &= \underbrace{(\mathrm{UP}, \mathrm{SO})}_{\text{up-parabola source}} \quad \text{if } \bar{x}_1 \in (-\infty, a_{111}),\\
(\bar{x}_1, a_{221}) &= \underbrace{(\mathrm{DP}, \mathrm{SO})}_{\text{down-parabola source}} \quad \text{if } \bar{x}_1 \in (a_{111}, \infty),
\end{aligned}
\tag{2.54}
$$

for $a_{110} < 0, a_{220} > 0$;

$$
\begin{aligned}
(\bar{x}_1, a_{221}) &= \underbrace{(\mathrm{UP}, \mathrm{SI})}_{\text{up-parabola sink}} \quad \text{if } \bar{x}_1 \in (-\infty, a_{111}),\\
(\bar{x}_1, a_{221}) &= \underbrace{(\mathrm{DP}, \mathrm{SI})}_{\text{down-parabola sink}} \quad \text{if } \bar{x}_1 \in (a_{111}, \infty),
\end{aligned}
\tag{2.55}
$$

for $a_{110} > 0, a_{220} > 0$;

$$
\begin{aligned}
(\bar{x}_1, a_{221}) &= \underbrace{(\mathrm{DP}, \mathrm{SI})}_{\text{down-parabola sink}} \quad \text{if } \bar{x}_1 \in (-\infty, a_{111}),\\
(\bar{x}_1, a_{221}) &= \underbrace{(\mathrm{UP}, \mathrm{SI})}_{\text{up-parabola sink}} \quad \text{if } \bar{x}_1 \in (a_{111}, \infty),
\end{aligned}
\tag{2.56}
$$

for $a_{110} < 0, a_{220} < 0$.

The infinite-equilibrium of $x_2^* = a_{121} = a_{222}$ is hyperbolic and hyperbolic-secant sources and sinks, as shown in Fig. 2.9. The infinite-equilibrium is the switching bifurcation for $a_{121} \in (a_{221}, a_{222})$ and $a_{121} > a_{222}$.

2.2.3 *Trumpet-Source and Sink Infinite-Equilibriums*

In this section, the third-order trumpet source and sink infinite-equilibrium will be discussed first, and the switching bifurcations are based the trumpet sink and sink infinite-equilibriums.

Consider a dynamical system for $a_{221} = a_{121}$ as

$$
\begin{aligned}
\dot{x}_1 &= a_{110}(x_1 - a_{111})(x_2 - a_{121})^2,\\
\dot{x}_2 &= a_{220}(x_2 - a_{221})^3,
\end{aligned}
\tag{2.81}
$$

and the corresponding first integral manifold for $a_{221} = a_{121}$ is

$$
\ln \frac{|x_1 - a_{111}|}{|x_{10} - a_{111}|} = \frac{a_{110}}{a_{220}} \ln \frac{|x_2 - a_{221}|}{|x_{20} - a_{221}|}. \tag{2.82}
$$

(i) For $a_{221} = a_{221}$, the equilibriums have the following properties.

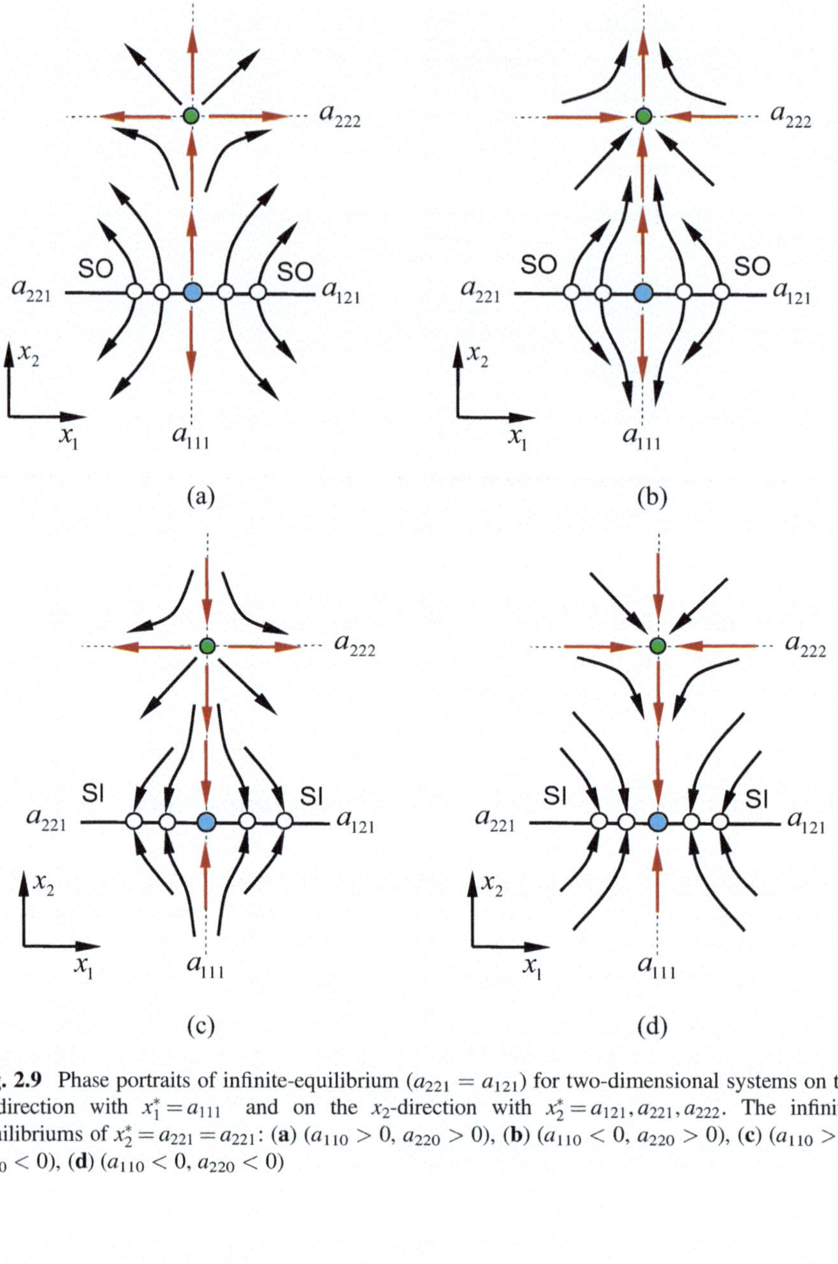

Fig. 2.9 Phase portraits of infinite-equilibrium ($a_{221} = a_{121}$) for two-dimensional systems on the x_1-direction with $x_1^* = a_{111}$ and on the x_2-direction with $x_2^* = a_{121}, a_{221}, a_{222}$. The infinite-equilibriums of $x_2^* = a_{221} = a_{221}$: (**a**) ($a_{110} > 0$, $a_{220} > 0$), (**b**) ($a_{110} < 0$, $a_{220} > 0$), (**c**) ($a_{110} > 0$, $a_{220} < 0$), (**d**) ($a_{110} < 0$, $a_{220} < 0$)

$$
\begin{aligned}
(a_{111}, a_{221}) &= \underbrace{(\mathrm{DT}:\mathrm{UT}, 3^{\mathrm{rd}}\mathrm{SO})}_{\text{third-order trumpet source}} \text{ for } a_{110} > 0, a_{220} > 0, \\
(a_{111}, a_{221}) &= \underbrace{(\mathrm{UT}:\mathrm{DT}, 3^{\mathrm{rd}}\mathrm{SO})}_{\text{third-order trumpet source}} \text{ for } a_{110} < 0, a_{220} > 0, \\
(a_{111}, a_{221}) &= \underbrace{(\mathrm{UT}:\mathrm{DT}, 3^{\mathrm{rd}}\mathrm{SI})}_{\text{third-order trumpet sink}} \text{ for } a_{110} > 0, a_{220} < 0, \\
(a_{111}, a_{221}) &= \underbrace{(\mathrm{DT}:\mathrm{UT}, 3^{\mathrm{rd}}\mathrm{SI})}_{\text{third-order trumpet sink}} \text{ for } a_{110} < 0, a_{220} < 0.
\end{aligned}
\tag{2.83}
$$

The infinite-equilibrium of $x_2^* = a_{121} = a_{221}$ is summarized as follows.

$$
\begin{aligned}
(\bar{x}_1, a_{221}) &= \underbrace{(\mathrm{DT}, 3^{\mathrm{rd}}\mathrm{SO})}_{\text{down-trumpet source}} \text{ if } \bar{x}_1 \in (-\infty, a_{111}), \\
(\bar{x}_1, a_{221}) &= \underbrace{(\mathrm{UT}, 3^{\mathrm{rd}}\mathrm{SO})}_{\text{up-trumpet source}} \text{ if } \bar{x}_1 \in (a_{111}, \infty),
\end{aligned}
\tag{2.84}
$$

for $a_{110} > 0, a_{220} > 0$;

$$
\begin{aligned}
(\bar{x}_1, a_{221}) &= \underbrace{(\mathrm{UT}, 3^{\mathrm{rd}}\mathrm{SO})}_{\text{up-trumpet source}} \text{ if } \bar{x}_1 \in (-\infty, a_{111}), \\
(\bar{x}_1, a_{221}) &= \underbrace{(\mathrm{DT}, 3^{\mathrm{rd}}\mathrm{SO})}_{\text{down-trumpet source}} \text{ if } \bar{x}_1 \in (a_{111}, \infty),
\end{aligned}
\tag{2.85}
$$

for $a_{110} < 0, a_{220} > 0$;

$$
\begin{aligned}
(\bar{x}_1, a_{221}) &= \underbrace{(\mathrm{UT}, 3^{\mathrm{rd}}\mathrm{SI})}_{\text{up-trumpet sink}} \text{ if } \bar{x}_1 \in (-\infty, a_{111}), \\
(\bar{x}_1, a_{221}) &= \underbrace{(\mathrm{DP}, 3^{\mathrm{rd}}\mathrm{SI})}_{\text{down-trumpet sink}} \text{ if } \bar{x}_1 \in (a_{111}, \infty),
\end{aligned}
\tag{2.86}
$$

for $a_{110} > 0, a_{220} < 0$;

$$
\begin{aligned}
(\bar{x}_1, a_{221}) &= \underbrace{(\mathrm{DT}, 3^{\mathrm{rd}}\mathrm{SI})}_{\text{down-trumpet sink}} \text{ if } \bar{x}_1 \in (-\infty, a_{111}), \\
(\bar{x}_1, a_{221}) &= \underbrace{(\mathrm{UT}, 3^{\mathrm{rd}}\mathrm{SI})}_{\text{up-trumpet sink}} \text{ if } \bar{x}_1 \in (a_{111}, \infty),
\end{aligned}
\tag{2.87}
$$

for $a_{110} < 0, a_{220} < 0$.

The infinite-equilibrium of $x_2^* = a_{121} = a_{222}$ are third-order up-trumpet and down-trumpet sources and sinks, as shown in Fig. 2.10. The infinite-equilibrium is the switching bifurcation for $a_{121} < a_{221}$ and $a_{121} > a_{221}$.

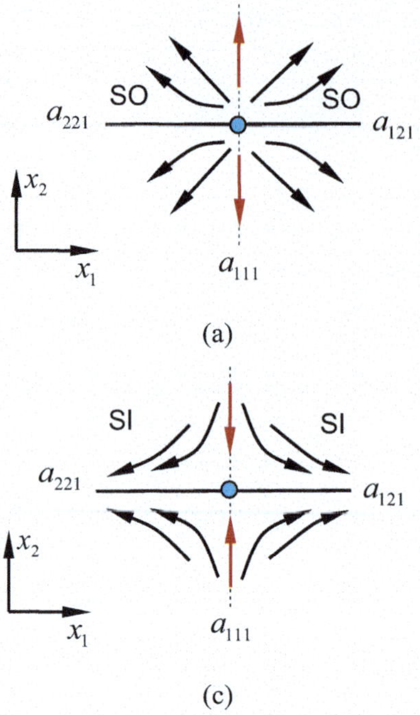

Fig. 2.10 Phase portraits ($a_{121} = a_{221}$) for two-dimensional systems on the x_1-direction with $x_1^* = a_{111}$ and on the x_2-direction with $x_2^* = a_{121}, a_{221}$. The infinite-equilibrium of $x_2^* = a_{121} = a_{221}$: (**a**) ($a_{110} > 0$, $a_{220} > 0$), (**b**) ($a_{110} < 0$, $a_{220} > 0$), (**c**) ($a_{110} > 0$, $a_{220} < 0$), (**d**) ($a_{110} < 0$, $a_{220} < 0$)

Chapter 3
Equilibrium Series and Switching Dynamics

In this chapter, equilibrium series with inflection-source and sink flows and the corresponding switching dynamics of the equilibrium series are discussed. The inflection-source and sink flows are for the appearing bifurcations of the connected hyperbolic and hyperbolic-secant flows. The parabolic source and sink are the switching bifurcations of the inflection-source flow with source and the inflection-sink flow with sink, respectively. The parabolic-secant source and sink are the switching bifurcations of the inflection-source and sink flows with saddles.

3.1 Equilibrium Series with Inflection-Source (Sink) Flows

Consider a dynamical system as

$$\begin{aligned} \dot{x}_1 &= a_{110}(x_1 - a_{111})(x_2 - a_{121})^2, \\ \dot{x}_2 &= a_{220}(x_2 - a_{221})(x_2 - a_{222})(x_2 - a_{223}), \end{aligned} \tag{3.1}$$

and the corresponding first integral manifold is

$$\ln \frac{|x_1 - a_{111}|}{|x_{10} - a_{111}|} = \frac{a_{110}}{a_{220}} \sum\nolimits_{s_1=1}^{3} \frac{(a_{22s_1} - a_{121})^2}{\prod_{s_2=1, s_2 \neq s_1}^{3} (a_{22s_1} - a_{22s_2})} \ln \frac{|x_2 - a_{22s_1}|}{|x_{20} - a_{22s_1}|}. \tag{3.2}$$

(i) For $a_{121} < a_{221}$ and $a_{121} \in (a_{222}, a_{223})$, the vertical equilibrium series of $x_1^* = a_{111}$ with $x_2^* = a_{121}, a_{221}, a_{222}, a_{223}$ have the following properties, as shown in Figs. 3.1 and 3.2. For the two cases, equilibriums in the series have the same dynamical behaviors. Just equilibrium locations in equilibrium series are different. The four

A. C. J. Luo, *Two-dimensional Self and Product Cubic Systems, Vol. I*,
https://doi.org/10.1007/978-3-031-57096-4_3

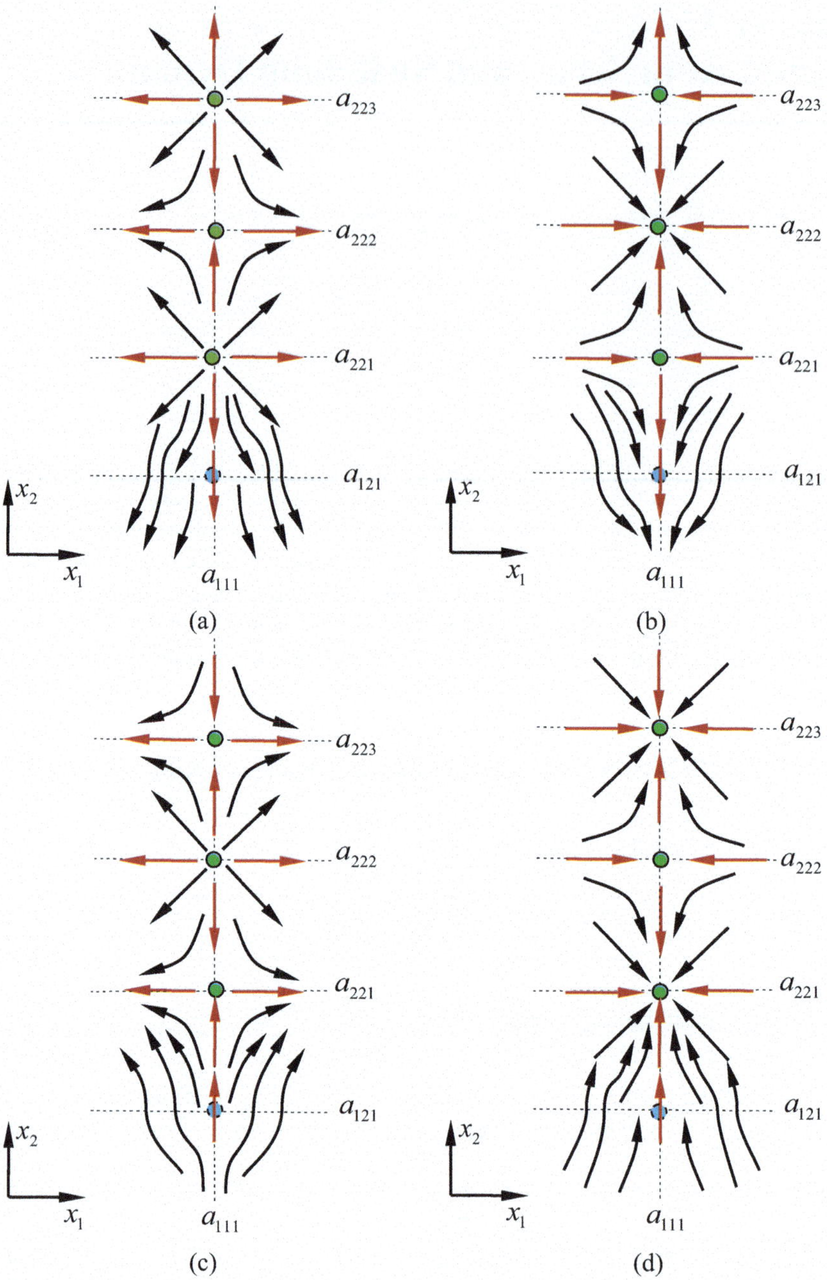

Fig. 3.1 Phase portraits ($a_{121} < a_{221}$) for two-dimensional systems on the x_1-direction with $x_1^* = a_{111}$ and on the x_2-direction with $x_2^* = a_{121}, a_{221}, a_{222}$. The equilibriums flows: (**a**) ($a_{110} > 0$, $a_{220} > 0$), (**b**) ($a_{110} < 0$, $a_{220} > 0$), (**c**) ($a_{110} > 0$, $a_{220} < 0$), (**d**) ($a_{110} < 0$, $a_{220} < 0$)

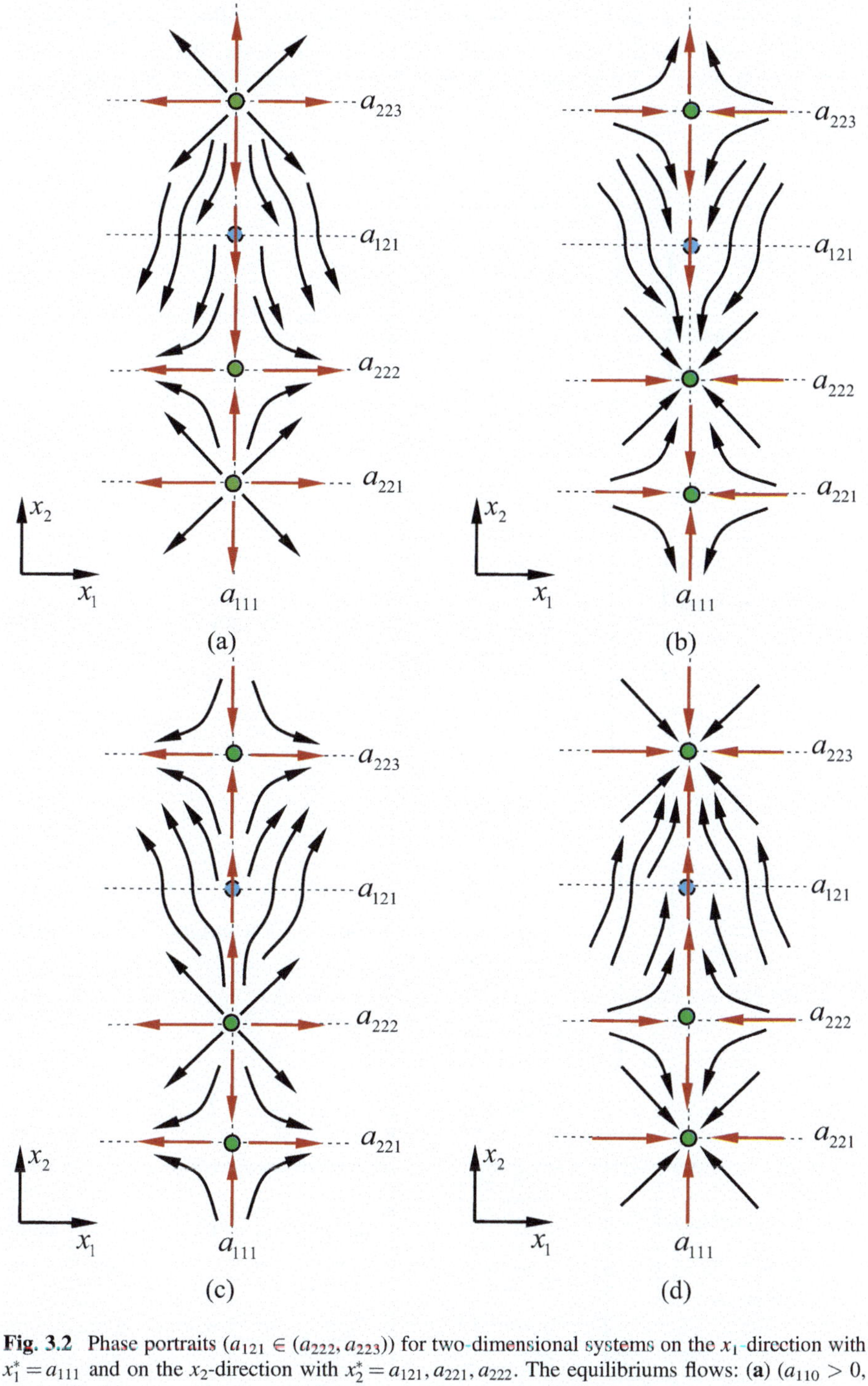

Fig. 3.2 Phase portraits ($a_{121} \in (a_{222}, a_{223})$) for two-dimensional systems on the x_1-direction with $x_1^* = a_{111}$ and on the x_2-direction with $x_2^* = a_{121}, a_{221}, a_{222}$. The equilibriums flows: (**a**) ($a_{110} > 0$, $a_{220} > 0$), (**b**) ($a_{110} < 0$, $a_{220} > 0$), (**c**) ($a_{110} > 0$, $a_{220} < 0$), (**d**) ($a_{110} < 0$, $a_{220} < 0$)

cases are (i) source, saddle, and source plus a negative inflection-source flow, (ii) saddle, sink, saddle with a negative inflection-sink flow, (iii) saddle, source, saddle with a positive inflection-source flow, and (iv) sink, saddle, and sink plus a positive inflection-sink flow.

$$\left\{\begin{array}{c}(a_{111},a_{223})\\(a_{111},a_{222})\\(a_{111},a_{221})\\(a_{111},a_{121})\end{array}\right\}=\left\{\begin{array}{c}\underbrace{(\mathrm{SO},\mathrm{SO})}_{\text{source}}\\\underbrace{(\mathrm{SO},\mathrm{SI})}_{\text{saddle}}\\\underbrace{(\mathrm{SO},\mathrm{SO})}_{\text{source}}\\\underbrace{({}_{\mathrm{II:DI}}\mathrm{SO},\mathrm{nF})}_{\text{inflection source flow }(-)}\end{array}\right\}\text{ for } a_{110}>0 \text{ and } a_{220}>0; \tag{3.3}$$

$$\left\{\begin{array}{c}(a_{111},a_{223})\\(a_{111},a_{222})\\(a_{111},a_{221})\\(a_{111},a_{121})\end{array}\right\}=\left\{\begin{array}{c}\underbrace{(\mathrm{SI},\mathrm{SO})}_{\text{saddle}}\\\underbrace{(\mathrm{SI},\mathrm{SI})}_{\text{sink}}\\\underbrace{(\mathrm{SI},\mathrm{SO})}_{\text{saddle}}\\\underbrace{({}_{\mathrm{DI:II}}\mathrm{SI},\mathrm{nF})}_{\text{inflection sink flow }(-)}\end{array}\right\}\text{ for } a_{110}<0 \text{ and } a_{220}>0; \tag{3.4}$$

$$\left\{\begin{array}{c}(a_{111},a_{223})\\(a_{111},a_{222})\\(a_{111},a_{221})\\(a_{111},a_{121})\end{array}\right\}=\left\{\begin{array}{c}\underbrace{(\mathrm{SO},\mathrm{SI})}_{\text{saddle}}\\\underbrace{(\mathrm{SO},\mathrm{SO})}_{\text{source}}\\\underbrace{(\mathrm{SO},\mathrm{SI})}_{\text{saddle}}\\\underbrace{({}_{\mathrm{DI:II}}\mathrm{SO},\mathrm{pF})}_{\text{inflection source flow }(+)}\end{array}\right\}\text{ for } a_{110}>0 \text{ and } a_{220}<0; \tag{3.5}$$

$$\left\{\begin{array}{c}(a_{111},a_{223})\\(a_{111},a_{222})\\(a_{111},a_{221})\\(a_{111},a_{121})\end{array}\right\}=\left\{\begin{array}{c}\underbrace{(\mathrm{SI},\mathrm{SI})}_{\text{sink}}\\\underbrace{(\mathrm{SI},\mathrm{SO})}_{\text{source}}\\\underbrace{(\mathrm{SI},\mathrm{SI})}_{\text{sink}}\\\underbrace{({}_{\mathrm{II:DI}}\mathrm{SI},\mathrm{pF})}_{\text{inflection sink flow }(+)}\end{array}\right\}\text{ for } a_{110}<0 \text{ and } a_{220}<0. \tag{3.6}$$

(ii) For $a_{121} \in (a_{221}, a_{222})$ and $a_{121} > a_{222}$, the equilibriums of $x_1^* = a_{111}$ with $x_2^* = a_{121}, a_{221}, a_{222}, a_{223}$ have the following properties, as shown in Figs. 3.3 and 3.4.

$$\left\{\begin{array}{l}(a_{111}, a_{223})\\(a_{111}, a_{222})\\(a_{111}, a_{121})\\(a_{111}, a_{221})\end{array}\right\} = \left\{\begin{array}{c}\underbrace{(\mathrm{SO}, \mathrm{SO})}_{\text{source}}\\\underbrace{(\mathrm{SO}, \mathrm{SI})}_{\text{saddle}}\\\underbrace{({}_{\mathrm{DI:II}}\mathrm{SO}, \mathrm{pF})}_{\text{inflection source flow }(+)}\\\underbrace{(\mathrm{SO}, \mathrm{SO})}_{\text{source}}\end{array}\right\} \text{ for } a_{110} > 0 \text{ and } a_{220} > 0; \tag{3.7}$$

$$\left\{\begin{array}{l}(a_{111}, a_{223})\\(a_{111}, a_{222})\\(a_{111}, a_{221})\\(a_{111}, a_{121})\end{array}\right\} = \left\{\begin{array}{c}\underbrace{(\mathrm{SI}, \mathrm{SO})}_{\text{saddle}}\\\underbrace{(\mathrm{SI}, \mathrm{SI})}_{\text{sink}}\\\underbrace{({}_{\mathrm{II:DI}}\mathrm{SI}, \mathrm{pF})}_{\text{inflection sink flow }(+)}\\\underbrace{(\mathrm{SI}, \mathrm{SO})}_{\text{saddle}}\end{array}\right\} \text{ for } a_{110} < 0 \text{ and } a_{220} > 0; \tag{3.8}$$

$$\left\{\begin{array}{l}(a_{111}, a_{223})\\(a_{111}, a_{222})\\(a_{111}, a_{221})\\(a_{111}, a_{121})\end{array}\right\} = \left\{\begin{array}{c}\underbrace{(\mathrm{SO}, \mathrm{SI})}_{\text{saddle}}\\\underbrace{(\mathrm{SO}, \mathrm{SO})}_{\text{source}}\\\underbrace{({}_{\mathrm{II:DI}}\mathrm{SO}, \mathrm{nF})}_{\text{inflection source flow }(-)}\\\underbrace{(\mathrm{SO}, \mathrm{SI})}_{\text{saddle}}\end{array}\right\} \text{ for } a_{110} > 0 \text{ and } a_{220} < 0; \tag{3.9}$$

$$\left\{\begin{array}{l}(a_{111}, a_{223})\\(a_{111}, a_{222})\\(a_{111}, a_{221})\\(a_{111}, a_{121})\end{array}\right\} = \left\{\begin{array}{c}\underbrace{(\mathrm{SI}, \mathrm{SI})}_{\text{sink}}\\\underbrace{(\mathrm{SI}, \mathrm{SO})}_{\text{source}}\\\underbrace{({}_{\mathrm{II:DI}}\mathrm{SI}, \mathrm{pF})}_{\text{inflection sink flow }(+)}\\\underbrace{(\mathrm{SI}, \mathrm{SI})}_{\text{sink}}\end{array}\right\} \text{ for } a_{110} < 0 \text{ and } a_{220} < 0. \tag{3.10}$$

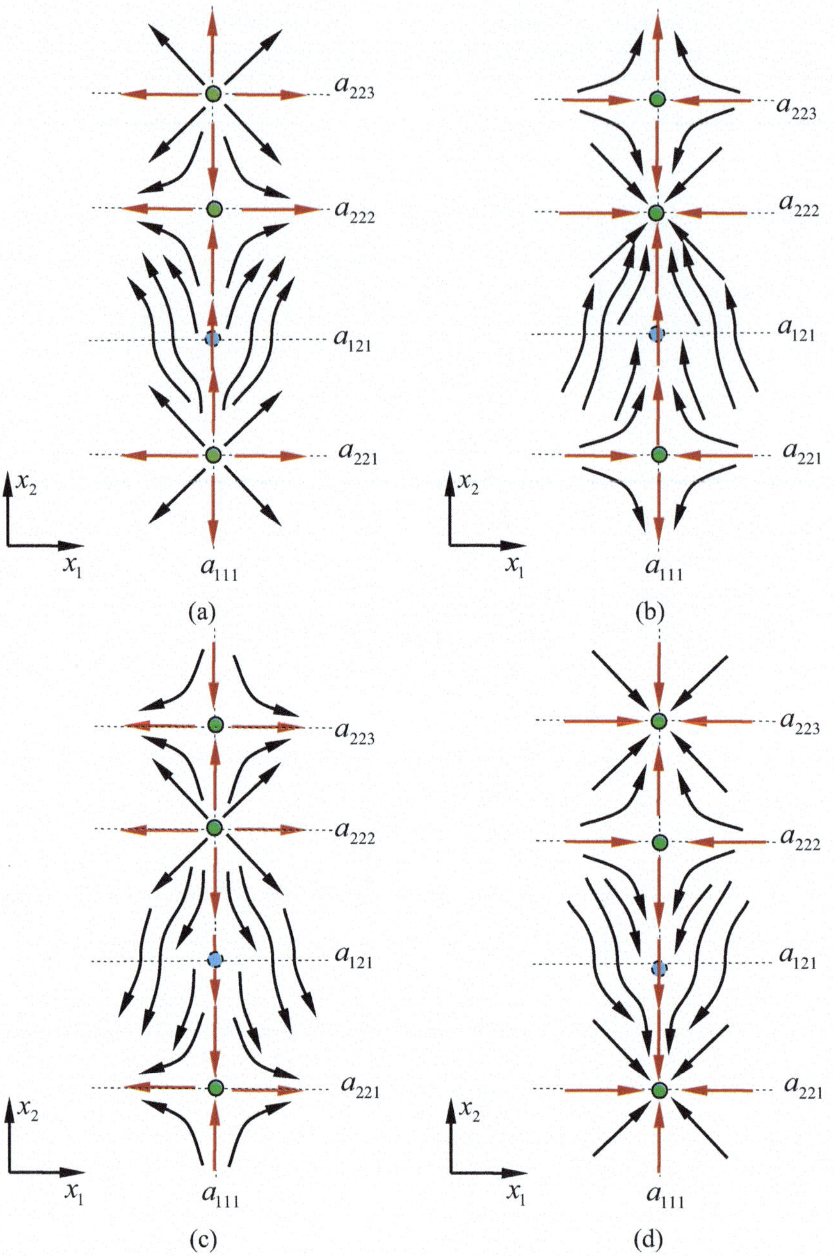

Fig. 3.3 Phase portraits ($a_{121} \in (a_{221}, a_{222})$) for two-dimensional systems on the x_1-direction with $x_1^* = a_{111}$ and on the x_2-direction with $x_2^* = a_{121}, a_{221}, a_{222}$. The equilibriums flows: (**a**) ($a_{110} > 0$, $a_{220} > 0$), (**b**) ($a_{110} < 0$, $a_{220} > 0$), (**c**) ($a_{110} > 0$, $a_{220} < 0$), (**d**) ($a_{110} < 0$, $a_{220} < 0$)

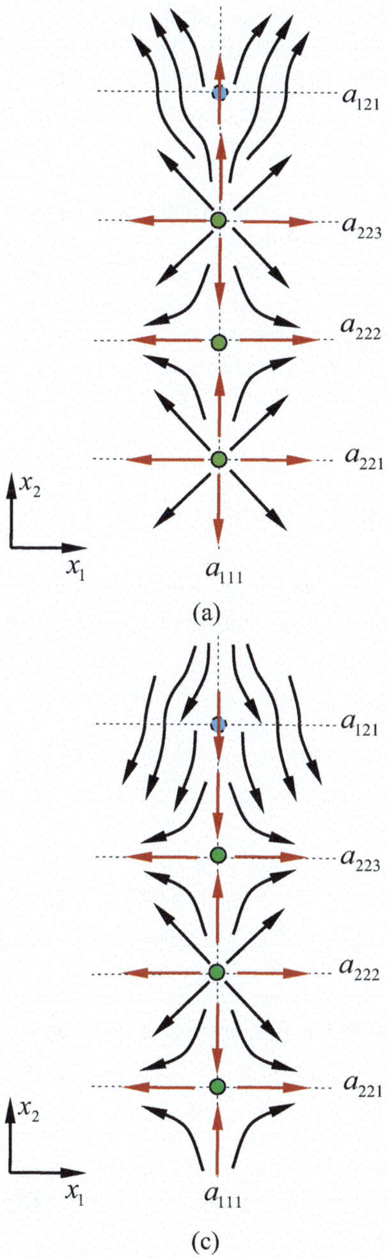

Fig. 3.4 Phase portraits ($a_{121} > a_{223}$) for two-dimensional systems on the x_1-direction with $x_1^* = a_{111}$ and on the x_2-direction with $x_2^* = a_{121}, a_{221}, a_{222}$. The equilibriums flows: (**a**) ($a_{110} > 0$, $a_{220} > 0$), (**b**) ($a_{110} < 0$, $a_{220} > 0$), (**c**) ($a_{110} > 0$, $a_{220} < 0$), (**d**) ($a_{110} < 0$, $a_{220} < 0$)

For the two cases, equilibriums in the series have the same dynamical behaviors. Just equilibrium locations in equilibrium series are different. The four cases are (i) source, saddle, and source plus a *positive* inflection-source flow, (ii) saddle, sink, saddle with a *positive* inflection-sink flow, (iii) saddle, source, saddle with a *negative* inflection-source flow, and (iv) sink, saddle, and sink plus a *negative* inflection-sink flow.

For both cases of $a_{121} < a_{221}$ and $a_{121} \in (a_{222}, a_{223})$, the dynamical behaviors of equilibriums are identical. Only the locations of equilibriums are switched. Similarly, for both cases of $a_{121} \in (a_{221}, a_{222})$ and $a_{121} > a_{223}$, the dynamical behaviors of equilibriums are also identical, and only the locations of equilibriums are switched. For the four cases, the simple equilibriums have the same dynamical behaviors, but the *positive* and *negative* inflection-source and sink flows become the *negative* and *positive* inflection-source and sink flows, respectively.

3.2 Parabola-Source and Sink Infinite-Equilibriums

For the equilibrium series, the corresponding switching bifurcations will be discussed through the hyperbola-sink and source infinite-equilibriums with saddles, source, and sink.

(i) Consider a dynamical system for $a_{221} = a_{121}$ as

$$\begin{aligned}\dot{x}_1 &= a_{110}(x_1 - a_{111})(x_2 - a_{121})^2,\\ \dot{x}_2 &= a_{220}(x_2 - a_{221})(x_2 - a_{222})(x_2 - a_{223}),\end{aligned} \tag{3.12}$$

and the corresponding first integral manifold for $a_{22s_1} = a_{121}$ is

$$\ln\frac{|x_1 - a_{111}|}{|x_{10} - a_{111}|} = \frac{a_{110}}{a_{220}} \sum\nolimits_{s_2=1, s_2 \neq s_1}^{3} \frac{(a_{22s_2} - a_{121})}{(a_{22s_2} - a_{22s_3})} \ln\frac{|x_2 - a_{22s_2}|}{|x_{20} - a_{22s_2}|}. \tag{3.13}$$

(i) The equilibriums of $x_2^* = a_{221}, a_{222}, a_{121}$ with $x_1^* = a_{111}$ for $a_{221} = a_{121}$ are

$$\left\{\begin{array}{c}(a_{111}, a_{223})\\ (a_{111}, a_{222})\\ (a_{111}, a_{221})\end{array}\right\} = \left\{\begin{array}{c}\underbrace{(\text{SO}, \text{SO})}_{\text{source}}\\ \underbrace{(\text{SO}, \text{SI})}_{\text{saddle}}\\ \underbrace{(\text{DP}:\text{UP}, \text{SO})}_{\text{hyperbolic sink}}\end{array}\right\} \text{ for } a_{110} > 0, a_{220} > 0, \tag{3.14}$$

$$\left\{\begin{array}{l}(a_{111}, a_{223})\\(a_{111}, a_{222})\\(a_{111}, a_{221})\end{array}\right\} = \left\{\begin{array}{c}\underbrace{(\mathrm{SI}, \mathrm{SO})}_{\text{saddle}}\\\underbrace{(\mathrm{SI}, \mathrm{SI})}_{\text{sink}}\\\underbrace{(\mathrm{UP}:\mathrm{DP}, \mathrm{SO})}_{\text{hyperbolic-secant source}}\end{array}\right\} \text{ for } a_{110} < 0, a_{220} > 0, \tag{3.15}$$

$$\left\{\begin{array}{l}(a_{111}, a_{223})\\(a_{111}, a_{222})\\(a_{111}, a_{221})\end{array}\right\} = \left\{\begin{array}{c}\underbrace{(\mathrm{SO}, \mathrm{SI})}_{\text{saddle}}\\\underbrace{(\mathrm{SO}, \mathrm{SO})}_{\text{source}}\\\underbrace{(\mathrm{UP}:\mathrm{DP}, \mathrm{SI})}_{\text{hyperbolic-secant sink}}\end{array}\right\} \text{ for } a_{110} > 0, a_{220} < 0, \tag{3.16}$$

$$\left\{\begin{array}{l}(a_{111}, a_{223})\\(a_{111}, a_{222})\\(a_{111}, a_{221})\end{array}\right\} = \left\{\begin{array}{c}\underbrace{(\mathrm{SI}, \mathrm{SI})}_{\text{sink}}\\\underbrace{(\mathrm{SI}, \mathrm{SO})}_{\text{saddle}}\\\underbrace{(\mathrm{DP}:\mathrm{UP}, \mathrm{SI})}_{\text{hyperbolic sink}}\end{array}\right\} \text{ for } a_{110} < 0, a_{220} < 0. \tag{3.17}$$

and the infinite-equilibrium of $x_2^* = a_{121} = a_{221}$ is summarized as follows.

$$\begin{array}{l}(\bar{x}_1, a_{221}) = \underbrace{(\mathrm{DP}, \mathrm{SO})}_{\text{down-parabola source}} \text{ if } \bar{x}_1 \in (-\infty, a_{111}),\\(\bar{x}_1, a_{221}) = \underbrace{(\mathrm{UP}, \mathrm{SO})}_{\text{up-parabola source}} \text{ if } \bar{x}_1 \in (a_{111}, \infty),\\\text{for } a_{110} > 0, a_{220} > 0;\end{array} \tag{3.18}$$

$$\begin{array}{l}(\bar{x}_1, a_{221}) = \underbrace{(\mathrm{UP}, \mathrm{SO})}_{\text{up-parabola source}} \text{ if } \bar{x}_1 \in (-\infty, a_{111}),\\(\bar{x}_1, a_{221}) = \underbrace{(\mathrm{DP}, \mathrm{SO})}_{\text{down-parabola source}} \text{ if } \bar{x}_1 \in (a_{111}, \infty),\\\text{for } a_{110} < 0, a_{220} > 0;\end{array} \tag{3.19}$$

$$\begin{array}{l}(\bar{x}_1, a_{221}) = \underbrace{(\mathrm{UP}, \mathrm{SI})}_{\text{up-parabola sink}} \text{ if } \bar{x}_1 \in (-\infty, a_{111}),\\(\bar{x}_1, a_{221}) = \underbrace{(\mathrm{DP}, \mathrm{SI})}_{\text{down-parabola sink}} \text{ if } \bar{x}_1 \in (a_{111}, \infty),\\\text{for } a_{110} > 0, a_{220} < 0;\end{array} \tag{3.20}$$

$$(\bar{x}_1, a_{221}) = \underbrace{(\text{DP}, \text{SI})}_{\text{down-parabola sink}} \quad \text{if } \bar{x}_1 \in (-\infty, a_{111}),$$
$$(\bar{x}_1, a_{221}) = \underbrace{(\text{UP}, \text{SI})}_{\text{up-parabola sink}} \quad \text{if } \bar{x}_1 \in (a_{111}, \infty), \tag{3.21}$$
$$\text{for } a_{110} < 0, a_{220} < 0.$$

The infinite-equilibrium of $x_2^* = a_{121} = a_{221}$ for up-parabola and down-parabola source and sink is presented in Fig. 3.5. The infinite-equilibrium is for the switching bifurcation for the equilibriums for $a_{121} < a_{221}$ and $a_{121} \in (a_{221}, a_{222})$.

(ii) The equilibriums of $x_2^* = a_{221}, a_{222}, a_{121}$ with $x_1^* = a_{111}$ for $a_{222} = a_{121}$ are

$$\left\{\begin{matrix} (a_{111}, a_{223}) \\ (a_{111}, a_{222}) \\ (a_{111}, a_{221}) \end{matrix}\right\} = \left\{\begin{matrix} \underbrace{(\text{SO}, \text{SO})}_{\text{source}} \\ \underbrace{(\text{UP}:\text{DP}, \text{SI})}_{\text{hyperbolic-secant sink}} \\ \underbrace{(\text{SO}, \text{SO})}_{\text{source}} \end{matrix}\right\} \text{for } a_{110} > 0, a_{220} > 0, \tag{3.22}$$

$$\left\{\begin{matrix} (a_{111}, a_{223}) \\ (a_{111}, a_{222}) \\ (a_{111}, a_{221}) \end{matrix}\right\} = \left\{\begin{matrix} \underbrace{(\text{SI}, \text{SO})}_{\text{saddle}} \\ \underbrace{(\text{DP}:\text{UP}, \text{SI})}_{\text{hyperbolic sink}} \\ \underbrace{(\text{SI}, \text{SO})}_{\text{saddle}} \end{matrix}\right\} \text{for } a_{110} < 0, a_{220} > 0, \tag{3.23}$$

$$\left\{\begin{matrix} (a_{111}, a_{223}) \\ (a_{111}, a_{222}) \\ (a_{111}, a_{221}) \end{matrix}\right\} = \left\{\begin{matrix} \underbrace{(\text{SO}, \text{SI})}_{\text{saddle}} \\ \underbrace{(\text{DP}:\text{UP}, \text{SO})}_{\text{hyperbolic source}} \\ \underbrace{(\text{SO}, \text{SI})}_{\text{saddle}} \end{matrix}\right\} \text{for } a_{110} > 0, a_{220} < 0, \tag{3.24}$$

$$\left\{\begin{matrix} (a_{111}, a_{223}) \\ (a_{111}, a_{222}) \\ (a_{111}, a_{221}) \end{matrix}\right\} = \left\{\begin{matrix} \underbrace{(\text{SI}, \text{SI})}_{\text{sink}} \\ \underbrace{(\text{UP}:\text{DP}, \text{SO})}_{\text{hyperbolic-secant source}} \\ \underbrace{(\text{SI}, \text{SI})}_{\text{sink}} \end{matrix}\right\} \text{for } a_{110} < 0, a_{220} < 0. \tag{3.25}$$

and the infinite-equilibrium of $x_2^* = a_{121} = a_{222}$ is summarized as follows.

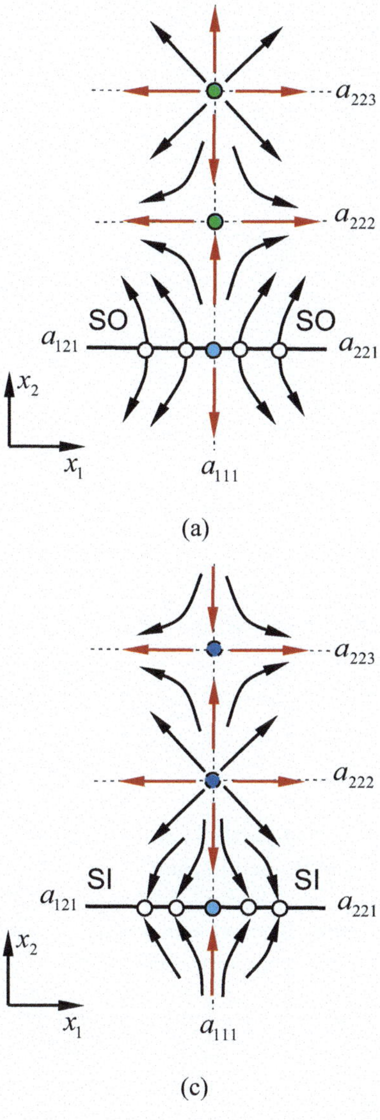

Fig. 3.5 Phase portraits ($a_{121} = a_{221}$) for two-dimensional systems on the x_1-direction with $x_1^* = a_{111}$ and on the x_2-direction with $x_2^* = a_{121}, a_{221}, a_{222}$. The infinite-equilibriums: (**a**) ($a_{110} > 0$, $a_{220} > 0$), (**b**) ($a_{110} < 0$, $a_{220} > 0$), (**c**) ($a_{110} > 0$, $a_{220} < 0$), (**d**) ($a_{110} < 0$, $a_{220} < 0$).

$$
\begin{aligned}
&(\bar{x}_1, a_{222}) = \underbrace{(\mathrm{UP}, \mathrm{SI})}_{\text{up-parabola sink}} \ \text{ if } \bar{x}_1 \in (-\infty, a_{111}),\\
&(\bar{x}_1, a_{222}) = \underbrace{(\mathrm{DP}, \mathrm{SI})}_{\text{down-parabola sink}} \ \text{ if } \bar{x}_1 \in (a_{111}, \infty),\\
&\text{for } a_{110} > 0, a_{220} > 0;
\end{aligned}
\tag{3.26}
$$

$$
\begin{aligned}
&(\bar{x}_1, a_{222}) = \underbrace{(\mathrm{DP}, \mathrm{SI})}_{\text{down-parabola sink}} \ \text{ if } \bar{x}_1 \in (-\infty, a_{111}),\\
&(\bar{x}_1, a_{222}) = \underbrace{(\mathrm{UP}, \mathrm{SI})}_{\text{up-parabola sink}} \ \text{ if } \bar{x}_1 \in (a_{111}, \infty),\\
&\text{for } a_{110} < 0, a_{220} > 0;
\end{aligned}
\tag{3.27}
$$

$$
\begin{aligned}
&(\bar{x}_1, a_{222}) = \underbrace{(\mathrm{DP}, \mathrm{SO})}_{\text{down-parabola source}} \ \text{ if } \bar{x}_1 \in (-\infty, a_{111}),\\
&(\bar{x}_1, a_{222}) = \underbrace{(\mathrm{UP}, \mathrm{SO})}_{\text{up-parabola source}} \ \text{ if } \bar{x}_1 \in (a_{111}, \infty),\\
&\text{for } a_{110} > 0, a_{220} < 0;
\end{aligned}
\tag{3.28}
$$

$$
\begin{aligned}
&(\bar{x}_1, a_{222}) = \underbrace{(\mathrm{UP}, \mathrm{SO})}_{\text{up-parabola source}} \ \text{ if } \bar{x}_1 \in (-\infty, a_{111}),\\
&(\bar{x}_1, a_{222}) = \underbrace{(\mathrm{DP}, \mathrm{SO})}_{\text{down-parabola source}} \ \text{ if } \bar{x}_1 \in (a_{111}, \infty),\\
&\text{for } a_{110} < 0, a_{220} < 0.
\end{aligned}
\tag{3.29}
$$

The infinite-equilibrium of $x_2^* = a_{121} = a_{222}$ for up-parabola and down-parabola source and sink is presented in Fig. 3.6. The infinite-equilibrium is the switching bifurcation for the equilibriums for $a_{121} \in (a_{221}, a_{222})$ and $a_{121} \in (a_{222}, a_{223})$. The properties of the infinite-equilibriums are different from $x_2^* = a_{121} = a_{221}$, and the equilibriums of (a_{111}, a_{221}) and (a_{111}, a_{223}) have the same properties.

(iii) The equilibriums of $x_2^* = a_{221}, a_{222}, a_{121}$ with $x_1^* = a_{111}$ for $a_{223} = a_{121}$ are

$$
\left\{\begin{array}{c}(a_{111}, a_{223})\\ (a_{111}, a_{222})\\ (a_{111}, a_{221})\end{array}\right\} = \left\{\begin{array}{c}\underbrace{(\mathrm{DP}:\mathrm{UP}, \mathrm{SO})}_{\text{hyperbolic source}}\\ \underbrace{(\mathrm{SO}, \mathrm{SI})}_{\text{saddle}}\\ \underbrace{(\mathrm{SO}, \mathrm{SO})}_{\text{source}}\end{array}\right\} \text{ for } a_{110} > 0, a_{220} > 0,
\tag{3.30}
$$

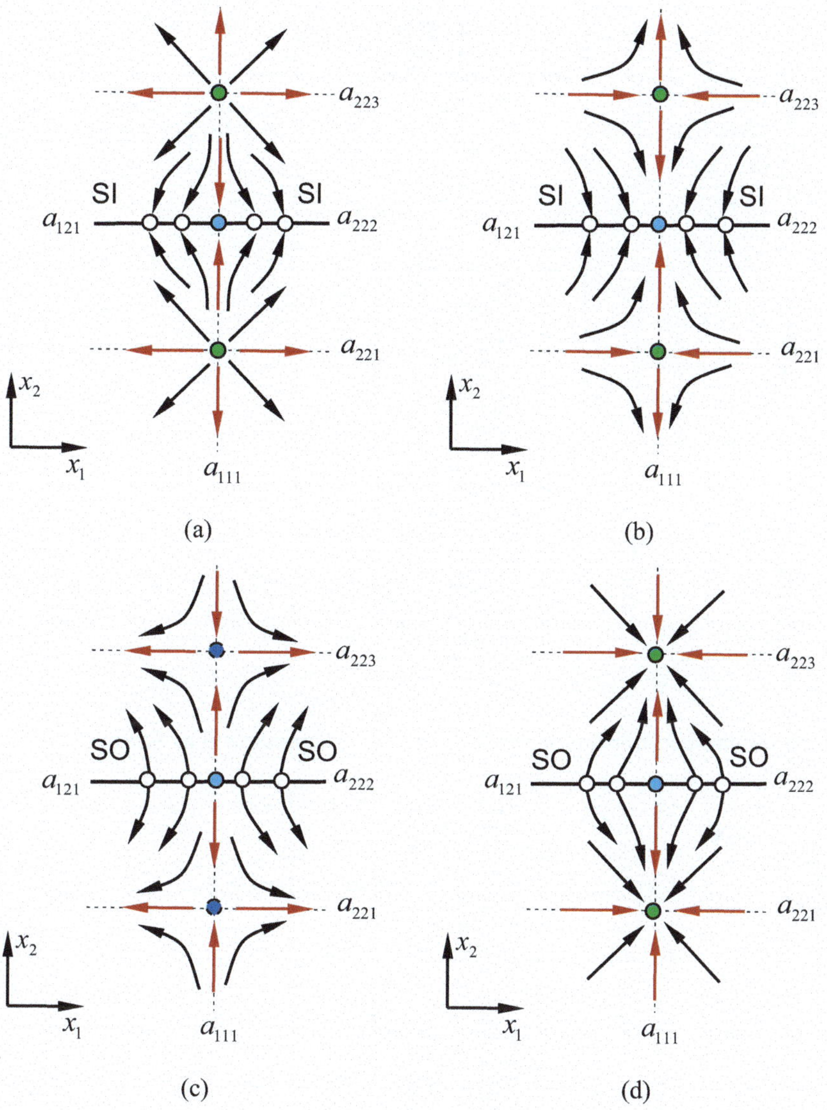

Fig. 3.6 Phase portraits ($a_{121} = a_{221}$) for two-dimensional systems on the x_1-direction with $x_1^* = a_{111}$ and on the x_2-direction with $x_2^* = a_{121}, a_{221}, a_{222}$. The infinite-equilibriums: (**a**) ($a_{110} > 0$, $a_{220} > 0$), (**b**) ($a_{110} < 0$, $a_{220} > 0$), (**c**) ($a_{110} > 0$, $a_{220} < 0$), (**d**) ($a_{110} < 0$, $a_{220} < 0$)

$$\left\{\begin{matrix}(a_{111},a_{223})\\(a_{111},a_{222})\\(a_{111},a_{221})\end{matrix}\right\}=\left\{\begin{matrix}\underbrace{(\mathrm{UP}:\mathrm{DP},\mathrm{SO})}_{\text{hyperbolic-secant source}}\\\underbrace{(\mathrm{SI},\mathrm{SI})}_{\text{sink}}\\\underbrace{(\mathrm{SI},\mathrm{SO})}_{\text{saddle}}\end{matrix}\right\}\text{ for } a_{110}<0, a_{220}>0, \tag{3.31}$$

$$\left\{\begin{matrix}(a_{111},a_{223})\\(a_{111},a_{222})\\(a_{111},a_{221})\end{matrix}\right\}=\left\{\begin{matrix}\underbrace{(\mathrm{UP}:\mathrm{DP},\mathrm{SI})}_{\text{hyperbolic-secant sink}}\\\underbrace{(\mathrm{SO},\mathrm{SO})}_{\text{source}}\\\underbrace{(\mathrm{SO},\mathrm{SI})}_{\text{saddle}}\end{matrix}\right\}\text{ for } a_{110}>0, a_{220}<0, \tag{3.32}$$

$$\left\{\begin{matrix}(a_{111},a_{223})\\(a_{111},a_{222})\\(a_{111},a_{221})\end{matrix}\right\}=\left\{\begin{matrix}\underbrace{(\mathrm{DP}:\mathrm{UP},\mathrm{SI})}_{\text{hyperbolic sink}}\\\underbrace{(\mathrm{SI},\mathrm{SO})}_{\text{saddle}}\\\underbrace{(\mathrm{SI},\mathrm{SI})}_{\text{sink}}\end{matrix}\right\}\text{ for } a_{110}<0, a_{220}<0. \tag{3.33}$$

and the infinite-equilibrium of $x_2^* = a_{121} = a_{223}$ is summarized as follows.

$$\begin{aligned}&(\bar{x}_1, a_{223}) = \underbrace{(\mathrm{DP},\mathrm{SO})}_{\text{down-parabola source}} \quad \text{if } \bar{x}_1 \in (-\infty, a_{111}),\\&(\bar{x}_1, a_{223}) = \underbrace{(\mathrm{UP},\mathrm{SO})}_{\text{up-parabola source}} \quad \text{if } \bar{x}_1 \in (a_{111}, \infty),\\&\text{for } a_{110}>0, a_{220}>0;\end{aligned} \tag{3.34}$$

$$\begin{aligned}&(\bar{x}_1, a_{223}) = \underbrace{(\mathrm{UP},\mathrm{SO})}_{\text{up-parabola source}} \quad \text{if } \bar{x}_1 \in (-\infty, a_{111}),\\&(\bar{x}_1, a_{223}) = \underbrace{(\mathrm{DP},\mathrm{SO})}_{\text{down-parabola source}} \quad \text{if } \bar{x}_1 \in (a_{111}, \infty),\\&\text{for } a_{110}<0, a_{220}>0;\end{aligned} \tag{3.35}$$

$$
\begin{aligned}
&(\bar{x}_1, a_{223}) = \underbrace{(\text{UP}, \text{SI})}_{\text{up-parabola sink}} \text{ if } \bar{x}_1 \in (-\infty, a_{111}),\\
&(\bar{x}_1, a_{223}) = \underbrace{(\text{DP}, \text{SI})}_{\text{down-parabola sink}} \text{ if } \bar{x}_1 \in (a_{111}, \infty),\\
&\text{for } a_{110} > 0, a_{220} < 0;
\end{aligned}
\tag{3.36}
$$

$$
\begin{aligned}
&(\bar{x}_1, a_{223}) = \underbrace{(\text{DP}, \text{SI})}_{\text{down-parabola sink}} \text{ if } \bar{x}_1 \in (-\infty, a_{111}),\\
&(\bar{x}_1, a_{223}) = \underbrace{(\text{UP}, \text{SI})}_{\text{up-parabola sink}} \text{ if } \bar{x}_1 \in (a_{111}, \infty),\\
&\text{for } a_{110} < 0, a_{220} < 0.
\end{aligned}
\tag{3.37}
$$

The infinite-equilibrium of $x_2^* = a_{121} = a_{223}$ for up-parabola and down-parabola source and sink is presented in Fig. 3.7. The infinite-equilibrium is for the switching bifurcation for the equilibriums for $a_{121} > a_{223}$ and $a_{121} \in (a_{222}, a_{223})$.

Compared to the infinite-equilibrium of $x_2^* = a_{121} = a_{221}$ with $x_2^* = a_{121} = a_{223}$, the infinite-equilibrium properties are unchanged. The properties at the equilibrium of (a_{111}, a_{223}) and (a_{111}, a_{221}) are switched for the two cases.

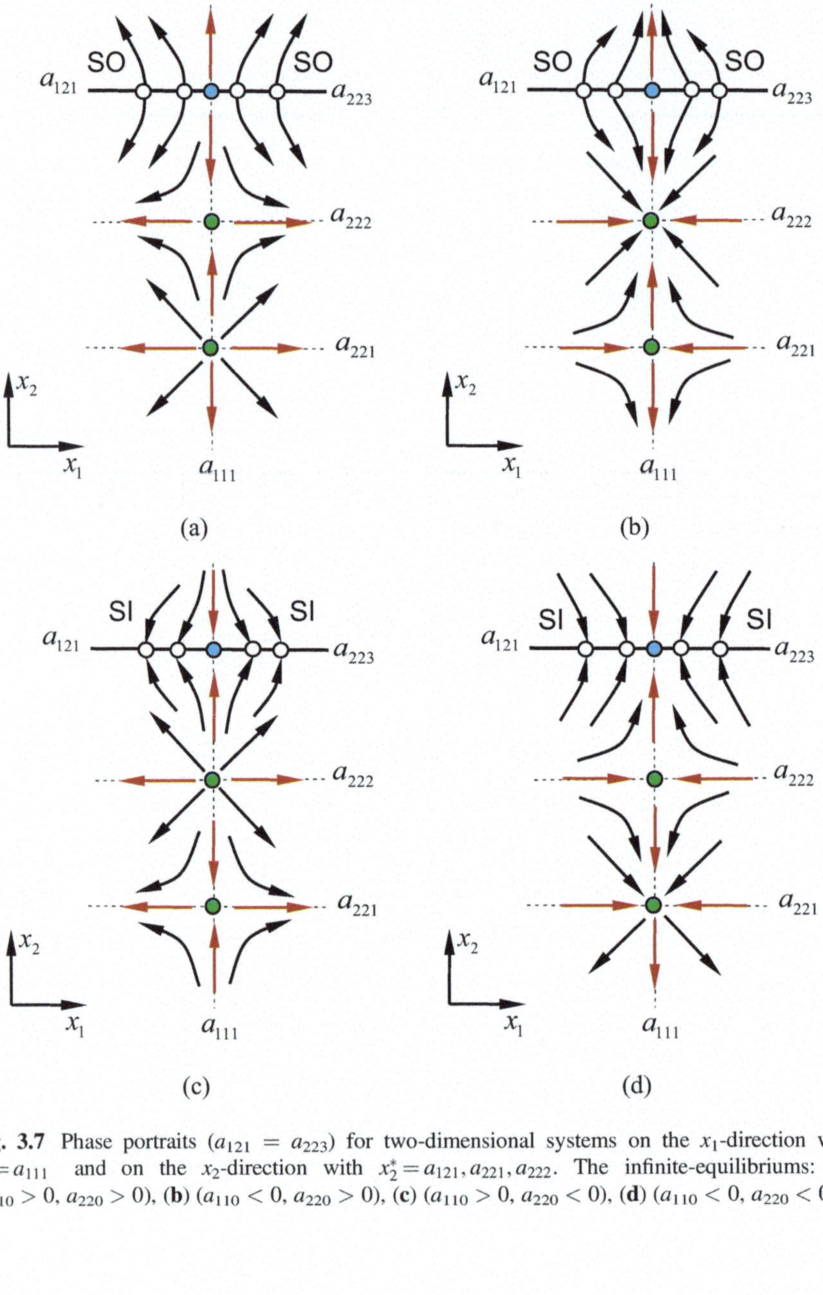

Fig. 3.7 Phase portraits ($a_{121} = a_{223}$) for two-dimensional systems on the x_1-direction with $x_1^* = a_{111}$ and on the x_2-direction with $x_2^* = a_{121}, a_{221}, a_{222}$. The infinite-equilibriums: (**a**) ($a_{110} > 0$, $a_{220} > 0$), (**b**) ($a_{110} < 0$, $a_{220} > 0$), (**c**) ($a_{110} > 0$, $a_{220} < 0$), (**d**) ($a_{110} < 0$, $a_{220} < 0$)

Chapter 4
Saddle-Nodes and Hyperbolic Flow Series

4.1 Equilibrium Series with Quadratic and Cubic Singularities

In this section, the connected hyperbolic and hyperbolic-secant flows with upper-saddle-sink and upper-saddle-source, lower-saddle-sink and lower-saddle-source plus sink, and source and saddle are presented. The saddle-sink and saddle-source are the appearing bifurcations for saddle and sink and for saddle and source, respectively. Based on the self-cubic vector fields, the third-order saddles, sink and source with hyperbolic and hyperbolic-secant flows are presented. All the equilibriums are on a line to form a series of equilibriums.

4.1.1 Equilibrium Series with Saddle-Nodes and Hyperbolic Flows

Consider a dynamical system as

$$\begin{aligned} \dot{x}_1 &= a_{110}(x_1 - a_{111})(x_2 - a_{121})(x_2 - a_{122}), \\ \dot{x}_2 &= a_{220}(x_2 - a_{22s_1})(x_2 - a_{22s_2})^2, \end{aligned} \tag{4.1}$$

and the corresponding first integral manifold is

A. C. J. Luo, *Two-dimensional Self and Product Cubic Systems, Vol. I*,
https://doi.org/10.1007/978-3-031-57096-4_4

$$
\begin{aligned}
\ln\frac{|x_1-a_{111}|}{|x_{10}-a_{111}|}=&\frac{a_{110}}{a_{220}}\Big\{\frac{\prod_{l_1=1}^{2}(a_{22s_1}-a_{12l_1})}{(a_{22s_1}-a_{22s_2})^2}\ln\frac{|x_2-a_{22s_1}|}{|x_{20}-a_{22s_1}|}\\
&+\Big[\frac{\sum_{l_1=1}^{2}(a_{22s_1}-a_{12l_1})}{(a_{22s_1}-a_{22s_1})}-\frac{\prod_{l_1=1}^{2}(a_{22s_2}-a_{12l_1})}{(a_{22s_2}-a_{22s_1})^2}\Big]\ln\frac{|x_2-a_{22s_2}|}{|x_{20}-a_{22s_2}|}\\
&-\frac{\prod_{l_1=1}^{2}(a_{22s_2}-a_{12l_1})}{(a_{22s_2}-a_{22s_1})}\Big(\frac{1}{x_2-a_{22s_2}}-\frac{1}{x_{20}-a_{22s_2}}\Big)\Big\}.
\end{aligned}
\tag{4.2}
$$

The following discussion focuses on phase portraits of $a_{22s_1}=a_{221}<a_{22s_2}=a_{222}$ through the first integral manifold. There are six cases:

- Case (i) $a_{121}<a_{122}<a_{221}<a_{222}$,
- Case (ii) $a_{221}<a_{121}<a_{122}<a_{222}$,
- Case (iii) $a_{221}<a_{222}<a_{121}<a_{122}$,
- Case (vi) $a_{121}<a_{221}<a_{122}<a_{222}$,
- Case (v) $a_{221}<a_{121}<a_{222}<a_{122}$,
- Case (vi) $a_{121}<a_{221}<a_{222}<a_{122}$.

Similarly, from the first integral manifold, readers can complete the phase portraits for $a_{22s_2}=a_{221}<a_{22s_1}=a_{222}$. The corresponding nonlinear dynamical behaviors can be discussed.

(i) For the case of$a_{122}<a_{221}<a_{222}$, the equilibrium series of $x_1^*=a_{111}$ with $x_2^*=a_{121},a_{122},a_{221}$ have the following properties, as shown in Fig. 4.1.

$$
\begin{Bmatrix}(a_{111},a_{222})\\(a_{111},a_{221})\\(a_{111},a_{122})\\(a_{111},a_{121})\end{Bmatrix}=\begin{Bmatrix}\underbrace{(\mathrm{SO},\mathrm{US})}_{\text{upper-saddle source}}\\\underbrace{(\mathrm{SO},\mathrm{SO})}_{\text{source}}\\\underbrace{(\mathrm{UP}:\mathrm{DP},\mathrm{nF})}_{\text{hyperbolic-secant flow }(-)}\\\underbrace{(\mathrm{DP}:\mathrm{UP},\mathrm{nF})}_{\text{hyperbolic flow }(-)}\end{Bmatrix}\text{ for }a_{110}>0\text{ and }a_{220}>0;
\tag{4.3}
$$

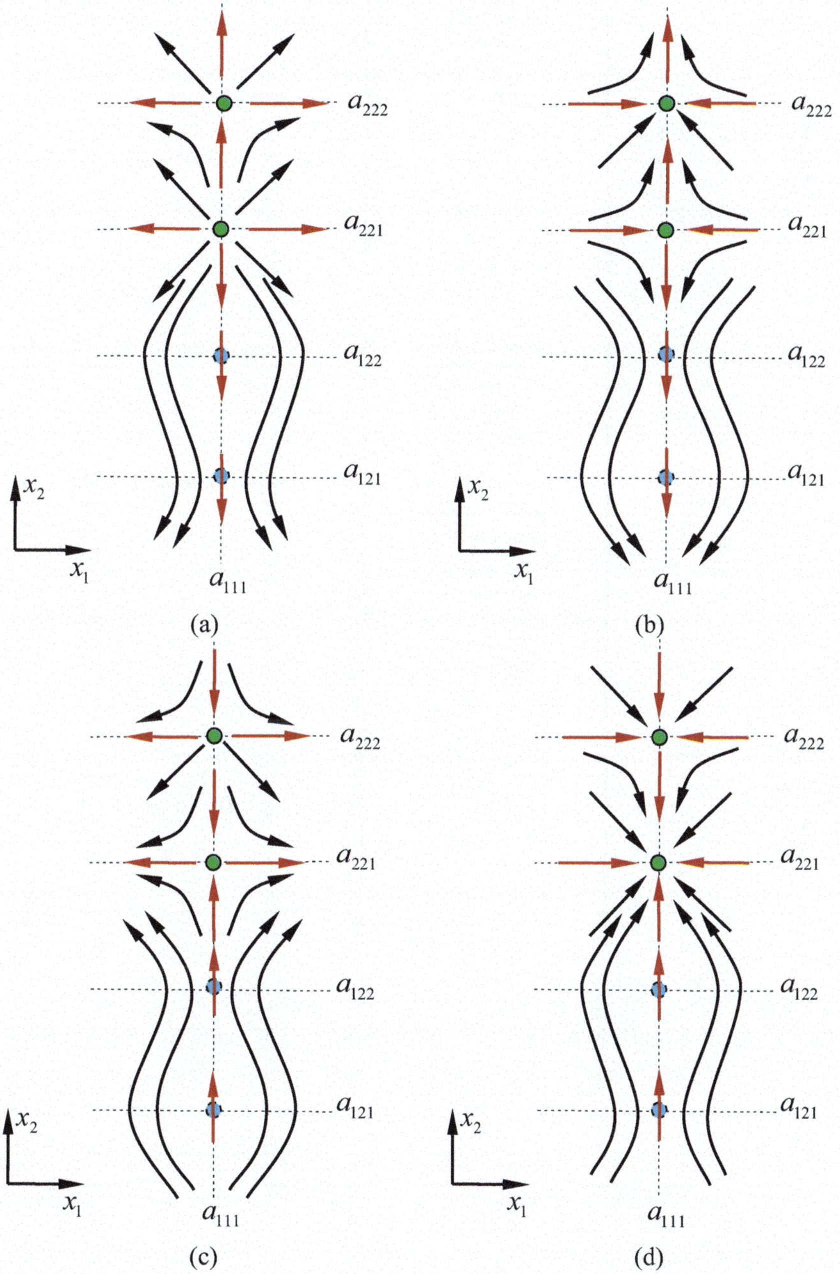

Fig. 4.1 Phase portraits ($a_{122} < a_{221}$) for two-dimensional systems on the x_1-direction with $x_1^* = a_{111}$ and on the x_2-direction with $x_2^* = a_{121}, a_{122}, a_{221}, a_{222}$. The equilibriums flows: (**a**) ($a_{110} > 0$, $a_{220} > 0$), (**b**) ($a_{110} < 0$, $a_{220} > 0$), (**c**) ($a_{110} > 0$, $a_{220} < 0$), (**d**) ($a_{110} < 0$, $a_{220} < 0$)

$$\left\{\begin{array}{c}(a_{111},a_{222})\\(a_{111},a_{221})\\(a_{111},a_{122})\\(a_{111},a_{121})\end{array}\right\}=\left\{\begin{array}{c}\underbrace{(\mathrm{SI},\mathrm{US})}_{\text{upper-saddle sink}}\\\underbrace{(\mathrm{SI},\mathrm{SO})}_{\text{saddle}}\\\underbrace{(\mathrm{DP}:\mathrm{UP},\mathrm{nF})}_{\text{hyperbolic flow }(-)}\\\underbrace{(\mathrm{UP}:\mathrm{DP},\mathrm{nF})}_{\text{hyperbolic-secant flow }(-)}\end{array}\right\}\text{ for }a_{110}<0\text{ and }a_{220}>0;\tag{4.4}$$

$$\left\{\begin{array}{c}(a_{111},a_{222})\\(a_{111},a_{221})\\(a_{111},a_{122})\\(a_{111},a_{121})\end{array}\right\}=\left\{\begin{array}{c}\underbrace{(\mathrm{SO},\mathrm{LS})}_{\text{lower-saddle source}}\\\underbrace{(\mathrm{SO},\mathrm{SI})}_{\text{saddle}}\\\underbrace{(\mathrm{DP}:\mathrm{UP},\mathrm{pF})}_{\text{hyperbolic flow }(+)}\\\underbrace{(\mathrm{UP}:\mathrm{DP},\mathrm{pF})}_{\text{hyperbolic-secant flow }(+)}\end{array}\right\}\text{ for }a_{110}>0\text{ and }a_{220}<0;\tag{4.5}$$

$$\left\{\begin{array}{c}(a_{111},a_{222})\\(a_{111},a_{221})\\(a_{111},a_{122})\\(a_{111},a_{121})\end{array}\right\}=\left\{\begin{array}{c}\underbrace{(\mathrm{SI},\mathrm{LS})}_{\text{lower-saddle sink}}\\\underbrace{(\mathrm{SI},\mathrm{SI})}_{\text{sink}}\\\underbrace{(\mathrm{UP}:\mathrm{DP},\mathrm{pF})}_{\text{hyperbolic-secant flow }(+)}\\\underbrace{(\mathrm{DP}:\mathrm{UP},\mathrm{pF})}_{\text{hyperbolic flow }(+)}\end{array}\right\}\text{ for }a_{110}<0\text{ and }a_{220}<0.\tag{4.6}$$

The saddle-source and saddle-sink with a simple equilibrium (sink, saddle, source) in the series of equilibriums are on top of connected hyperbolic and hyperbolic-secant flows. The saddle-nodes connected with saddle, source, and sink are observed.

(ii) For the case of $a_{221} < a_{121} < a_{122} < a_{222}$, the equilibrium series of $x_1^* = a_{111}$ with $x_2^* = a_{121}, a_{122}, a_{221}, a_{222}$ have the following properties, as shown in Fig. 4.2.

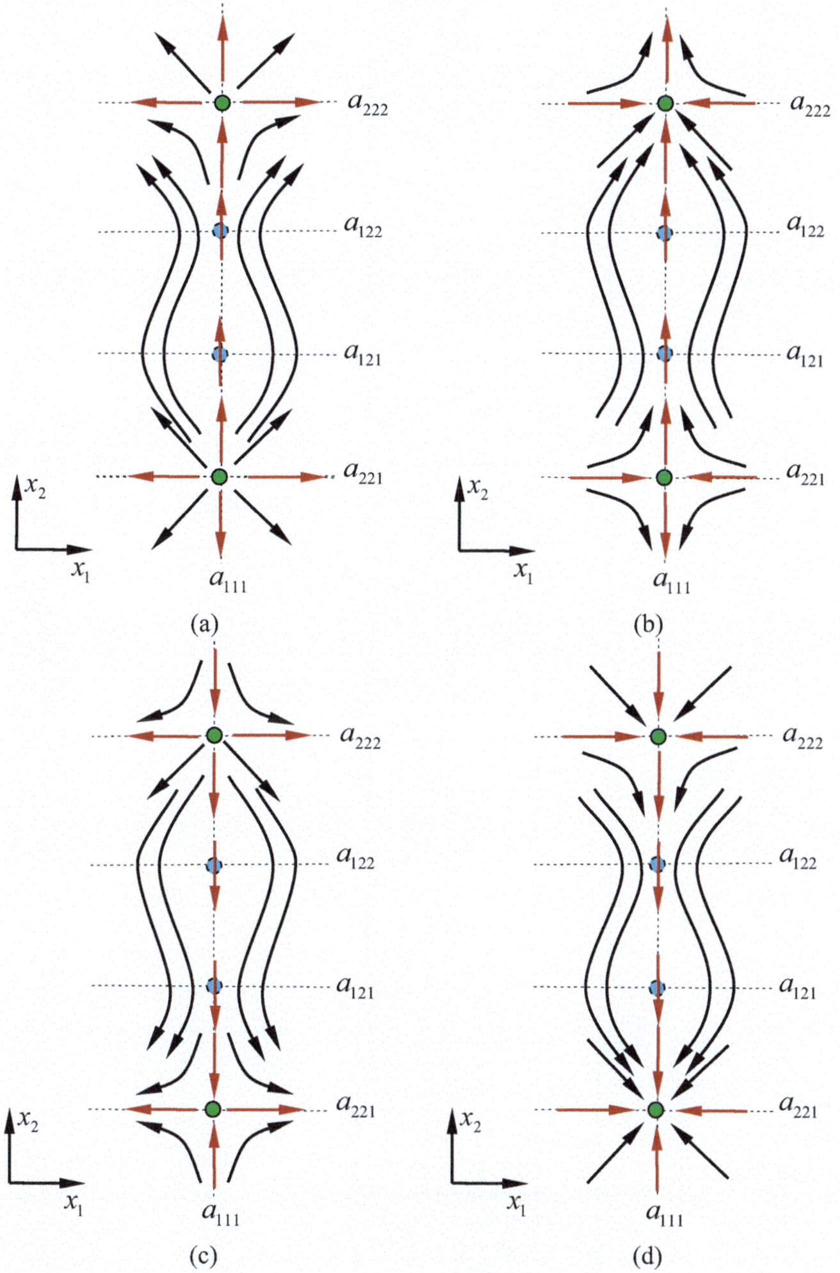

Fig. 4.2 Phase portraits ($a_{121}, a_{122} \in (a_{221}, a_{222})$) for two-dimensional systems on the x_1-direction with $x_1^* = a_{111}$ and on the x_2-direction with $x_2^* = a_{121}, a_{122}, a_{221}, a_{222}$. **(a)** ($a_{110} > 0$, $a_{220} > 0$), **(b)** ($a_{110} < 0$, $a_{220} > 0$), **(c)** ($a_{110} > 0$, $a_{220} < 0$), **(d)** ($a_{110} < 0$, $a_{220} < 0$)

$$\left.\begin{array}{c}(a_{111},a_{222})\\(a_{111},a_{122})\\(a_{111},a_{121})\\(a_{111},a_{221})\end{array}\right\} = \left\{\begin{array}{c}\underbrace{(\mathrm{SO},\mathrm{US})}_{\text{upper-saddle source}}\\ \underbrace{(\mathrm{DP}:\mathrm{UP},\mathrm{pF})}_{\text{hyperbolic flow }(+)}\\ \underbrace{(\mathrm{UP}:\mathrm{DP},\mathrm{pF})}_{\text{hyperbolic-secant flow }(+)}\\ \underbrace{(\mathrm{SO},\mathrm{SO})}_{\text{source}}\end{array}\right\} \text{for } a_{110}>0 \text{ and } a_{220}>0; \tag{4.7}$$

$$\left.\begin{array}{c}(a_{111},a_{222})\\(a_{111},a_{122})\\(a_{111},a_{121})\\(a_{111},a_{221})\end{array}\right\} = \left\{\begin{array}{c}\underbrace{(\mathrm{SI},\mathrm{US})}_{\text{upper-saddle sink}}\\ \underbrace{(\mathrm{UP}:\mathrm{DP},\mathrm{pF})}_{\text{hyperbolic-secant flow }(+)}\\ \underbrace{(\mathrm{DP}:\mathrm{UP},\mathrm{pF})}_{\text{hyperbolic flow }(+)}\\ \underbrace{(\mathrm{SI},\mathrm{SO})}_{\text{saddle}}\end{array}\right\} \text{for } a_{110}<0 \text{ and } a_{220}>0; \tag{4.8}$$

$$\left.\begin{array}{c}(a_{111},a_{222})\\(a_{111},a_{122})\\(a_{111},a_{121})\\(a_{111},a_{221})\end{array}\right\} = \left\{\begin{array}{c}\underbrace{(\mathrm{SO},\mathrm{LS})}_{\text{lower-saddle source}}\\ \underbrace{(\mathrm{UP}:\mathrm{DP},\mathrm{nF})}_{\text{hyperbolic-secant flow }(-)}\\ \underbrace{(\mathrm{DP}:\mathrm{UP},\mathrm{nF})}_{\text{hyperbolic flow }(-)}\\ \underbrace{(\mathrm{SO},\mathrm{SI})}_{\text{saddle}}\end{array}\right\} \text{for } a_{110}>0 \text{ and } a_{220}<0; \tag{4.9}$$

$$\left.\begin{array}{c}(a_{111},a_{222})\\(a_{111},a_{122})\\(a_{111},a_{121})\\(a_{111},a_{221})\end{array}\right\} = = \left\{\begin{array}{c}\underbrace{(\mathrm{SI},\mathrm{LS})}_{\text{lower-saddle sink}}\\ \underbrace{(\mathrm{DP}:\mathrm{UP},\mathrm{nF})}_{\text{hyperbolic flow }(-)}\\ \underbrace{(\mathrm{UP}:\mathrm{DP},\mathrm{nF})}_{\text{hyperbolic-secant flow }(-)}\\ \underbrace{(\mathrm{SI},\mathrm{SI})}_{\text{sink}}\end{array}\right\} \text{for } a_{110}<0 \text{ and } a_{220}<0. \tag{4.10}$$

The saddle-source and saddle-sink with a simple equilibrium (sink, saddle, source) in the series of equilibriums are separated by the connected hyperbolic and hyperbolic-secant flows.

(iii) For the case of $a_{221} < a_{222} < a_{121}$, the equilibrium series of $x_1^* = a_{111}$ with $x_2^* = a_{121}, a_{122}, a_{221}$ have the following properties, as shown in Fig. 4.3.

$$\left\{\begin{array}{c} (a_{111}, a_{122}) \\ (a_{111}, a_{121}) \\ (a_{111}, a_{222}) \\ (a_{111}, a_{221}) \end{array}\right\} = \left\{\begin{array}{c} \underbrace{(\mathrm{DP}:\mathrm{UP},\mathrm{pF})}_{\text{hyperbolic flow }(+)} \\ \underbrace{(\mathrm{UP}:\mathrm{DP},\mathrm{pF})}_{\text{hyperbolic-secant flow }(+)} \\ \underbrace{(\mathrm{SO},\mathrm{US})}_{\text{upper-saddle source}} \\ \underbrace{(\mathrm{SO},\mathrm{SO})}_{\text{source}} \end{array}\right\} \text{ for } a_{110} > 0 \text{ and } a_{220} > 0; \tag{4.11}$$

$$\left\{\begin{array}{c} (a_{111}, a_{122}) \\ (a_{111}, a_{121}) \\ (a_{111}, a_{222}) \\ (a_{111}, a_{221}) \end{array}\right\} = \left\{\begin{array}{c} \underbrace{(\mathrm{UP}:\mathrm{DP},\mathrm{pF})}_{\text{hyperbolic-secant flow }(+)} \\ \underbrace{(\mathrm{DP}:\mathrm{UP},\mathrm{pF})}_{\text{hyperbolic flow }(+)} \\ \underbrace{(\mathrm{SI},\mathrm{US})}_{\text{upper-saddle sink}} \\ \underbrace{(\mathrm{SI},\mathrm{SO})}_{\text{saddle}} \end{array}\right\} \text{ for } a_{110} < 0 \text{ and } a_{220} > 0; \tag{4.12}$$

$$\left\{\begin{array}{c} (a_{111}, a_{122}) \\ (a_{111}, a_{121}) \\ (a_{111}, a_{222}) \\ (a_{111}, a_{221}) \end{array}\right\} = \left\{\begin{array}{c} \underbrace{(\mathrm{UP}:\mathrm{DP},\mathrm{nF})}_{\text{hyperbolic-secant flow }(-)} \\ \underbrace{(\mathrm{DP}:\mathrm{UP},\mathrm{nF})}_{\text{hyperbolic flow }(-)} \\ \underbrace{(\mathrm{SO},\mathrm{LS})}_{\text{lower-saddle source}} \\ \underbrace{(\mathrm{SO},\mathrm{SI})}_{\text{saddle}} \end{array}\right\} \text{ for } a_{110} > 0 \text{ and } a_{220} < 0; \tag{4.13}$$

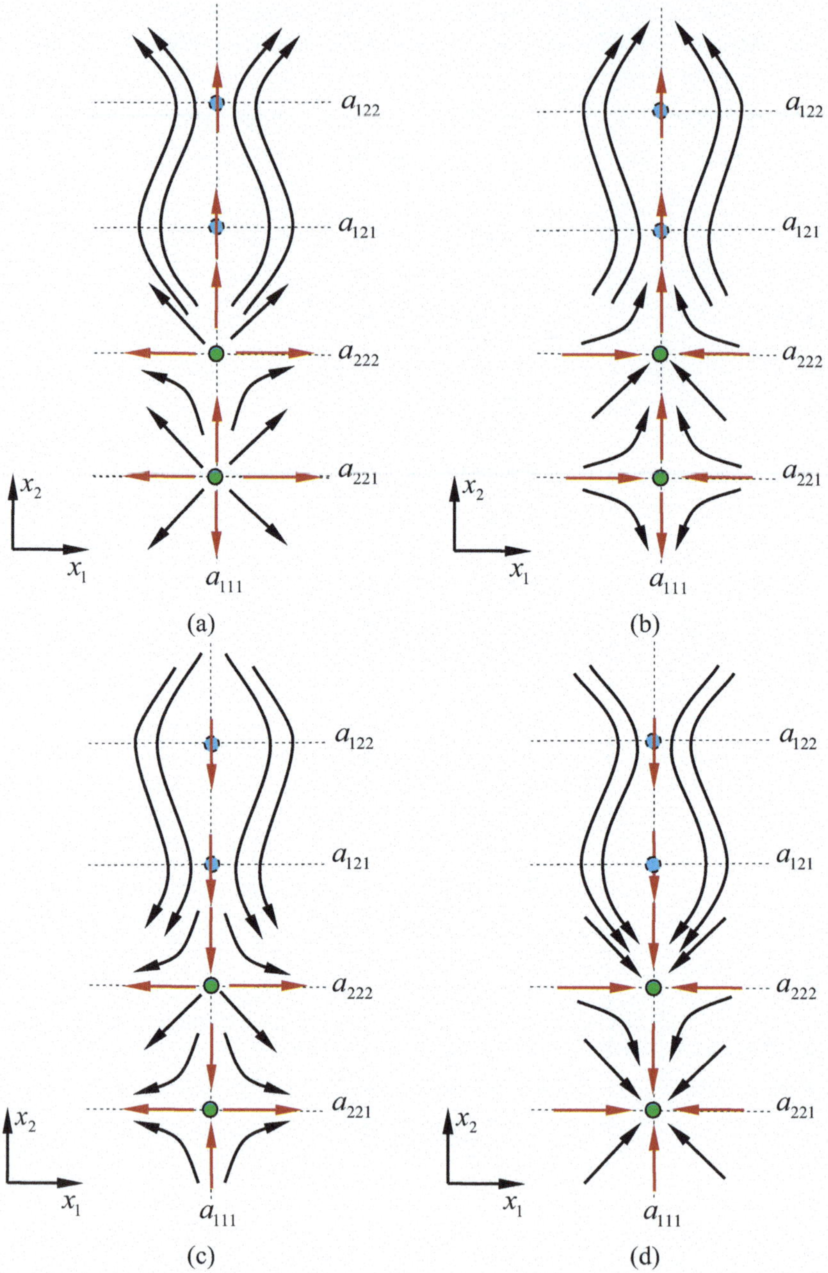

Fig. 4.3 Phase portraits ($a_{222} \in (a_{221}, a_{221})$) for two-dimensional systems on the x_1-direction with $x_1^* = a_{111}$ and on the x_2-direction with $x_2^* = a_{121}, a_{122}, a_{221}, a_{222}$. (**a**) ($a_{110} > 0$, $a_{220} > 0$), (**b**) ($a_{110} < 0$, $a_{220} > 0$), (**c**) ($a_{110} > 0$, $a_{220} < 0$), (**d**) ($a_{110} < 0$, $a_{220} < 0$)

$$\left\{\begin{array}{c}(a_{111},a_{122})\\(a_{111},a_{121})\\(a_{111},a_{222})\\(a_{111},a_{221})\end{array}\right\}=\left\{\begin{array}{c}\underbrace{(\mathrm{DP}:\mathrm{UP},\mathrm{nF})}_{\text{hyperbolic flow }(-)}\\\underbrace{(\mathrm{UP}:\mathrm{DP},\mathrm{nF})}_{\text{hyperbolic-secant flow }(-)}\\\underbrace{(\mathrm{SI},\mathrm{LS})}_{\text{lower-saddle sink}}\\\underbrace{(\mathrm{SI},\mathrm{SI})}_{\text{sink}}\end{array}\right\}\text{ for }a_{110}<0\text{ and }a_{220}<0. \tag{4.14}$$

The saddle-source and saddle-sink with a simple equilibrium (sink, saddle, source) in the series of equilibriums are on the bottom of hyperbolic and hyperbolic-secant flows.

(iv) For $a_{221}<a_{121}<a_{222}<a_{122}$, the equilibriums of $x_2^*=a_{121},a_{122},a_{221},a_{222}$ with $x_1^*=a_{111}$ have the following properties.

$$\left\{\begin{array}{c}(a_{111},a_{122})\\(a_{111},a_{222})\\(a_{111},a_{121})\\(a_{111},a_{221})\end{array}\right\}=\left\{\begin{array}{c}\underbrace{(\mathrm{DP}:\mathrm{UP},\mathrm{pF})}_{\text{hyperbolic flow }(+)}\\\underbrace{(\mathrm{SI},\mathrm{US})}_{\text{upper-saddle source}}\\\underbrace{(\mathrm{UP}:\mathrm{DP},\mathrm{pF})}_{\text{hyperbolic-secant flow }(+)}\\\underbrace{(\mathrm{SO},\mathrm{SO})}_{\text{source}}\end{array}\right\}\text{ for }a_{110}>0\text{ and }a_{220}>0; \tag{4.15}$$

$$\left\{\begin{array}{c}(a_{111},a_{122})\\(a_{111},a_{222})\\(a_{111},a_{121})\\(a_{111},a_{221})\end{array}\right\}=\left\{\begin{array}{c}\underbrace{(\mathrm{UP}:\mathrm{DP},\mathrm{pF})}_{\text{hyperbolic-secant flow }(+)}\\\underbrace{(\mathrm{SO},\mathrm{US})}_{\text{upper-saddle source}}\\\underbrace{(\mathrm{DP}:\mathrm{UP},\mathrm{pF})}_{\text{hyperbolic flow }(+)}\\\underbrace{(\mathrm{SI},\mathrm{SO})}_{\text{saddle}}\end{array}\right\}\text{ for }a_{110}<0\text{ and }a_{220}>0; \tag{4.16}$$

$$\left\{\begin{array}{c}(a_{111},a_{122})\\(a_{111},a_{222})\\(a_{111},a_{121})\\(a_{111},a_{221})\end{array}\right\}=\left\{\begin{array}{c}\underbrace{(\mathrm{UP}:\mathrm{DP},\mathrm{nF})}_{\text{hyperbolic-secant flow }(-)}\\\underbrace{(\mathrm{SI},\mathrm{LS})}_{\text{lower-saddle sink}}\\\underbrace{(\mathrm{DP}:\mathrm{UP},\mathrm{nF})}_{\text{hyperbolic flow }(-)}\\\underbrace{(\mathrm{SO},\mathrm{SI})}_{\text{saddle}}\end{array}\right\}\text{ for }a_{110}>0\text{ and }a_{220}<0; \tag{4.17}$$

$$\left\{\begin{array}{c}(a_{111},a_{122})\\(a_{111},a_{222})\\(a_{111},a_{121})\\(a_{111},a_{221})\end{array}\right\}=\left\{\begin{array}{c}\underbrace{(\mathrm{DP}:\mathrm{UP},\mathrm{nF})}_{\text{hyperbolic flow }(-)}\\\underbrace{(\mathrm{SO},\mathrm{LS})}_{\text{lower-saddle source}}\\\underbrace{(\mathrm{UP}:\mathrm{DP},\mathrm{nF})}_{\text{hyperbolic-secant flow }(-)}\\\underbrace{(\mathrm{SI},\mathrm{SI})}_{\text{sink}}\end{array}\right\}\text{ for }a_{110}<0\text{ and }a_{220}<0, \tag{4.18}$$

as shown in Fig. 4.4. The saddle-source and saddle-sink with a simple equilibrium (sink, source, and saddle) are separated by the hyperbolic and hyperbolic-secant flows. The hyperbolic or hyperbolic-secant flows are on the top of equilibrium series.

(v) For $a_{121} < a_{221} < a_{122} < a_{222}$, the equilibriums of $x_2^* = a_{121}, a_{122}, a_{221}, a_{222}$ with $x_1^* = a_{111}$ have the following properties.

$$\left\{\begin{array}{c}(a_{111},a_{222})\\(a_{111},a_{122})\\(a_{111},a_{221})\\(a_{111},a_{121})\end{array}\right\}=\left\{\begin{array}{c}\underbrace{(\mathrm{SO},\mathrm{US})}_{\text{upper-saddle source}}\\\underbrace{(\mathrm{DP}:\mathrm{UP},\mathrm{pF})}_{\text{hyperbolic flow }(+)}\\\underbrace{(\mathrm{SI},\mathrm{SO})}_{\text{saddle}}\\\underbrace{(\mathrm{DP}:\mathrm{UP},\mathrm{nF})}_{\text{hyperbolic flow }(-)}\end{array}\right\}\text{ for }a_{110}>0\text{ and }a_{220}>0; \tag{4.19}$$

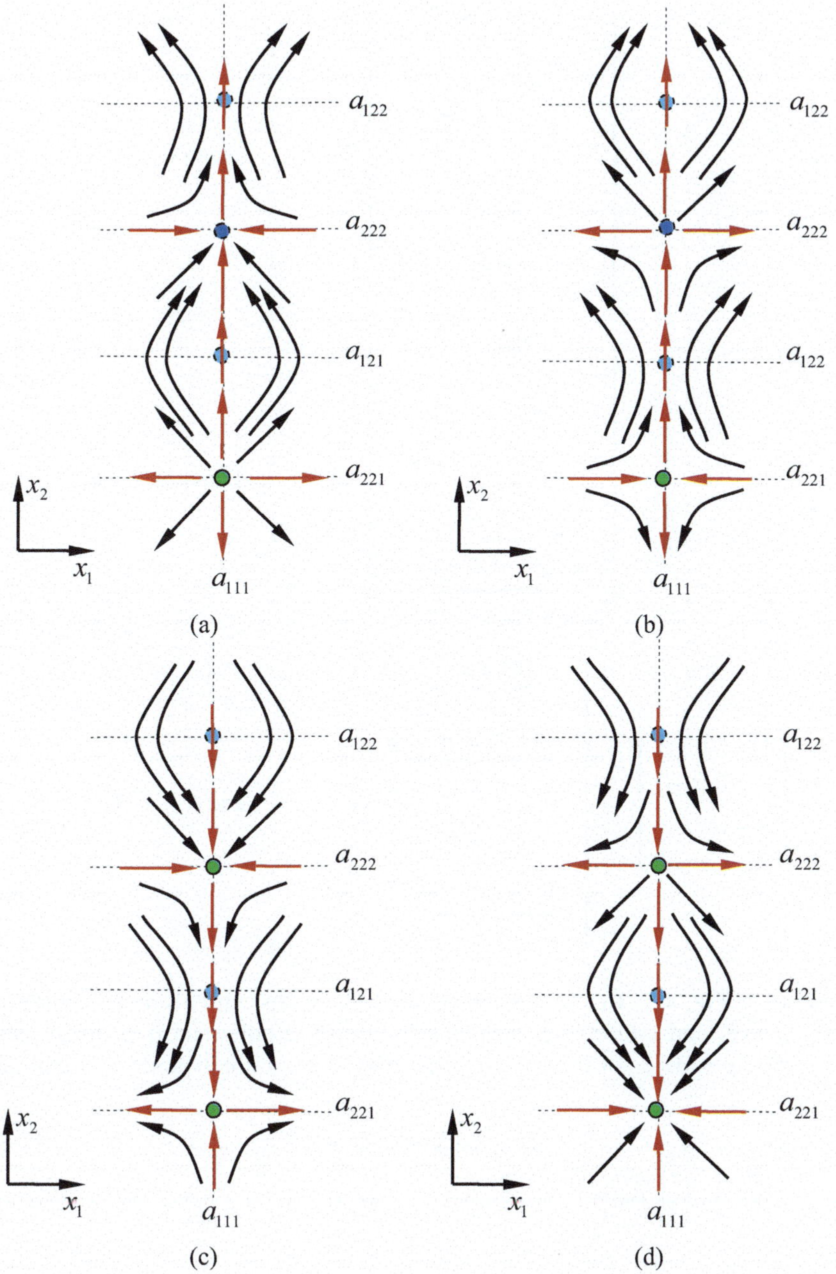

Fig. 4.4 Phase portraits ($a_{221} < a_{121} < a_{222} < a_{122}$) for two-dimensional systems on the x_1-direction with $x_1^* = a_{111}$ and on the x_2-direction with $x_2^* = a_{121}, a_{122}, a_{221}, a_{222}$. (**a**) ($a_{110} > 0$, $a_{220} > 0$), (**b**) ($a_{110} < 0$, $a_{220} > 0$), (**c**) ($a_{110} > 0$, $a_{220} < 0$), (**d**) ($a_{110} < 0$, $a_{220} < 0$)

$$\left\{\begin{array}{c}(a_{111},a_{222})\\(a_{111},a_{122})\\(a_{111},a_{221})\\(a_{111},a_{121})\end{array}\right\}=\left\{\begin{array}{c}\underbrace{(\mathrm{SI},\mathrm{US})}_{\text{upper-saddle source}}\\\underbrace{(\mathrm{UP}:\mathrm{DP},\mathrm{pF})}_{\text{hyperbolic-secant flow }(+)}\\\underbrace{(\mathrm{SO},\mathrm{SO})}_{\text{source}}\\\underbrace{(\mathrm{UP}:\mathrm{DP},\mathrm{nF})}_{\text{hyperbolic-secant flow }(-)}\end{array}\right\}\text{ for }a_{110}<0\text{ and }a_{220}>0; \tag{4.20}$$

$$\left\{\begin{array}{c}(a_{111},a_{222})\\(a_{111},a_{122})\\(a_{111},a_{221})\\(a_{111},a_{121})\end{array}\right\}=\left\{\begin{array}{c}\underbrace{(\mathrm{SO},\mathrm{LS})}_{\text{lower-saddle source}}\\\underbrace{(\mathrm{UP}:\mathrm{DP},\mathrm{nF})}_{\text{hyperbolic-secant flow }(-)}\\\underbrace{(\mathrm{SI},\mathrm{SI})}_{\text{sink}}\\\underbrace{(\mathrm{UP}:\mathrm{DP},\mathrm{pF})}_{\text{hyperbolic-secant flow }(+)}\end{array}\right\}\text{ for }a_{110}>0\text{ and }a_{220}<0; \tag{4.21}$$

$$\left\{\begin{array}{c}(a_{111},a_{222})\\(a_{111},a_{122})\\(a_{111},a_{221})\\(a_{111},a_{121})\end{array}\right\}=\left\{\begin{array}{c}\underbrace{(\mathrm{SI},\mathrm{LS})}_{\text{lower-saddle sink}}\\\underbrace{(\mathrm{DP}:\mathrm{UP},\mathrm{nF})}_{\text{hyperbolic flow }(-)}\\\underbrace{(\mathrm{SO},\mathrm{SI})}_{\text{saddle}}\\\underbrace{(\mathrm{DP}:\mathrm{UP},\mathrm{pF})}_{\text{hyperbolic flow }(+)}\end{array}\right\}\text{ for }a_{110}<0\text{ and }a_{220}<0, \tag{4.22}$$

as shown in Fig. 4.5. The saddle-source and saddle-sink with a simple equilibrium (sink, source, and saddle) are separated by the hyperbolic and hyperbolic-secant flows. The hyperbolic or hyperbolic-secant flows are on the bottom of equilibrium series.

(vi) For $a_{121} < a_{221} < a_{222} < a_{122}$, the equilibriums of $x_2^* = a_{121}, a_{122}, a_{221}, a_{222}$ with $x_1^* = a_{111}$ have the following properties.

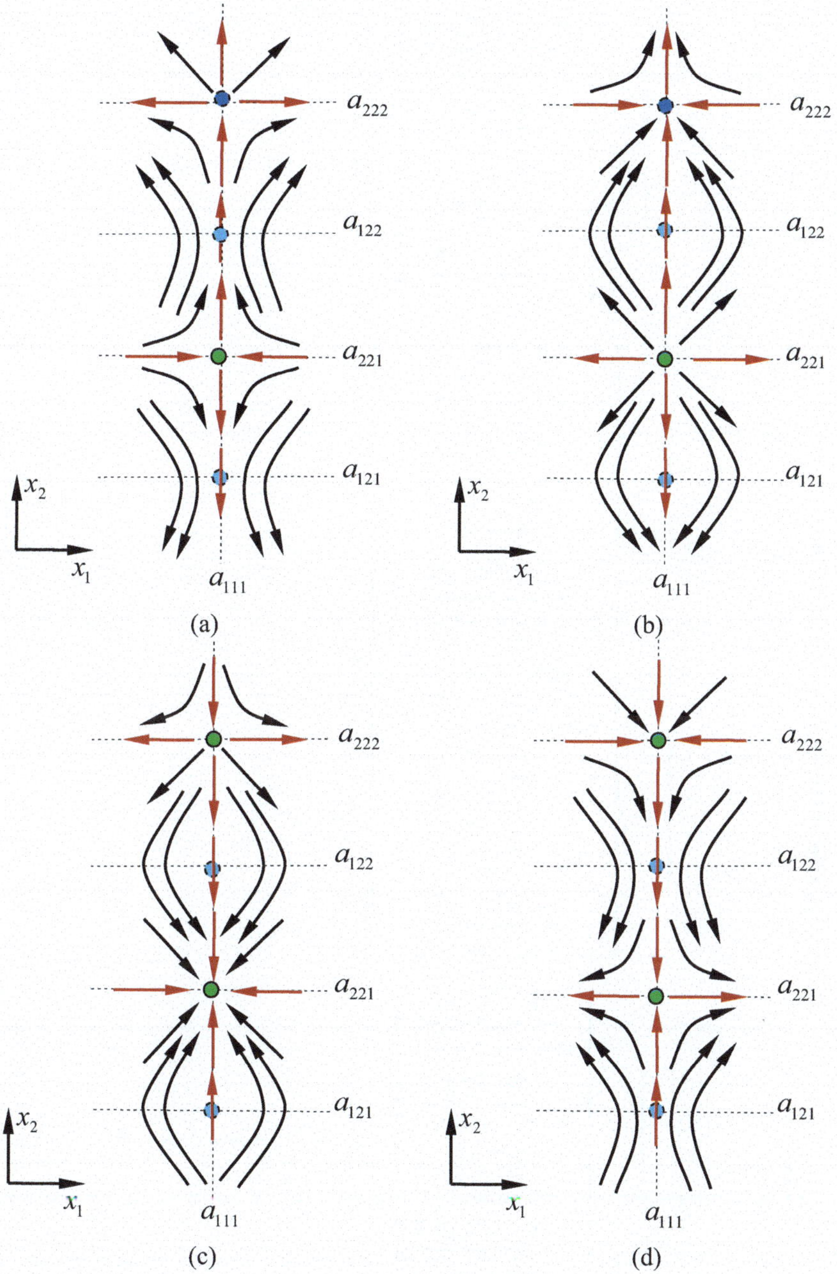

Fig. 4.5 Phase portraits ($a_{121} < a_{221} < a_{122} < a_{222}$) for two-dimensional systems on the x_1-direction with $x_1^* = a_{111}$ and on the x_2-direction with $x_2^* = a_{121}, a_{122}, a_{221}, a_{222}$. **(a)** ($a_{110} > 0$, $a_{220} > 0$), **(b)** ($a_{110} < 0$, $a_{220} > 0$), **(c)** ($a_{110} > 0$, $a_{220} < 0$), **(d)** ($a_{110} < 0$, $a_{220} < 0$)

$$\left\{\begin{array}{c}(a_{111},a_{122})\\(a_{111},a_{222})\\(a_{111},a_{221})\\(a_{111},a_{121})\end{array}\right\}=\left\{\begin{array}{c}\underbrace{(\mathrm{DP}:\mathrm{UP},\mathrm{pF})}_{\text{hyperbolic flow }(+)}\\\underbrace{(\mathrm{SI},\mathrm{US})}_{\text{upper-saddle sink}}\\\underbrace{(\mathrm{SI},\mathrm{SO})}_{\text{saddle}}\\\underbrace{(\mathrm{DP}:\mathrm{UP},\mathrm{nF})}_{\text{hyperbolic flow }(-)}\end{array}\right\}\text{ for } a_{110}>0 \text{ and } a_{220}>0; \tag{4.23}$$

$$\left\{\begin{array}{c}(a_{111},a_{122})\\(a_{111},a_{222})\\(a_{111},a_{221})\\(a_{111},a_{121})\end{array}\right\}=\left\{\begin{array}{c}\underbrace{(\mathrm{UP}:\mathrm{DP},\mathrm{pF})}_{\text{hyperbolic-secant flow }(+)}\\\underbrace{(\mathrm{SO},\mathrm{US})}_{\text{upper-saddle source}}\\\underbrace{(\mathrm{SO},\mathrm{SO})}_{\text{source}}\\\underbrace{(\mathrm{UP}:\mathrm{DP},\mathrm{nF})}_{\text{hyperbolic-secant flow }(-)}\end{array}\right\}\text{ for } a_{110}<0 \text{ and } a_{220}>0; \tag{4.24}$$

$$\left\{\begin{array}{c}(a_{111},a_{122})\\(a_{111},a_{222})\\(a_{111},a_{221})\\(a_{111},a_{121})\end{array}\right\}=\left\{\begin{array}{c}\underbrace{(\mathrm{UP}:\mathrm{DP},\mathrm{nF})}_{\text{hyperbolic-secant flow }(-)}\\\underbrace{(\mathrm{SI},\mathrm{LS})}_{\text{lower-saddle sink}}\\\underbrace{(\mathrm{SI},\mathrm{SI})}_{\text{sink}}\\\underbrace{(\mathrm{UP}:\mathrm{DP},\mathrm{pF})}_{\text{hyperbolic-secant flow }(+)}\end{array}\right\}\text{ for } a_{110}>0 \text{ and } a_{220}<0; \tag{4.25}$$

$$\left\{\begin{array}{c}(a_{111},a_{122})\\(a_{111},a_{222})\\(a_{111},a_{221})\\(a_{111},a_{121})\end{array}\right\}=\left\{\begin{array}{c}\underbrace{(\mathrm{DP}:\mathrm{UP},\mathrm{nF})}_{\text{hyperbolic flow }(-)}\\\underbrace{(\mathrm{SO},\mathrm{LS})}_{\text{lower-saddle source}}\\\underbrace{(\mathrm{SO},\mathrm{SI})}_{\text{saddle}}\\\underbrace{(\mathrm{DP}:\mathrm{UP},\mathrm{pF})}_{\text{hyperbolic flow }(+)}\end{array}\right\}\text{ for } a_{110}<0 \text{ and } a_{220}<0, \tag{4.26}$$

as shown in Fig. 4.6. The saddle-source and saddle-sink with a simple equilibrium (sink, source, and saddle) are separated by the hyperbolic and hyperbolic-secant flows. The hyperbolic and hyperbolic-secant flows are on the top and bottom of equilibrium series.

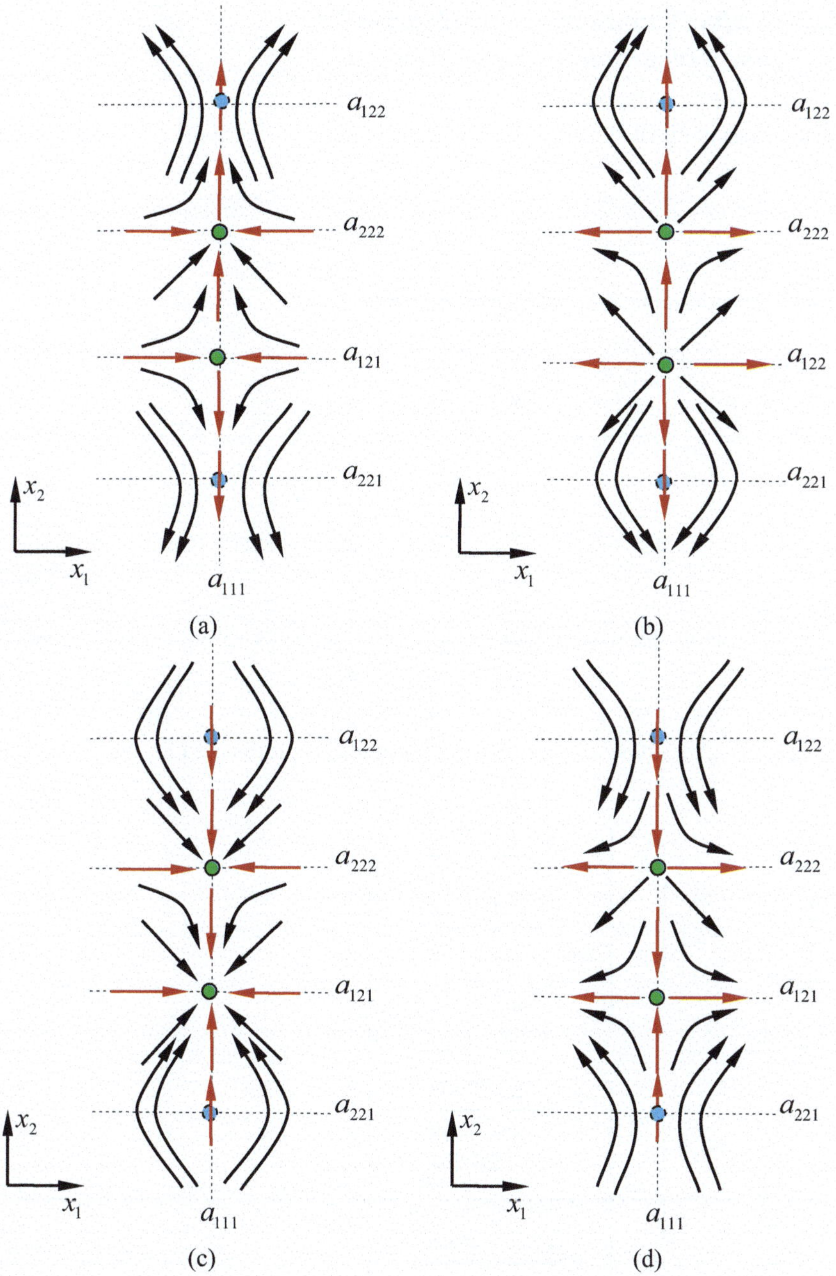

Fig. 4.6 Phase portraits ($a_{121} < a_{221} < a_{222} < a_{122}$) for two-dimensional systems on the x_1-direction with $x_1^* = a_{111}$ and on the x_2-direction with $x_2^* = a_{121}, a_{122}, a_{221}, a_{222}$. **(a)** ($a_{110} > 0$, $a_{220} > 0$), **(b)** ($a_{110} < 0$, $a_{220} > 0$), **(c)** ($a_{110} > 0$, $a_{220} < 0$), **(d)** ($a_{110} < 0$, $a_{220} < 0$)

4.1.2 Cubic Singularity with Hyperbolic and Hyperbolic-Secant Flows

Consider a dynamical system as

$$\begin{aligned}\dot{x}_1 &= a_{110}(x_1 - a_{111})(x_2 - a_{121})(x_2 - a_{122}),\\ \dot{x}_2 &= a_{220}(x_2 - a_{221})^3,\end{aligned} \tag{4.27}$$

and the corresponding first integral manifold is

$$\begin{aligned}\ln\frac{|x_1 - a_{111}|}{|x_{10} - a_{111}|} = \frac{a_{110}}{a_{220}}\Bigg\{ &\ln\frac{|x_2 - a_{221}|}{|x_{20} - a_{221}|}\\ &-\sum\nolimits_{l_1=1}^{2}(a_{221} - a_{12l_1})\Big(\frac{1}{x_2 - a_{221}} - \frac{1}{x_{20} - a_{221}}\Big)\\ &-\frac{1}{2}\prod\nolimits_{l_1=1}^{2}(a_{221} - a_{12l_1})\left[\frac{1}{(x_2 - a_{221})^2} - \frac{1}{(x_{20} - a_{221})^2}\right]\Bigg\}.\end{aligned} \tag{4.28}$$

The following discussion focuses on phase portraits through the first integral manifold.

(i) For the case of $a_{122} < a_{221}$, the equilibrium series of $x_1^* = a_{111}$ with $x_2^* = a_{121}, a_{122}, a_{221}$ have the following properties, as shown in Fig. 4.7.

$$\begin{Bmatrix}(a_{111}, a_{221})\\ (a_{111}, a_{122})\\ (a_{111}, a_{121})\end{Bmatrix} = \begin{Bmatrix}\underbrace{(\text{SO}, 3^{\text{rd}}\text{SO})}_{\text{third-order source}}\\ \underbrace{(\text{UP : DP, nF})}_{\text{hyperbolic-secant flow }(-)}\\ \underbrace{(\text{DP : UP, nF})}_{\text{hyperbolic flow }(-)}\end{Bmatrix} \text{ for } a_{110} > 0 \text{ and } a_{220} > 0; \tag{4.29}$$

$$\begin{Bmatrix}(a_{111}, a_{221})\\ (a_{111}, a_{122})\\ (a_{111}, a_{121})\end{Bmatrix} = \begin{Bmatrix}\underbrace{(\text{SI}, 3^{\text{rd}}\text{SO})}_{\text{third-order saddle}}\\ \underbrace{(\text{UP : DP, nF})}_{\text{hyperbolic-secant flow }(-)}\\ \underbrace{(\text{DP : UP, nF})}_{\text{hyperbolic flow }(-)}\end{Bmatrix} \text{ for } a_{110} < 0 \text{ and } a_{220} > 0; \tag{4.30}$$

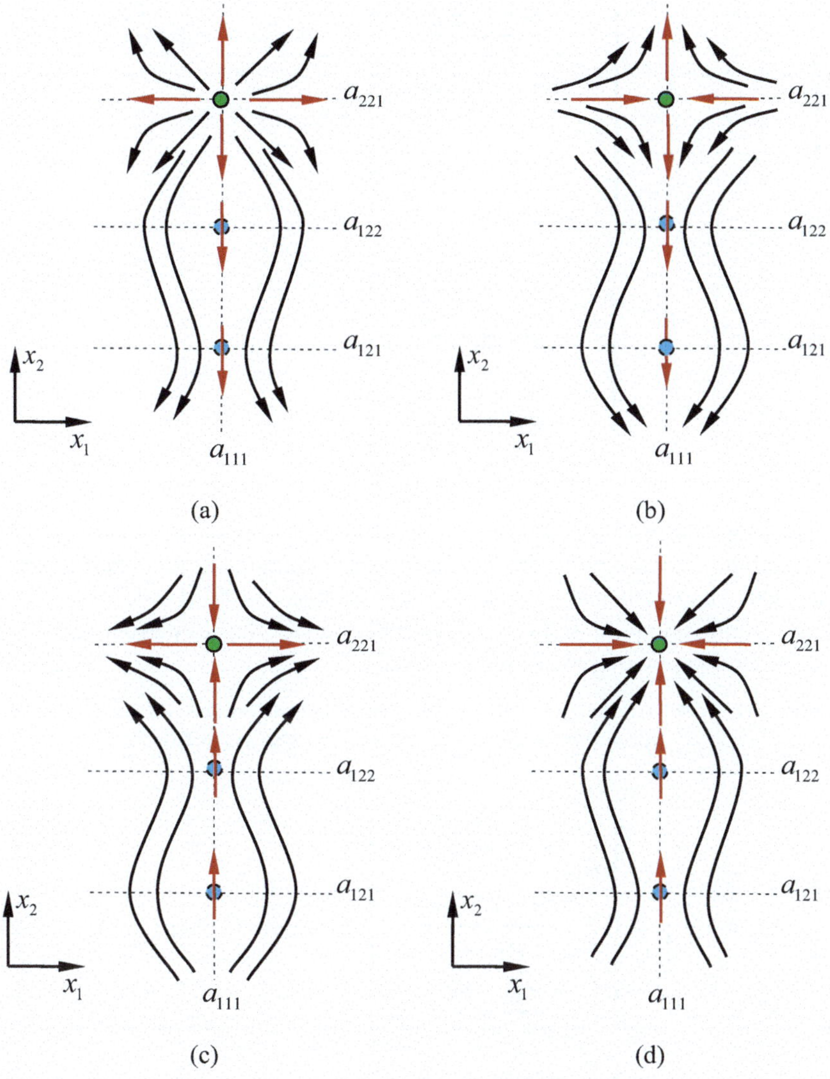

Fig. 4.7 Phase portraits ($a_{122} < a_{221}$) for two-dimensional systems on the x_1-direction with $x_1^* = a_{111}$ and on the x_2-direction with $x_2^* = a_{121}, a_{122}, a_{221}, a_{222}$. The equilibriums flows: (**a**) ($a_{110} > 0$, $a_{220} > 0$), (**b**) ($a_{110} < 0$, $a_{220} > 0$), (**c**) ($a_{110} > 0$, $a_{220} < 0$), (**d**) ($a_{110} < 0$, $a_{220} < 0$)

$$\left\{\begin{array}{c}(a_{111},a_{221})\\(a_{111},a_{122})\\(a_{111},a_{121})\end{array}\right\}=\left\{\begin{array}{c}\underbrace{(\mathrm{SO},3^{\mathrm{rd}}\mathrm{SI})}_{\text{third-order saddle}}\\\underbrace{(\mathrm{UP}:\mathrm{DP},\mathrm{pF})}_{\text{hyperbolic-secant flow }(+)}\\\underbrace{(\mathrm{DP}:\mathrm{UP},\mathrm{nF})}_{\text{hyperbolic flow }(+)}\end{array}\right\}\text{ for } a_{110}>0 \text{ and } a_{220}<0; \tag{4.31}$$

$$\left\{\begin{array}{c}(a_{111},a_{221})\\(a_{111},a_{122})\\(a_{111},a_{121})\end{array}\right\}=\left\{\begin{array}{c}\underbrace{(\mathrm{SI},3^{\mathrm{rd}}\mathrm{SI})}_{\text{third-order sink}}\\\underbrace{(\mathrm{UP}:\mathrm{DP},\mathrm{pF})}_{\text{hyperbolic-secant flow }(+)}\\\underbrace{(\mathrm{DP}:\mathrm{UP},\mathrm{pF})}_{\text{hyperbolic flow }(+)}\end{array}\right\}\text{ for } a_{110}<0 \text{ and } a_{220}<0. \tag{4.32}$$

One of the third-order saddle, source, and sink are on top of connected hyperbolic and hyperbolic-secant flows. The two third-order saddles are the bifurcations for saddle, sink, and saddle and for saddle, source, saddle. The third-order source is the appearing bifurcations of source, saddle, and source. The third-order sink is the appearing bifurcation of sink, saddle, and sink.

(ii) For the case of $a_{122} < a_{221}$, the equilibrium series of $x_1^* = a_{111}$ with $x_2^* = a_{121}, a_{122}, a_{221}$ have the following properties, as shown in Fig. 4.8.

$$\left\{\begin{array}{c}(a_{111},a_{221})\\(a_{111},a_{122})\\(a_{111},a_{121})\end{array}\right\}=\left\{\begin{array}{c}\underbrace{(\mathrm{UP}:\mathrm{DP},\mathrm{pF})}_{\text{hyperbolic-secant flow }(+)}\\\underbrace{(\mathrm{DP}:\mathrm{UP},\mathrm{nF})}_{\text{hyperbolic flow }(+)}\\\underbrace{(\mathrm{SO},3^{\mathrm{rd}}\mathrm{SO})}_{\text{third-order source}}\end{array}\right\}\text{ for } a_{110}>0 \text{ and } a_{220}>0; \tag{4.33}$$

$$\left\{\begin{array}{c}(a_{111},a_{221})\\(a_{111},a_{122})\\(a_{111},a_{121})\end{array}\right\}=\left\{\begin{array}{c}\underbrace{(\mathrm{UP}:\mathrm{DP},\mathrm{pF})}_{\text{hyperbolic-secant flow }(+)}\\\underbrace{(\mathrm{DP}:\mathrm{UP},\mathrm{pF})}_{\text{hyperbolic flow }(+)}\\\underbrace{(\mathrm{SI},3^{\mathrm{rd}}\mathrm{SO})}_{\text{third-order saddle}}\end{array}\right\}\text{ for } a_{110}<0 \text{ and } a_{220}>0; \tag{4.34}$$

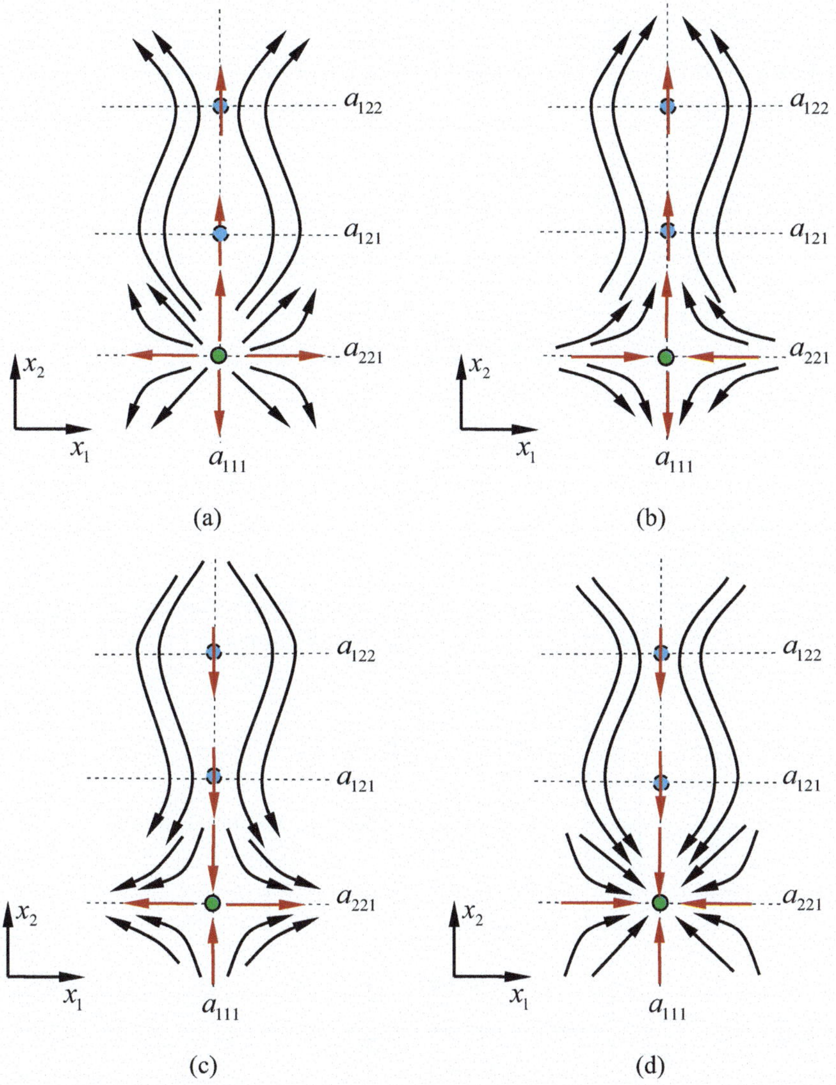

Fig. 4.8 Phase portraits ($a_{221} < a_{121}$) for two-dimensional systems on the x_1-direction with $x_1^* = a_{111}$ and on the x_2-direction with $x_2^* = a_{121}, a_{122}, a_{221}, a_{222}$. The equilibriums flows: (**a**) ($a_{110} > 0$, $a_{220} > 0$), (**b**) ($a_{110} < 0$, $a_{220} > 0$), (**c**) ($a_{110} > 0$, $a_{220} < 0$), (**d**) ($a_{110} < 0$, $a_{220} < 0$)

$$\left\{\begin{array}{c}(a_{111},a_{221})\\(a_{111},a_{122})\\(a_{111},a_{121})\end{array}\right\}=\left\{\begin{array}{c}\underbrace{(\mathrm{UP}:\mathrm{DP},\mathrm{nF})}_{\text{hyperbolic-secant flow }(-)}\\\underbrace{(\mathrm{DP}:\mathrm{UP},\mathrm{nF})}_{\text{hyperbolic flow }(-)}\\\underbrace{(\mathrm{SO},3^{\mathrm{rd}}\mathrm{SI})}_{\text{third-order saddle}}\end{array}\right\}\text{ for }a_{110}>0\text{ and }a_{220}<0;\qquad(4.35)$$

$$\left\{\begin{array}{c}(a_{111},a_{221})\\(a_{111},a_{122})\\(a_{111},a_{121})\end{array}\right\}=\left\{\begin{array}{c}\underbrace{(\mathrm{UP}:\mathrm{DP},\mathrm{nF})}_{\text{hyperbolic-secant flow }(-)}\\\underbrace{(\mathrm{DP}:\mathrm{UP},\mathrm{nF})}_{\text{hyperbolic flow }(-)}\\\underbrace{(\mathrm{SI},3^{\mathrm{rd}}\mathrm{SI})}_{\text{third-order sink}}\end{array}\right\}\text{ for }a_{110}<0\text{ and }a_{220}<0.\qquad(4.36)$$

One of the third-order saddle, source, and sink are on bottom of the connected hyperbolic and hyperbolic-secant flows.

(iii) For the case of $a_{122}\in(a_{221},a_{222})$, the equilibrium series of $x_1^*=a_{111}$ with $x_2^*=a_{121},a_{122},a_{221}$ have the following properties, as shown in Fig. 4.9.

$$\left\{\begin{array}{c}(a_{111},a_{221})\\(a_{111},a_{122})\\(a_{111},a_{121})\end{array}\right\}=\left\{\begin{array}{c}\underbrace{(\mathrm{DP}:\mathrm{UP},\mathrm{pF})}_{\text{hyperbolic flow }(+)}\\\underbrace{(\mathrm{SI},3^{\mathrm{rd}}\mathrm{SO})}_{\text{third-order saddle}}\\\underbrace{(\mathrm{DP}:\mathrm{UP},\mathrm{nF})}_{\text{hyperbolic flow }(-)}\end{array}\right\}\text{ for }a_{110}>0\text{ and }a_{220}>0;\qquad(4.37)$$

$$\left\{\begin{array}{c}(a_{111},a_{221})\\(a_{111},a_{122})\\(a_{111},a_{121})\end{array}\right\}=\left\{\begin{array}{c}\underbrace{(\mathrm{UP}:\mathrm{DP},\mathrm{pF})}_{\text{hyperbolic-secant flow }(+)}\\\underbrace{(\mathrm{SO},3^{\mathrm{rd}}\mathrm{SO})}_{\text{third-order source}}\\\underbrace{(\mathrm{UP}:\mathrm{DP},\mathrm{nF})}_{\text{hyperbolic-secant flow }(-)}\end{array}\right\}\text{ for }a_{110}<0\text{ and }a_{220}>0;\qquad(4.38)$$

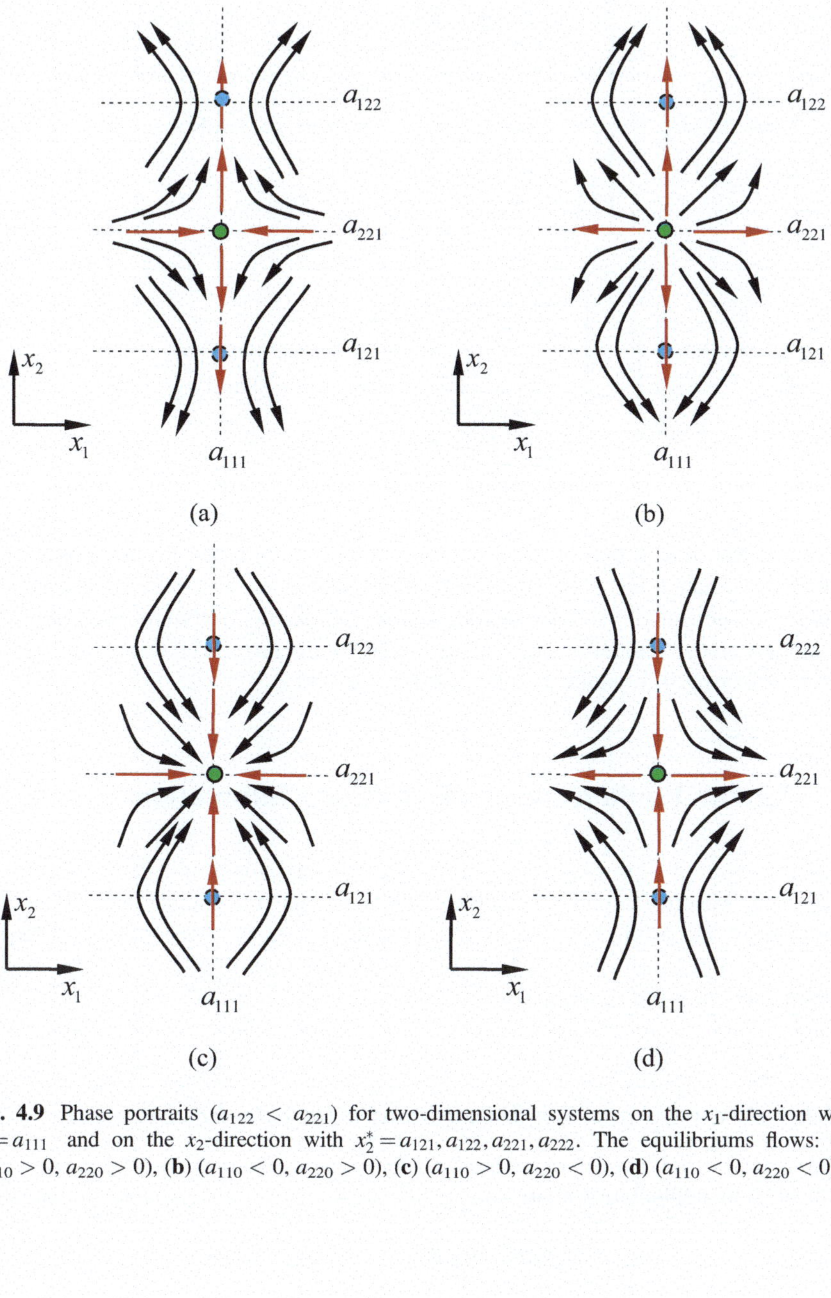

Fig. 4.9 Phase portraits ($a_{122} < a_{221}$) for two-dimensional systems on the x_1-direction with $x_1^* = a_{111}$ and on the x_2-direction with $x_2^* = a_{121}, a_{122}, a_{221}, a_{222}$. The equilibriums flows: (**a**) ($a_{110} > 0$, $a_{220} > 0$), (**b**) ($a_{110} < 0$, $a_{220} > 0$), (**c**) ($a_{110} > 0$, $a_{220} < 0$), (**d**) ($a_{110} < 0$, $a_{220} < 0$)

$$\left\{\begin{array}{l}(a_{111},a_{221})\\(a_{111},a_{122})\\(a_{111},a_{121})\end{array}\right\}=\left\{\begin{array}{c}\underbrace{(\mathrm{UP}:\mathrm{DP},\mathrm{nF})}_{\text{hyperbolic-secant flow }(-)}\\\underbrace{(\mathrm{SI},3^{\mathrm{rd}}\mathrm{SI})}_{\text{third-order sink}}\\\underbrace{(\mathrm{UP}:\mathrm{DP},\mathrm{pF})}_{\text{hyperbolic-secant flow }(+)}\end{array}\right\}\text{for } a_{110}>0 \text{ and } a_{220}<0; \tag{4.39}$$

$$\left\{\begin{array}{l}(a_{111},a_{221})\\(a_{111},a_{122})\\(a_{111},a_{121})\end{array}\right\}=\left\{\begin{array}{c}\underbrace{(\mathrm{DP}:\mathrm{UP},\mathrm{nF})}_{\text{hyperbolic flow }(-)}\\\underbrace{(\mathrm{SO},3^{\mathrm{rd}}\mathrm{SI})}_{\text{third-order saddle}}\\\underbrace{(\mathrm{DP}:\mathrm{UP},\mathrm{pF})}_{\text{hyperbolic flow }(+)}\end{array}\right\}\text{for } a_{110}<0 \text{ and } a_{220}<0. \tag{4.40}$$

One of the third-order saddle, source, and sink is between the hyperbolic and hyperbolic-secant flows. The two third-order saddles are also the bifurcations for saddle, sink, and saddle and for saddle, source, and saddle. The third-order source is for the appearing bifurcations of source, saddle, and source. The third-order sink is for the appearing bifurcation of sink, saddle, and sink.

4.2 Infinite-Equilibriums and Switching Bifurcations

In this section, the infinite-equilibriums for the switching of saddle-source and saddle-sink with the hyperbolic and hyperbolic-secant flows and for the switching of a simple equilibrium (sink, source, and saddle) with the hyperbolic and hyperbolic-secant flows are discussed. Thus, such infinite-equilibriums are the corresponding switching bifurcations.

4.2.1 Down-Up and Up-Down Saddle Infinite-Equilibriums

Consider a dynamical system as

$$\begin{aligned}\dot{x}_1&=a_{110}(x_1-a_{111})(x_2-a_{121})(x_2-a_{122}),\\\dot{x}_2&=a_{220}(x_2-a_{22s_1})(x_2-a_{22s_2})^2.\end{aligned} \tag{4.41}$$

For $a_{12l_1} = a_{22s_2}(l_1, l_2 \in \{1,2\})$, the corresponding first integral manifold is

$$\ln \frac{|x_1 - a_{111}|}{|x_{10} - a_{111}|} = \frac{a_{110}}{a_{220}} \left[\frac{(a_{22s_1} - a_{12l_2})}{(a_{22s_1} - a_{22s_2})} \ln \frac{|x_2 - a_{22s_1}|}{|x_{20} - a_{22s_1}|} + \frac{(a_{22s_2} - a_{12l_2})}{(a_{22s_2} - a_{22s_1})} \ln \frac{|x_2 - a_{22s_2}|}{|x_{20} - a_{22s_2}|} \right]. \tag{4.42}$$

(i) The equilibriums of $x_2^* = a_{121}, a_{122}, a_{221}, a_{222}$ with $x_1^* = a_{111}$ for $a_{121} = a_{22s_2} = a_{222}$ have the following properties.

$$\left\{ \begin{array}{l} (a_{111}, a_{122}) \\ (a_{111}, a_{222}) \\ (a_{111}, a_{221}) \end{array} \right\} = \left\{ \begin{array}{c} \underbrace{(\mathrm{DP}:\mathrm{UP},\mathrm{pF})}_{\text{hyperbolic flow }(+)} \\ \underbrace{(\mathrm{DU}:\mathrm{UD},\mathrm{US})}_{\text{down-up-to-up-down upper-saddle}} \\ \underbrace{(\mathrm{SO},\mathrm{SO})}_{\text{source}} \end{array} \right\} \text{ for } a_{110} > 0 \text{ and } a_{220} > 0; \tag{4.43}$$

$$\left\{ \begin{array}{l} (a_{111}, a_{122}) \\ (a_{111}, a_{222}) \\ (a_{111}, a_{221}) \end{array} \right\} = \left\{ \begin{array}{c} \underbrace{(\mathrm{UP}:\mathrm{DP},\mathrm{pF})}_{\text{hyperbolic-secant flow }(+)} \\ \underbrace{(\mathrm{UD}:\mathrm{DU},\mathrm{US})}_{\text{up-down-to-down-up upper-saddle}} \\ \underbrace{(\mathrm{SI},\mathrm{SO})}_{\text{saddle}} \end{array} \right\} \text{ for } a_{110} < 0 \text{ and } a_{220} > 0; \tag{4.44}$$

$$\left\{ \begin{array}{l} (a_{111}, a_{122}) \\ (a_{111}, a_{222}) \\ (a_{111}, a_{221}) \end{array} \right\} = \left\{ \begin{array}{c} \underbrace{(\mathrm{UP}:\mathrm{DP},\mathrm{nF})}_{\text{hyperbolic-secant flow }(-)} \\ \underbrace{(\mathrm{UD}:\mathrm{DU},\mathrm{LS})}_{\text{up-down-to-down-up lower-saddle}} \\ \underbrace{(\mathrm{SO},\mathrm{SI})}_{\text{saddle}} \end{array} \right\} \text{ for } a_{110} > 0 \text{ and } a_{220} < 0; \tag{4.45}$$

$$\left\{ \begin{array}{l} (a_{111}, a_{122}) \\ (a_{111}, a_{222}) \\ (a_{111}, a_{221}) \end{array} \right\} = \left\{ \begin{array}{c} \underbrace{(\mathrm{DP}:\mathrm{UP},\mathrm{nF})}_{\text{hyperbolic flow }(-)} \\ \underbrace{(\mathrm{DU}:\mathrm{UD},\mathrm{LS})}_{\text{down-up-to-up-down lower-saddle}} \\ \underbrace{(\mathrm{SI},\mathrm{SI})}_{\text{source}} \end{array} \right\} \text{ for } a_{110} < 0 \text{ and } a_{220} < 0. \tag{4.46}$$

The infinite-equilibriums of $x_2^* = a_{121} = a_{221}$ are

$$\begin{aligned}
&(\bar{x}_1, a_{222}) = \underbrace{(\text{DU}, \text{US})}_{\text{down-up upper-saddle}} \quad \text{if } \bar{x} \in (-\infty, a_{111}),\\
&(\bar{x}_1, a_{222}) = \underbrace{(\text{UD}, \text{US})}_{\text{up-down upper-saddle}} \quad \text{if } \bar{x} \in (a_{111}, \infty),\\
&\text{for } a_{110} > 0 \text{ and } a_{220} > 0;
\end{aligned} \tag{4.47}$$

$$\begin{aligned}
&(\bar{x}_1, a_{222}) = \underbrace{(\text{UD}, \text{US})}_{\text{up-down upper-saddle}} \quad \text{if } \bar{x} \in (-\infty, a_{111}),\\
&(\bar{x}_1, a_{222}) = \underbrace{(\text{DU}, \text{US})}_{\text{down-up upper-saddle}} \quad \text{if } \bar{x} \in (a_{111}, \infty),\\
&\text{for } a_{110} < 0 \text{ and } a_{220} > 0;
\end{aligned} \tag{4.48}$$

$$\begin{aligned}
&(\bar{x}_1, a_{222}) = \underbrace{(\text{UD}, \text{LS})}_{\text{up-down lower-saddle}} \quad \text{if } \bar{x} \in (-\infty, a_{111}),\\
&(\bar{x}_1, a_{222}) = \underbrace{(\text{DU}, \text{LS})}_{\text{down-up lower-saddle}} \quad \text{if } \bar{x} \in (a_{111}, \infty),\\
&\text{for } a_{110} > 0 \text{ and } a_{220} < 0;
\end{aligned} \tag{4.49}$$

$$\begin{aligned}
&(\bar{x}_1, a_{222}) = \underbrace{(\text{DU}, \text{LS})}_{\text{down-up lower-saddle}} \quad \text{if } \bar{x} \in (-\infty, a_{111}),\\
&(\bar{x}_1, a_{222}) = \underbrace{(\text{UD}, \text{LS})}_{\text{up-down lower-saddle}} \quad \text{if } \bar{x} \in (a_{111}, \infty),\\
&\text{for } a_{110} < 0 \text{ and } a_{220} < 0.
\end{aligned} \tag{4.50}$$

The above infinite-equilibriums are shown in Fig. 4.10. The hyperbolic-flow or hyperbolic-secant flows are on the upper sides of the up-down, down-up upper-saddle and lower-saddle infinite-equilibriums.

(ii) The equilibriums of $x_2^* = a_{121}, a_{122}, a_{221}, a_{222}$ with $x_1^* = a_{111}$ for $a_{122} = a_{22s_2} = a_{222}$ are

$$\left\{\begin{matrix} (a_{111}, a_{222}) \\ (a_{111}, a_{121}) \\ (a_{111}, a_{221}) \end{matrix}\right\} = \left\{\begin{matrix} \underbrace{(\text{UD}:\text{DU}, \text{US})}_{\text{up-down-to-down-up upper-saddle}} \\ \underbrace{(\text{UP}:\text{DP}, \text{pF})}_{\text{hyperbolic-secant flow } (+)} \\ \underbrace{(\text{SO}, \text{SO})}_{\text{source}} \end{matrix}\right\} \text{ for } a_{110} > 0 \text{ and } a_{220} > 0; \tag{4.51}$$

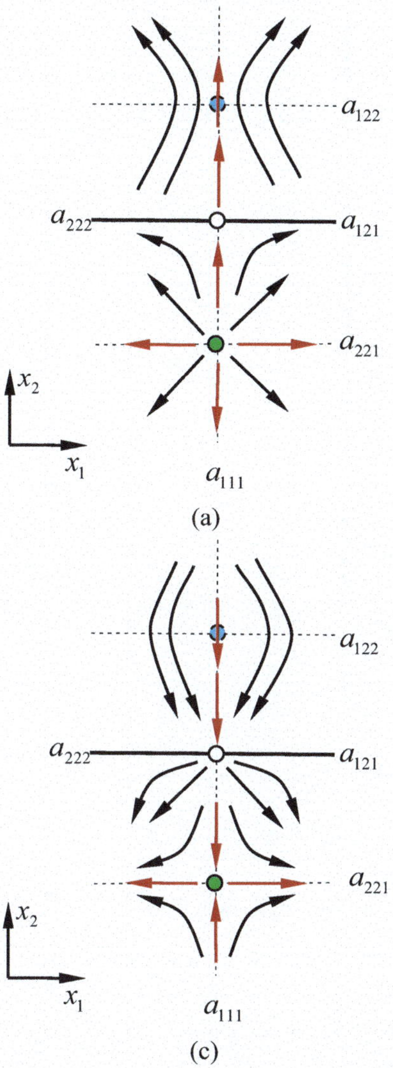

Fig. 4.10 Phase portraits ($a_{121} = a_{222}$) for two-dimensional systems on the x_1-direction with $x_1^* = a_{111}$ and on the x_2-direction with $x_2^* = a_{121}, a_{122}, a_{221}, a_{222}$. The equilibriums flows: **(a)** ($a_{110} > 0$, $a_{220} > 0$), **(b)** ($a_{110} < 0$, $a_{220} > 0$), **(c)** ($a_{110} > 0$, $a_{220} < 0$), **(d)** ($a_{110} < 0$, $a_{220} < 0$)

$$\left\{\begin{array}{l}(a_{111},a_{222})\\(a_{111},a_{121})\\(a_{111},a_{221})\end{array}\right\}=\left\{\begin{array}{c}\underbrace{(\mathrm{DU}:\mathrm{UD},\mathrm{US})}_{\text{down-up-to-up-down upper-saddle}}\\\underbrace{(\mathrm{DP}:\mathrm{UP},\mathrm{pF})}_{\text{hyperbolic flow }(+)}\\\underbrace{(\mathrm{SO},\mathrm{SI})}_{\text{saddle}}\end{array}\right\}\text{ for }a_{110}<0\text{ and }a_{220}>0;\quad(4.52)$$

$$\left\{\begin{array}{l}(a_{111},a_{222})\\(a_{111},a_{121})\\(a_{111},a_{221})\end{array}\right\}=\left\{\begin{array}{c}\underbrace{(\mathrm{DU}:\mathrm{UD},\mathrm{LS})}_{\text{down-up-to-up-down lower-saddle}}\\\underbrace{(\mathrm{DP}:\mathrm{UP},\mathrm{nF})}_{\text{hyperbolic flow }(-)}\\\underbrace{(\mathrm{SI},\mathrm{SO})}_{\text{saddle}}\end{array}\right\}\text{ for }a_{110}>0\text{ and }a_{220}<0;\quad(4.53)$$

$$\left\{\begin{array}{l}(a_{111},a_{222})\\(a_{111},a_{121})\\(a_{111},a_{221})\end{array}\right\}=\left\{\begin{array}{c}\underbrace{(\mathrm{UD}:\mathrm{DU},\mathrm{LS})}_{\text{up-down-to-down-up lower-saddle}}\\\underbrace{(\mathrm{UP}:\mathrm{DP},\mathrm{nF})}_{\text{hyperbolic-secant flow }(-)}\\\underbrace{(\mathrm{SI},\mathrm{SI})}_{\text{sink}}\end{array}\right\}\text{ for }a_{110}<0\text{ and }a_{220}<0.\quad(4.54)$$

The infinite-equilibriums of $x_2^*=a_{122}=a_{222}$ are

$$\begin{array}{ll}(\bar{x}_1,a_{222})=\underbrace{(\mathrm{UD},\mathrm{US})}_{\text{up-down upper-saddle}} & \text{if } \bar{x}\in(-\infty,a_{111}),\\(\bar{x}_1,a_{222})=\underbrace{(\mathrm{DU},\mathrm{US})}_{\text{down-up upper-saddle}} & \text{if } \bar{x}\in(a_{111},\infty),\\\text{for } a_{110}>0 \text{ and } a_{220}>0; &\end{array}\quad(4.55)$$

$$\begin{array}{ll}(\bar{x}_1,a_{222})=\underbrace{(\mathrm{DU},\mathrm{US})}_{\text{down-up upper-saddle}} & \text{if } \bar{x}\in(-\infty,a_{111}),\\(\bar{x}_1,a_{222})=\underbrace{(\mathrm{UD},\mathrm{US})}_{\text{up-down upper-saddle}} & \text{if } \bar{x}\in(a_{111},\infty),\\\text{for } a_{110}<0 \text{ and } a_{220}>0; &\end{array}\quad(4.56)$$

$$
\begin{aligned}
(\bar{x}_1, a_{222}) &= \underbrace{(\mathrm{DU}, \mathrm{LS})}_{\text{down-up lower-saddle}} \quad \text{if } \bar{x} \in (-\infty, a_{111}), \\
(\bar{x}_1, a_{222}) &= \underbrace{(\mathrm{UD}, \mathrm{LS})}_{\text{up-down lower-saddle}} \quad \text{if } \bar{x} \in (a_{111}, \infty), \\
&\text{for } a_{110} > 0 \text{ and } a_{220} < 0;
\end{aligned}
\tag{4.57}
$$

$$
\begin{aligned}
(\bar{x}_1, a_{222}) &= \underbrace{(\mathrm{UD}, \mathrm{LS})}_{\text{up-down lower-saddle}} \quad \text{if } \bar{x} \in (-\infty, a_{111}), \\
(\bar{x}_1, a_{222}) &= \underbrace{(\mathrm{DU}, \mathrm{LS})}_{\text{down-up lower-saddle}} \quad \text{if } \bar{x} \in (a_{111}, \infty), \\
&\text{for } a_{110} < 0 \text{ and } a_{220} < 0.
\end{aligned}
\tag{4.58}
$$

The corresponding infinite-equilibriums are sketched in Fig. 4.11. The hyperbolic-flow or hyperbolic-secant flows are on the lower sides of the up-down, down-up upper-saddle and lower-saddle infinite-equilibriums. The infinite-equilibriums are for the switching bifurcations of the saddle-sink and saddle-source with hyperbolic and hyperbolic-secant flows.

For $a_{22s_2} = a_{221}$ and $a_{22s_1} = a_{222}$, the two cases can be discussed similarly, which can be done by readers as an exercise.

4.2.2 *Inflection-Source and Sink Infinite-Equilibriums*

Consider a dynamical system as

$$
\begin{aligned}
\dot{x}_1 &= a_{110}(x_1 - a_{111})(x_2 - a_{121})(x_2 - a_{122}), \\
\dot{x}_2 &= a_{220}(x_2 - a_{22s_1})(x_2 - a_{22s_2})^2.
\end{aligned}
\tag{4.59}
$$

For $a_{12l_1} = a_{22s_1} (l_1, l_2 \in \{1, 2\})$, the corresponding first integral manifold is

$$
\begin{aligned}
\ln \frac{|x_1 - a_{111}|}{|x_{10} - a_{111}|} = \frac{a_{110}}{a_{220}} \Bigg[& \frac{2(a_{22s_2} - a_{12l_2})}{(a_{22s_2} - a_{22s_1})} \ln \frac{|x_2 - a_{22s_2}|}{|x_{20} - a_{22s_2}|} \\
& - \frac{(a_{22s_2} - a_{12l_2})}{(a_{22s_2} - a_{22s_1})^2} (\frac{1}{x_2 - a_{22s_2}} - \frac{1}{x_{20} - a_{22s_2}}) \Bigg].
\end{aligned}
\tag{4.60}
$$

(i) The equilibriums of $x_2^* = a_{121}, a_{122}, a_{221}, a_{222}$ with $x_1^* = a_{111}$ for $a_{121} = a_{22s_1} = a_{221}$ have the following properties.

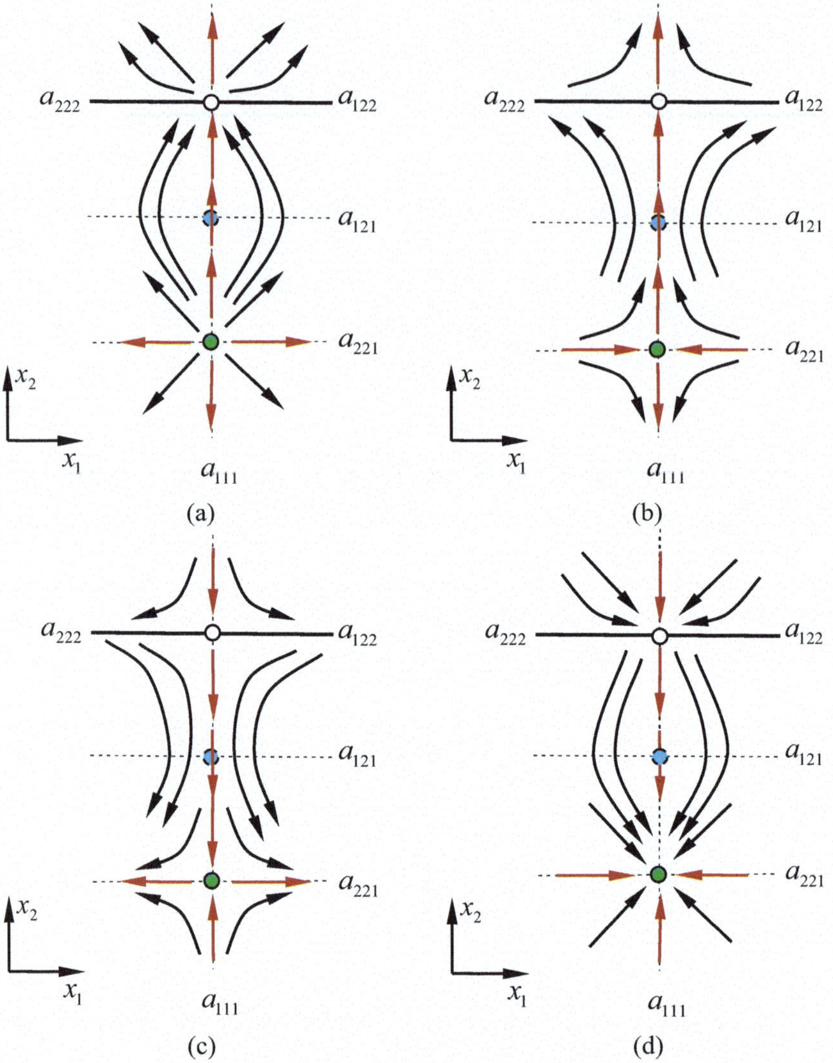

Fig. 4.11 Phase portraits ($a_{122} = a_{222}$) for two-dimensional systems on the x_1-direction with $x_1^* = a_{111}$ and on the x_2-direction with $x_2^* = a_{121}, a_{122}, a_{221}, a_{222}$. The equilibriums flows: (**a**) ($a_{110} > 0$, $a_{220} > 0$), (**b**) ($a_{110} < 0$, $a_{220} > 0$), (**c**) ($a_{110} > 0$, $a_{220} < 0$), (**d**) ($a_{110} < 0$, $a_{220} < 0$)

$$\left\{\begin{array}{l}(a_{111}, a_{122})\\(a_{111}, a_{222})\\(a_{111}, a_{221})\end{array}\right\} = \left\{\begin{array}{c}\underbrace{(\mathrm{DP}:\mathrm{UP},\mathrm{pF})}_{\text{hyperbolic flow }(+)}\\\underbrace{(\mathrm{SI},\mathrm{US})}_{\text{upper-saddle sink}}\\\underbrace{(\mathrm{II}:\mathrm{DI},\mathrm{SO})}_{\text{inflection source}}\end{array}\right\} \text{ for } a_{110} > 0 \text{ and } a_{220} > 0; \tag{4.61}$$

$$\left\{\begin{array}{l}(a_{111}, a_{122})\\(a_{111}, a_{222})\\(a_{111}, a_{221})\end{array}\right\} = \left\{\begin{array}{c}\underbrace{(\mathrm{UP}:\mathrm{DP},\mathrm{pF})}_{\text{hyperbolic-secant flow }(+)}\\\underbrace{(\mathrm{SO},\mathrm{US})}_{\text{upper-saddle sink}}\\\underbrace{(\mathrm{DI}:\mathrm{II},\mathrm{SO})}_{\text{inflection source}}\end{array}\right\} \text{ for } a_{110} < 0 \text{ and } a_{220} > 0; \tag{4.62}$$

$$\left\{\begin{array}{l}(a_{111}, a_{122})\\(a_{111}, a_{222})\\(a_{111}, a_{221})\end{array}\right\} = \left\{\begin{array}{c}\underbrace{(\mathrm{UP}:\mathrm{DP},\mathrm{nF})}_{\text{hyperbolic-secant flow }(-)}\\\underbrace{(\mathrm{SI},\mathrm{LS})}_{\text{lower-saddle source}}\\\underbrace{(\mathrm{DI}:\mathrm{II},\mathrm{SI})}_{\text{inflection sink}}\end{array}\right\} \text{ for } a_{110} > 0 \text{ and } a_{220} < 0; \tag{4.63}$$

$$\left\{\begin{array}{l}(a_{111}, a_{122})\\(a_{111}, a_{222})\\(a_{111}, a_{221})\end{array}\right\} = \left\{\begin{array}{c}\underbrace{(\mathrm{DP}:\mathrm{UP},\mathrm{nF})}_{\text{hyperbolic flow }(-)}\\\underbrace{(\mathrm{SO},\mathrm{LS})}_{\text{lower-saddle source}}\\\underbrace{(\mathrm{II}:\mathrm{DI},\mathrm{SI})}_{\text{inflection sink}}\end{array}\right\} \text{ for } a_{110} < 0 \text{ and } a_{220} < 0. \tag{4.64}$$

The inflection infinite-equilibriums of $x_2^* = a_{121} = a_{221}$ are

$$\begin{aligned}(\bar{x}_1, a_{221}) &= \underbrace{(\mathrm{II},\mathrm{SO})}_{\text{increasing-inflection source}} \quad \text{if } \bar{x} \in (-\infty, a_{111}),\\(\bar{x}_1, a_{221}) &= \underbrace{(\mathrm{DI},\mathrm{SO})}_{\text{decreasing-inflection source}} \quad \text{if } \bar{x} \in (a_{111}, \infty),\end{aligned} \tag{4.65}$$

for $a_{110} > 0$ and $a_{220} > 0$;

$$\begin{aligned}
&(\bar{x}_1, a_{221}) = \underbrace{(\mathrm{DI}, \mathrm{SO})}_{\text{decreasing-inflection source}} \quad \text{if } \bar{x} \in (-\infty, a_{111}),\\
&(\bar{x}_1, a_{221}) = \underbrace{(\mathrm{II}, \mathrm{SO})}_{\text{increasing-inflection source}} \quad \text{if } \bar{x} \in (a_{111}, \infty),\\
&\text{for } a_{110} < 0 \text{ and } a_{220} > 0;
\end{aligned} \tag{4.66}$$

$$\begin{aligned}
&(\bar{x}_1, a_{221}) = \underbrace{(\mathrm{DI}, \mathrm{SI})}_{\text{decreasing-inflection sink}} \quad \text{if } \bar{x} \in (-\infty, a_{111}),\\
&(\bar{x}_1, a_{221}) = \underbrace{(\mathrm{II}, \mathrm{SI})}_{\text{increasing-inflection sink}} \quad \text{if } \bar{x} \in (a_{111}, \infty),\\
&\text{for } a_{110} > 0 \text{ and } a_{220} < 0;
\end{aligned} \tag{4.67}$$

$$\begin{aligned}
&(\bar{x}_1, a_{221}) = \underbrace{(\mathrm{II}, \mathrm{SI})}_{\text{increasing-inflection sink}} \quad \text{if } \bar{x} \in (-\infty, a_{111}),\\
&(\bar{x}_1, a_{221}) = \underbrace{(\mathrm{DI}, \mathrm{SI})}_{\text{decreasing-inflection sink}} \quad \text{if } \bar{x} \in (a_{111}, \infty),\\
&\text{for } a_{110} < 0 \text{ and } a_{220} < 0.
\end{aligned} \tag{4.68}$$

The above inflection infinite-equilibriums are shown in Fig. 4.12. The hyperbolic-flow or hyperbolic-secant flows are on the upper sides of the inflection-source and sink infinite-equilibriums.

(ii) The equilibriums of $x_2^* = a_{121}, a_{122}, a_{221}, a_{222}$ with $x_1^* = a_{111}$ for $a_{122} = a_{22s_1} = a_{221}$ have the following properties.

$$\left\{\begin{matrix} (a_{111}, a_{222}) \\ (a_{111}, a_{221}) \\ (a_{111}, a_{121}) \end{matrix}\right\} = \left\{\begin{matrix} \underbrace{(\mathrm{SO}, \mathrm{US})}_{\text{upper-saddle source}} \\ \underbrace{(\mathrm{DI} : \mathrm{II}, \mathrm{SO})}_{\text{inflection source}} \\ \underbrace{(\mathrm{DP} : \mathrm{UP}, \mathrm{nF})}_{\text{hyperbolic flow } (-)} \end{matrix}\right\} \text{ for } a_{110} > 0 \text{ and } a_{220} > 0; \tag{4.69}$$

$$\left\{\begin{matrix} (a_{111}, a_{222}) \\ (a_{111}, a_{221}) \\ (a_{111}, a_{121}) \end{matrix}\right\} = \left\{\begin{matrix} \underbrace{(\mathrm{SI}, \mathrm{US})}_{\text{upper-saddle sink}} \\ \underbrace{(\mathrm{II} : \mathrm{DI}, \mathrm{SO})}_{\text{inflection source}} \\ \underbrace{(\mathrm{UP} : \mathrm{DP}, \mathrm{nF})}_{\text{hyperbolic-secant flow } (-)} \end{matrix}\right\} \text{ for } a_{110} < 0 \text{ and } a_{220} > 0; \tag{4.70}$$

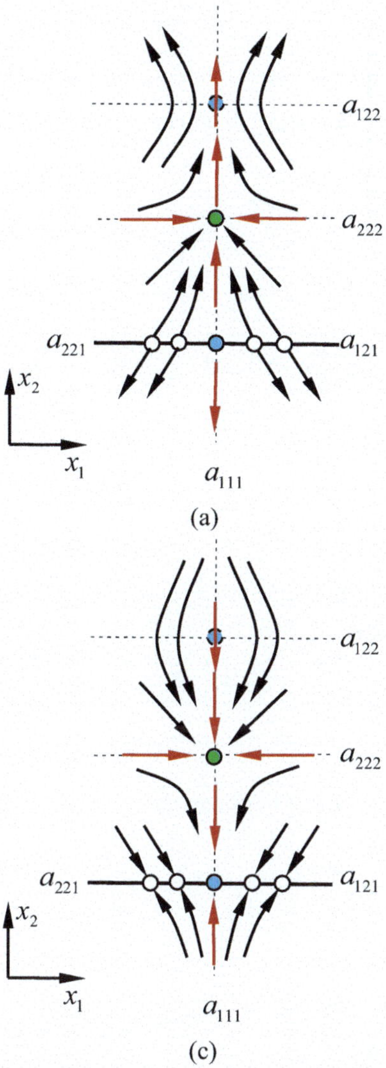

Fig. 4.12 Phase portraits ($a_{121} = a_{221}$) for two-dimensional systems on the x_1-direction with $x_1^* = a_{111}$ and on the x_2-direction with $x_2^* = a_{121}, a_{122}, a_{221}, a_{222}$. The equilibriums flows: (**a**) ($a_{110} > 0$, $a_{220} > 0$), (**b**) ($a_{110} < 0$, $a_{220} > 0$), (**c**) ($a_{110} > 0$, $a_{220} < 0$), (**d**) ($a_{110} < 0$, $a_{220} < 0$)

$$\left\{\begin{array}{l}(a_{111}, a_{222}) \\ (a_{111}, a_{221}) \\ (a_{111}, a_{121})\end{array}\right\} = \left\{\begin{array}{c}\underbrace{(\mathrm{SO}, \mathrm{LS})}_{\text{lower-saddle source}} \\ \underbrace{(\mathrm{II} : \mathrm{DI}, \mathrm{SI})}_{\text{inflection sink}} \\ \underbrace{(\mathrm{UP} : \mathrm{DP}, \mathrm{pF})}_{\text{hyperbolic-secant flow } (+)}\end{array}\right\} \text{ for } a_{110} > 0 \text{ and } a_{220} < 0; \tag{4.71}$$

$$\left\{\begin{array}{l}(a_{111}, a_{222}) \\ (a_{111}, a_{221}) \\ (a_{111}, a_{121})\end{array}\right\} = \left\{\begin{array}{c}\underbrace{(\mathrm{SI}, \mathrm{LS})}_{\text{lower-saddle sink}} \\ \underbrace{(\mathrm{DI} : \mathrm{II}, \mathrm{SI})}_{\text{inflection sink}} \\ \underbrace{(\mathrm{DP} : \mathrm{UP}, \mathrm{pF})}_{\text{hyperbolic flow } (+)}\end{array}\right\} \text{ for } a_{110} < 0 \text{ and } a_{220} < 0. \tag{4.72}$$

The inflection infinite-equilibriums of $x_2^* = a_{122} = a_{221}$ are

$$\begin{aligned}
(\bar{x}_1, a_{221}) &= \underbrace{(\mathrm{DI}, \mathrm{SO})}_{\text{decreasing-inflection source}} \quad \text{if } \bar{x} \in (-\infty, a_{111}), \\
(\bar{x}_1, a_{221}) &= \underbrace{(\mathrm{II}, \mathrm{SO})}_{\text{increasing-inflection source}} \quad \text{if } \bar{x} \in (a_{111}, \infty), \\
&\text{for } a_{110} > 0 \text{ and } a_{220} > 0;
\end{aligned} \tag{4.73}$$

$$\begin{aligned}
(\bar{x}_1, a_{221}) &= \underbrace{(\mathrm{II}, \mathrm{SO})}_{\text{increasing-inflection source}} \quad \text{if } \bar{x} \in (-\infty, a_{111}), \\
(\bar{x}_1, a_{221}) &= \underbrace{(\mathrm{DI}, \mathrm{SO})}_{\text{decreasing-inflection source}} \quad \text{if } \bar{x} \in (a_{111}, \infty), \\
&\text{for } a_{110} < 0 \text{ and } a_{220} > 0;
\end{aligned} \tag{4.74}$$

$$\begin{aligned}
(\bar{x}_1, a_{221}) &= \underbrace{(\mathrm{II}, \mathrm{SI})}_{\text{increasing-inflection sink}} \quad \text{if } \bar{x} \in (-\infty, a_{111}), \\
(\bar{x}_1, a_{221}) &= \underbrace{(\mathrm{DI}, \mathrm{SI})}_{\text{decreasing-inflection sink}} \quad \text{if } \bar{x} \in (a_{111}, \infty), \\
&\text{for } a_{110} > 0 \text{ and } a_{220} < 0;
\end{aligned} \tag{4.75}$$

$$\begin{aligned}
(\bar{x}_1, a_{221}) &= \underbrace{(\mathrm{DI}, \mathrm{SI})}_{\text{decreasing-inflection sink}} \quad \text{if } \bar{x} \in (-\infty, a_{111}), \\
(\bar{x}_1, a_{221}) &= \underbrace{(\mathrm{II}, \mathrm{SI})}_{\text{increasing-inflection sink}} \quad \text{if } \bar{x} \in (a_{111}, \infty), \\
&\text{for } a_{110} < 0 \text{ and } a_{220} < 0.
\end{aligned} \tag{4.76}$$

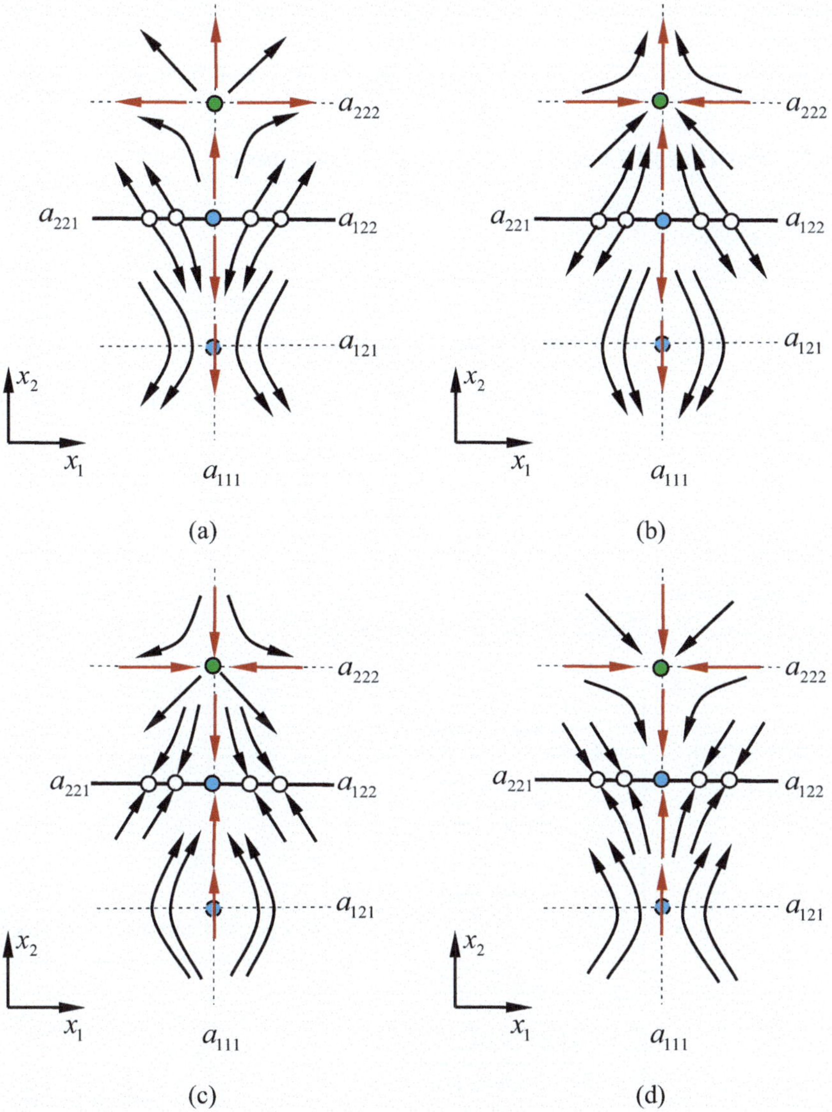

Fig. 4.13 Phase portraits ($a_{121} = a_{221}$) for two-dimensional systems on the x_1-direction with $x_1^* = a_{111}$ and on the x_2-direction with $x_2^* = a_{121}, a_{122}, a_{221}, a_{222}$. The equilibriums flows: (**a**) ($a_{110} > 0$, $a_{220} > 0$), (**b**) ($a_{110} < 0$, $a_{220} > 0$), (**c**) ($a_{110} > 0$, $a_{220} < 0$), (**d**) ($a_{110} < 0$, $a_{220} < 0$)

The corresponding inflection infinite-equilibriums are sketched in Fig. 4.13. The hyperbolic-flow or hyperbolic-secant flows are on the lower sides of the up-down, down-up upper-saddle and lower-saddle infinite-equilibriums. The infinite-equilibriums are for the switching bifurcations of the saddle-sink and saddle-source with hyperbolic and hyperbolic-secant flows.

Similarly, the phase portraits for $a_{121}=a_{22s_1}=a_{222}$ and $a_{122}=a_{22s_1}=a_{222}$ can be completed.

4.2.3 *Paralleled Double Switching Bifurcations*

Consider a dynamical system as

$$\begin{aligned}\dot{x}_1&=a_{110}(x_1-a_{111})(x_2-a_{121})(x_2-a_{122}),\\ \dot{x}_2&=a_{220}(x_2-a_{22s_1})(x_2-a_{22s_2})^2.\end{aligned} \tag{4.77}$$

For $a_{12l_1}=a_{22s_1}$ and $a_{12l_2}=a_{22s_2}$ ($l_1,\ l_2\in\{1,2\}$), the corresponding first integral manifold is

$$\ln\frac{|\,x_1-a_{111}\,|}{|\,x_{10}-a_{111}\,|}=\frac{a_{110}}{a_{220}}\ln\frac{|\,x_2-a_{22s_2}\,|}{|\,x_{20}-a_{22s_2}\,|} \tag{4.78}$$

(i) The equilibriums of $x_2^*=a_{121},a_{122},a_{221},a_{222}$ with $x_1^*=a_{111}$ for $a_{122}=a_{22s_2}=a_{222}$ and $a_{121}=a_{22s_1}=a_{221}$ have the following properties.

$$\left\{\begin{matrix}(a_{111},a_{222})\\(a_{111},a_{221})\end{matrix}\right\}=\left\{\begin{matrix}\underbrace{(\text{UD}:\text{DU},\text{US})}_{\text{up-down-to-down-up upper-saddle}}\\ \underbrace{(\text{II}:\text{DI},\text{SO})}_{\text{inflection source}}\end{matrix}\right\}\text{ for }a_{110}>0\text{ and }a_{220}>0; \tag{4.79}$$

$$\left\{\begin{matrix}(a_{111},a_{222})\\(a_{111},a_{221})\end{matrix}\right\}=\left\{\begin{matrix}\underbrace{(\text{DU}:\text{UD},\text{US})}_{\text{down-up-to-up-down upper-saddle}}\\ \underbrace{(\text{DI}:\text{II},\text{SO})}_{\text{inflection source}}\end{matrix}\right\}\text{ for }a_{110}<0\text{ and }a_{220}>0; \tag{4.80}$$

$$\left\{\begin{matrix}(a_{111},a_{222})\\(a_{111},a_{221})\end{matrix}\right\}=\left\{\begin{matrix}\underbrace{(\text{DU}:\text{UD},\text{LS})}_{\text{down-up-to-up-down lower-saddle}}\\ \underbrace{(\text{DI}:\text{II},\text{SI})}_{\text{inflection sink}}\end{matrix}\right\}\text{ for }a_{110}>0\text{ and }a_{220}<0; \tag{4.81}$$

$$\left\{\begin{matrix}(a_{111},a_{222})\\(a_{111},a_{221})\end{matrix}\right\}=\left\{\begin{matrix}\underbrace{(\text{UD}:\text{DU},\text{LS})}_{\text{up-down-to-down-up lower-saddle}}\\ \underbrace{(\text{II}:\text{DI},\text{SI})}_{\text{inflection sink}}\end{matrix}\right\}\text{ for }a_{110}<0\text{ and }a_{220}<0. \tag{4.82}$$

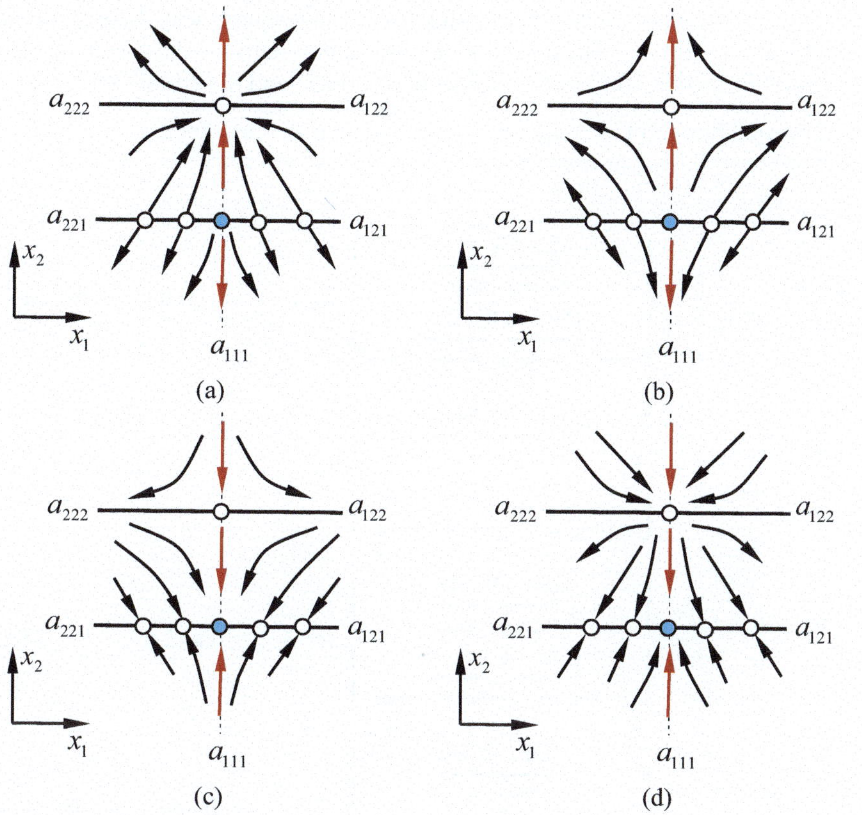

Fig. 4.14 Phase portraits ($a_{122} = a_{221}$, $a_{221} = a_{121}$) for two-dimensional systems on the x_1-direction with $x_1^* = a_{111}$ and on the x_2-direction with$x_2^* = a_{121}, a_{122}, a_{221}, a_{222}$. The equilibriums flows: **(a)** ($a_{110} > 0$, $a_{220} > 0$), **(b)** ($a_{110} < 0$, $a_{220} > 0$), **(c)** ($a_{110} > 0$, $a_{220} < 0$) , **(d)** ($a_{110} < 0$, $a_{220} < 0$)

Table 4.1 Inflection and down-up and up-down infinite-equilibrium ($a_{121} = a_{22s_1} = a_{221}$, $a_{122} = a_{22s_2} = a_{222}$)

		$\bar{x}_1$ for $(\bar{x}_1, a_{221})$		$\bar{x}_1$ for $(\bar{x}_1, a_{222})$	
Fig. 4.14	(a_{110}, a_{220})	$(-\infty, a_{111})$	(a_{111}, ∞)	$(-\infty, a_{111})$	(a_{111}, ∞)
(a)	(+,+)	(II,SO)	(DI,SO)	(UD,US)	(DU,US)
(b)	(−,+)	(DI,SO)	(II,SO)	(DU,US)	(UD,US)
(c)	(+,−)	(DI,SI)	(II,SI)	(UD,LS)	(UD,LS)
(d)	(−,−)	(II,SI)	(DI,SI)	(UD,LS)	(DU,LS)

The infinite-equilibriums of $x_2^* = a_{121} = a_{22s_1} = a_{221}$ and $x_2^* = a_{122} = a_{22s_2} = a_{222}$ are tabulated in Table 4.1. The inflection, up-down and down-up infinite-equilibriums are sketched in Fig. 4.14. The up-down and down-up infinite-equilibriums are for the switching bifurcations of the saddle-sink and saddle-source with hyperbolic and hyperbolic-secant flows. The inflection infinite-equilibriums are the switching bifurcations of the saddle, sink, and source with hyperbolic and hyperbolic-secant flows.

(ii) The equilibriums of $x_2^* = a_{121}, a_{122}, a_{221}, a_{222}$ with $x_1^* = a_{111}$ for $a_{122} = a_{22s_1} = a_{222}$ and $a_{121} = a_{22s_2} = a_{221}$ have the following properties.

$$\left\{\begin{matrix}(a_{111}, a_{222})\\(a_{111}, a_{221})\end{matrix}\right\} = \left\{\begin{matrix}\underbrace{(\text{DI : II, SO})}_{\text{inflection source}}\\\underbrace{(\text{UD : DU, LS})}_{\text{up-down-to-down-up lower-saddle}}\end{matrix}\right\} \text{ for } a_{110} > 0 \text{ and } a_{220} > 0; \quad (4.83)$$

$$\left\{\begin{matrix}(a_{111}, a_{222})\\(a_{111}, a_{221})\end{matrix}\right\} = \left\{\begin{matrix}\underbrace{(\text{II : DI, SO})}_{\text{inflection source}}\\\underbrace{(\text{DU : UD, LS})}_{\text{down-up-to-up-down lower-saddle}}\end{matrix}\right\} \text{ for } a_{110} < 0 \text{ and } a_{220} > 0; \quad (4.84)$$

$$\left\{\begin{matrix}(a_{111}, a_{222})\\(a_{111}, a_{221})\end{matrix}\right\} = \left\{\begin{matrix}\underbrace{(\text{II : DI, SI})}_{\text{inflection sink}}\\\underbrace{(\text{DU : UD, US})}_{\text{down-up-to-up-down upper-saddle}}\end{matrix}\right\} \text{ for } a_{110} > 0 \text{ and } a_{220} < 0; \quad (4.85)$$

$$\left\{\begin{matrix}(a_{111}, a_{222})\\(a_{111}, a_{221})\end{matrix}\right\} = \left\{\begin{matrix}\underbrace{(\text{DI : II, SI})}_{\text{inflection sink}}\\\underbrace{(\text{UD : DU, US})}_{\text{up-down-to-down-up upper-saddle}}\end{matrix}\right\} \text{ for } a_{110} < 0 \text{ and } a_{220} < 0. \quad (4.86)$$

The infinite-equilibriums of $x_2^* = a_{121} = a_{22s_2} = a_{221}$ and $x_2^* = a_{122} = a_{22s_1} = a_{222}$ are tabulated in Table 4.2. The inflection, up-down and down-up infinite-equilibriums are sketched in Fig. 4.15. The up-down and down-up infinite-equilibriums are for the switching bifurcations of the saddle-sink and saddle-source with hyperbolic and hyperbolic-secant flows. The inflection infinite-equilibriums are the switching bifurcations of the saddle, sink, and source with hyperbolic and hyperbolic-secant flows.

Table 4.2 Inflection and down-up and up-down infinite-equilibrium ($a_{121} = a_{22s_2} = a_{221}$, $a_{122} = a_{22s_1} = a_{222}$)

		$\bar{x}_1$ for $(\bar{x}_1, a_{222})$		$\bar{x}_1$ for $(\bar{x}_1, a_{221})$	
Fig. 4.15	(a_{110}, a_{220})	$(-\infty, a_{111})$	(a_{111}, ∞)	$(-\infty, a_{111})$	(a_{111}, ∞)
(a)	(+,+)	(DI,SO)	(II,SO)	(UD,LS)	(DU,LS)
(b)	(−,+)	(II,SO)	(DI,SO)	(DU,LS)	(UD,LS)
(c)	(+,−)	(II,SI)	(DI,SI)	(DU,US)	(UD,US)
(d)	(−,−)	(DI,SI)	(II,SI)	(UD,US)	(DU,US)

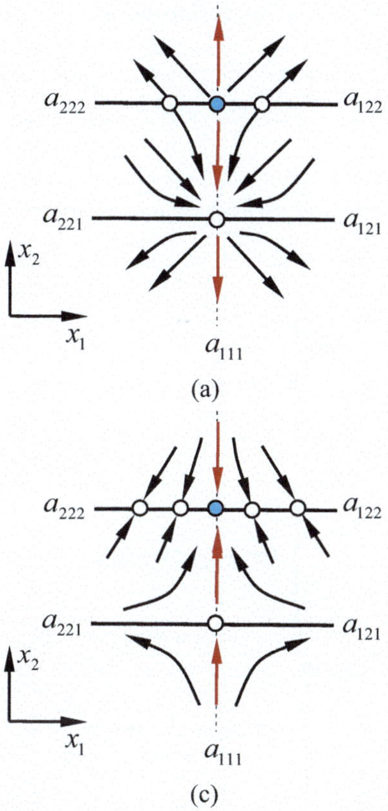

Fig. 4.15 Phase portraits ($a_{122} = a_{222}$, $a_{221} = a_{121}$) for two-dimensional systems on the x_1-direction with $x_1^* = a_{111}$ and on the x_2-direction with $x_2^* = a_{121}, a_{122}, a_{221}, a_{222}$. The equilibriums flows: **(a)** ($a_{110} > 0$, $a_{220} > 0$), **(b)** ($a_{110} < 0$, $a_{220} > 0$), **(c)** ($a_{110} > 0$, $a_{220} < 0$), **(d)** ($a_{110} < 0$, $a_{220} < 0$)

4.2.4 *Increasing and Decreasing-Concave Switching Bifurcations*

Consider a dynamical system as

$$\begin{aligned} \dot{x}_1 &= a_{110}(x_1 - a_{111})(x_2 - a_{121})(x_2 - a_{122}), \\ \dot{x}_2 &= a_{220}(x_2 - a_{221})^3. \end{aligned} \tag{4.87}$$

For $a_{12l_1} = a_{221}$ ($l_1, l_2 \in \{1,2\}$), the corresponding first integral manifold is

$$\ln \frac{|x_1 - a_{111}|}{|x_{10} - a_{111}|} = \frac{a_{110}}{a_{220}} \left[\ln \frac{|x_2 - a_{221}|}{|x_{20} - a_{221}|} - (a_{221} - a_{12l_1})(\frac{1}{x_2 - a_{221}} - \frac{1}{x_{20} - a_{221}}) \right]. \tag{4.88}$$

(i) The equilibriums of $x_2^* = a_{121}, a_{122}, a_{221}$ with $x_1^* = a_{111}$ for $a_{121} = a_{221}$ have the following properties.

$$\left\{ \begin{array}{l} (a_{111}, a_{122}) \\ (a_{111}, a_{221}) \end{array} \right\} = \left\{ \begin{array}{c} \underbrace{(\mathrm{DP}:\mathrm{UP},\mathrm{pF})}_{\text{hyperbolic flow } (+)} \\ \underbrace{(\mathrm{IC}:\mathrm{DC},\mathrm{SO})}_{\text{increasing-to-decreasing concave source}} \end{array} \right\} \text{ for } a_{110} > 0 \text{ and } a_{220} > 0; \tag{4.89}$$

$$\left\{ \begin{array}{l} (a_{111}, a_{122}) \\ (a_{111}, a_{221}) \end{array} \right\} = \left\{ \begin{array}{c} \underbrace{(\mathrm{UP}:\mathrm{DP},\mathrm{pF})}_{\text{hyperbolic-secant flow } (+)} \\ \underbrace{(\mathrm{DC}:\mathrm{IC},\mathrm{SO})}_{\text{decreasing-to-increasing concave source}} \end{array} \right\} \text{ for } a_{110} < 0 \text{ and } a_{220} > 0; \tag{4.90}$$

$$\left\{ \begin{array}{l} (a_{111}, a_{122}) \\ (a_{111}, a_{221}) \end{array} \right\} = \left\{ \begin{array}{c} \underbrace{(\mathrm{UP}:\mathrm{DP},\mathrm{nF})}_{\text{hyperbolic-secant flow } (-)} \\ \underbrace{(\mathrm{DC}:\mathrm{IC},\mathrm{SI})}_{\text{decreasing-to-increasing concave sink}} \end{array} \right\} \text{ for } a_{110} > 0 \text{ and } a_{220} < 0; \tag{4.91}$$

$$\left\{ \begin{array}{l} (a_{111}, a_{122}) \\ (a_{111}, a_{221}) \end{array} \right\} = \left\{ \begin{array}{c} \underbrace{(\mathrm{DP}:\mathrm{UP},\mathrm{nF})}_{\text{hyperbolic flow } (-)} \\ \underbrace{(\mathrm{IC}:\mathrm{DC},\mathrm{SI})}_{\text{increasing-to-decreasing concave sink}} \end{array} \right\} \text{ for } a_{110} < 0 \text{ and } a_{220} < 0. \tag{4.92}$$

The infinite-equilibrium of $x_2^* = a_{121} = a_{221}$ is tabulated in Table 4.3. The decreasing and increasing-concave source and sink infinite-equilibriums are sketched in Fig. 4.16. The decreasing and increasing-concave infinite-equilibriums are for the switching bifurcations of the third-order saddle, source, and sink with hyperbolic and hyperbolic-secant flows.

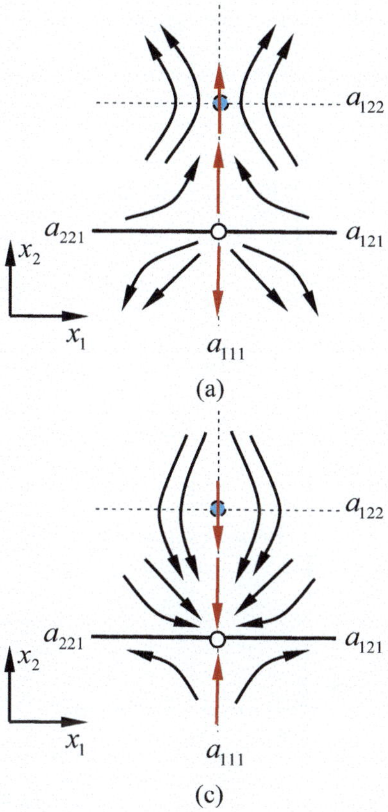

Fig. 4.16 Phase portraits ($a_{121} = a_{221}$) for two-dimensional systems on the x_1-direction with $x_1^* = a_{111}$ and on the x_2-direction with $x_2^* = a_{121}, a_{122}, a_{221}, a_{222}$. The equilibriums flows: (**a**) ($a_{110} > 0$, $a_{220} > 0$), (**b**) ($a_{110} < 0$, $a_{220} > 0$), (**c**) ($a_{110} > 0$, $a_{220} < 0$), (**d**) ($a_{110} < 0$, $a_{220} < 0$)

Table 4.3 Increasing and decreasing concave source and sink infinite-equilibriums ($a_{122} = a_{221}$)

	$(\bar{x}_1, a_{221})$	$\bar{x}_1$	
Fig. 4.16	(a_{110}, a_{220})	$(-\infty, a_{111})$	(a_{111}, ∞)
(a)	(+,+)	(DC,SO)	(IC,SO)
(b)	(−,+)	(IC,SO)	(DC,SO)
(c)	(+,−)	(IC,SI)	(DC,SI)
(d)	(−,−)	(DC,SI)	(IC,SI)

(ii) The equilibriums of $x_2^* = a_{121}, a_{122}, a_{221}$ with $x_1^* = a_{111}$ for $a_{122} = a_{221}$ have the following properties.

$$\left\{\begin{matrix} (a_{111}, a_{221}) \\ (a_{111}, a_{121}) \end{matrix}\right\} = \left\{\begin{matrix} \underbrace{(\text{DC}:\text{IC}, \text{SO})}_{\text{decreasing-to-increasing concave source}} \\ \underbrace{(\text{DP}:\text{UP}, \text{nF})}_{\text{hyperbolic flow }(-)} \end{matrix}\right\} \text{ for } a_{110} > 0 \text{ and } a_{220} > 0; \tag{4.89}$$

$$\left\{\begin{matrix} (a_{111}, a_{221}) \\ (a_{111}, a_{121}) \end{matrix}\right\} = \left\{\begin{matrix} \underbrace{(\text{IC}:\text{DC}, \text{SO})}_{\text{increasing-to-decreasing concave source}} \\ \underbrace{(\text{UP}:\text{DP}, \text{nF})}_{\text{hyperbolic-secant flow }(-)} \end{matrix}\right\} \text{ for } a_{110} < 0 \text{ and } a_{220} > 0; \tag{4.90}$$

$$\left\{\begin{matrix} (a_{111}, a_{221}) \\ (a_{111}, a_{121}) \end{matrix}\right\} = \left\{\begin{matrix} \underbrace{(\text{IC}:\text{DC}, \text{SI})}_{\text{increasing-to-decreasing concave sink}} \\ \underbrace{(\text{UP}:\text{DP}, \text{pF})}_{\text{hyperbolic-secant flow }(+)} \end{matrix}\right\} \text{ for } a_{110} > 0 \text{ and } a_{220} < 0; \tag{4.91}$$

$$\left\{\begin{matrix} (a_{111}, a_{221}) \\ (a_{111}, a_{121}) \end{matrix}\right\} = \left\{\begin{matrix} \underbrace{(\text{DC}:\text{IC}, \text{SI})}_{\text{decreasing-to-increasing concave sink}} \\ \underbrace{(\text{DP}:\text{UP}, \text{pF})}_{\text{hyperbolic flow }(+)} \end{matrix}\right\} \text{ for } a_{110} < 0 \text{ and } a_{220} < 0. \tag{4.92}$$

The infinite-equilibrium of $x_2^* = a_{122} = a_{221}$ is tabulated in Table 4.4. The decreasing and increasing concave source and sink infinite-equilibriums are sketched in Fig. 4.17.

Table 4.4 Increasing and decreasing concave source and sink infinite-equilibriums ($a_{122} = a_{221}$)

	$(\overline{x}_1, a_{221})$	$\overline{x}_1$	
Fig. 4.17	(a_{110}, a_{220})	$(-\infty, a_{111})$	(a_{111}, ∞)
(a)	(+,+)	(DC,SO)	(IC,SO)
(b)	(−,+)	(IC,SO)	(DC,SO)
(c)	(+,−)	(IC,SI)	(DC,SI)
(d)	(−,−)	(DC,SI)	(IC,SI)

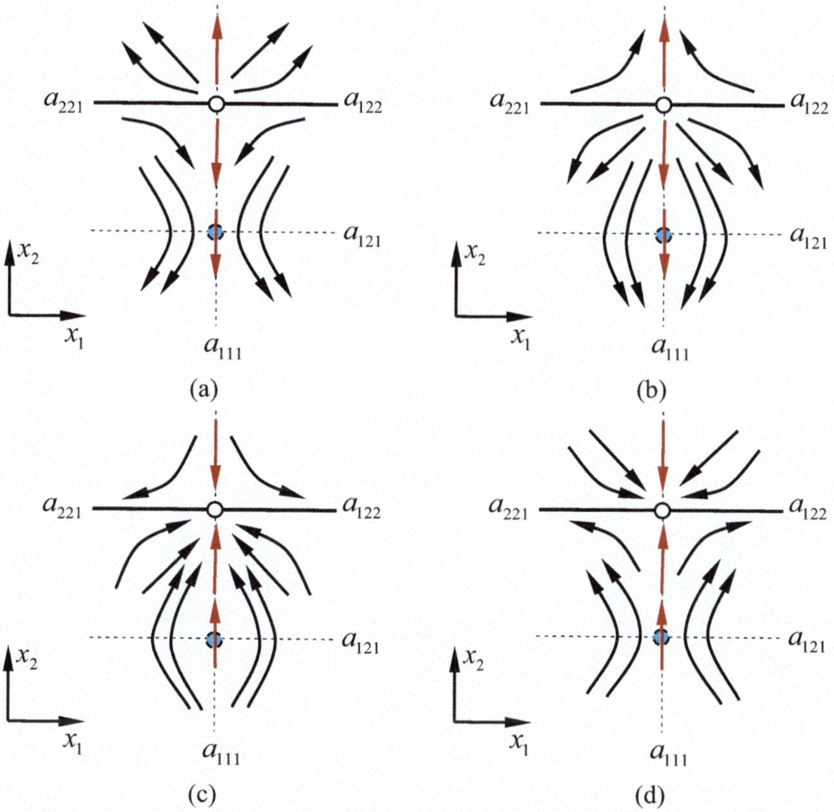

Fig. 4.17 Phase portraits ($a_{122} = a_{221}$) for two-dimensional systems on the x_1-direction with $x_1^* = a_{111}$ and on the x_2-direction with $x_2^* = a_{121}, a_{122}, a_{221}, a_{222}$. The equilibriums flows: (**a**) ($a_{110} > 0$, $a_{220} > 0$), (**b**) ($a_{110} < 0$, $a_{220} > 0$), (**c**) ($a_{110} > 0$, $a_{220} < 0$), (**d**) ($a_{110} < 0$, $a_{220} < 0$)

Chapter 5
Simple Equilibrium Series and Switching Dynamics

In this chapter, simple equilibrium series with connected hyperbolic flows are presented, and the corresponding switching dynamics are discussed through the inflection-source and sink infinite-equilibriums. The inflection-source is the switching bifurcation for a hyperbolic-secant flow with a connected source and for a hyperbolic flow with a connected saddle. The inflection-sink is the switching bifurcation for a hyperbolic-secant flow with connected sink and for a hyperbolic flow with a connected saddle. The double switching bifurcations are generated by the two inflection-source and sink infinite-equilibriums.

5.1 Equilibrium Series with Connected Hyperbolic Flows

Consider a dynamical system as

$$\begin{aligned}\dot{x}_1 &= a_{110}(x_1 - a_{111})(x_2 - a_{121})(x_2 - a_{122}),\\ \dot{x}_2 &= a_{220}(x_2 - a_{221})(x_2 - a_{222})(x_2 - a_{223}),\end{aligned} \tag{5.1}$$

and the corresponding first integral manifold is

$$\ln\frac{|x_1 - a_{111}|}{|x_{10} - a_{111}|} = \frac{a_{110}}{a_{220}}\sum\nolimits_{s_1=1}^{3}\frac{\prod_{l_2=1}^{2}(a_{22s_1} - a_{12l_2})}{\prod_{s_2=1,s_2\neq s_1}^{3}(a_{22s_1} - a_{22s_2})}\ln\frac{|x_2 - a_{22s_1}|}{|x_{20} - a_{22s_1}|}. \tag{5.2}$$

(i) For $a_{122} < a_{221}$, the equilibriums of $x_1^* = a_{111}$ with $x_2^* = a_{121}, a_{122}, a_{221}, a_{222}, a_{223}$ have the following properties, as shown in Fig. 5.1. The saddle and source or sink with hyperbolic and hyperbolic-secant flows exist on the series of equilibriums. The hyperbolic and hyperbolic-secant flows with directrix are connected together on the bottom. The saddle and source or sink on the top are from the

A. C. J. Luo, *Two-dimensional Self and Product Cubic Systems, Vol. I*,
https://doi.org/10.1007/978-3-031-57096-4_5

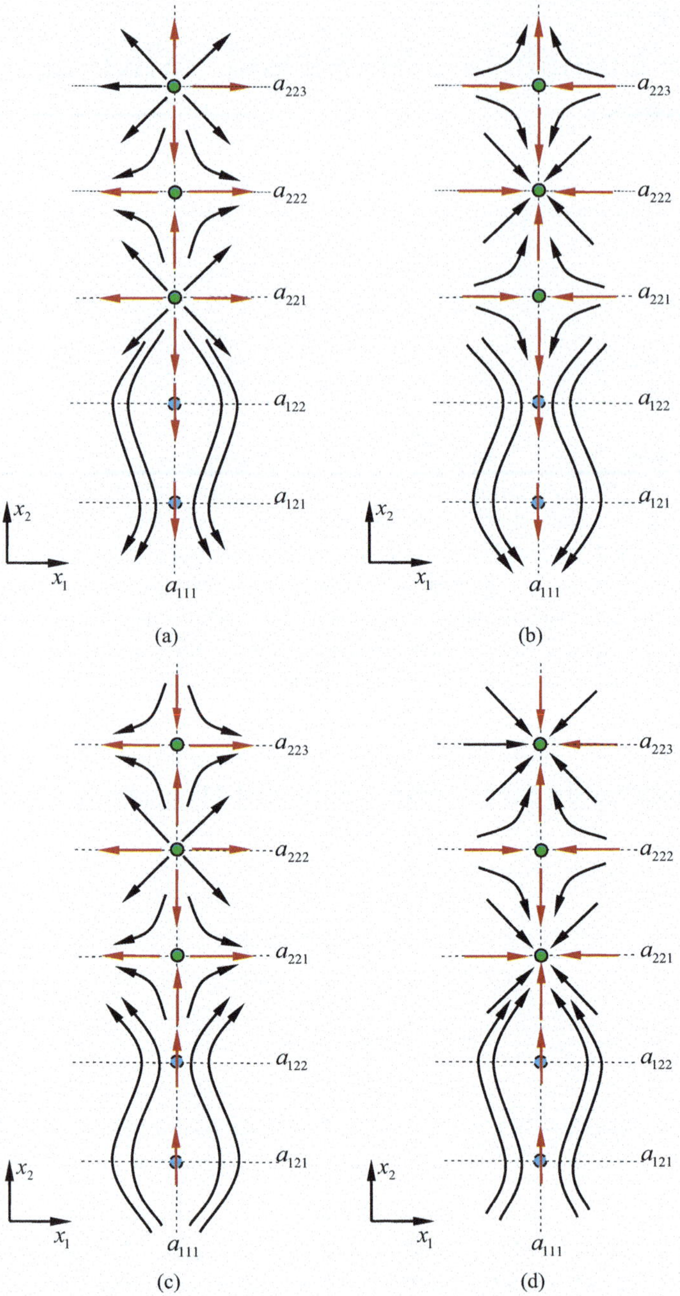

Fig. 5.1 Phase portraits ($a_{122} < a_{221}$) for two-dimensional systems on the x_1-direction with $x_1^* = a_{111}$ and on the x_2-direction with $x_2^* = a_{121}, a_{122}, a_{221}, a_{222}, a_{223}$. The equilibriums flows: (**a**) ($a_{110} > 0$, $a_{220} > 0$), (**b**) ($a_{110} < 0$, $a_{220} > 0$), (**c**) ($a_{110} > 0$, $a_{220} < 0$), (**d**) ($a_{110} < 0$, $a_{220} < 0$)

Table 5.1 Equilibriums on the line of $x_1^* = a_{111}$ for $a_{121} < a_{122} \in (a_{221}, a_{222})$

Fig.5.1	(a_{111}, x_2^*)	x_2^*				
	(a_{110}, a_{220})	a_{121}	a_{122}	a_{221}	a_{222}	a_{223}
(a)	(+,+)	(DP:UP,nF)	(UP:DP,nF)	(SO,SO)	(SO,SI)	(SO,SO)
(b)	(–,+)	(UP:DP,nF)	(DP:UP,nF)	(SI,SO)	(SI,SI)	(SI,SO)
(c)	(+,–)	(UP:DP,pF)	(DP:UP,pF)	(SO,SI)	(SO,SO)	(SO,SI)
(d)	(–,–)	(DP:UP,pF)	(UP:DP,pF)	(SI,SI)	(SI,SO)	(SI,SI)

appearing bifurcation of saddle–saddle or saddle–sink. The properties of the equilibriums on the line of $x_1^* = a_{111}$ are listed in Table 5.1.

(ii) For $a_{221} < a_{121} < a_{122} < a_{222}$, the equilibriums of $x_1^* = a_{111}$ with $x_2^* = a_{121}$, a_{122}, a_{221}, a_{222}, a_{223} have the following properties, as shown in Fig. 5.2. Compared to the case of $a_{122} < a_{221}$, the saddle, source, or sink with hyperbolic and hyperbolic-secant flows exist on the series of equilibriums. Only the hyperbolic and hyperbolic-secant flows are switched. The properties of the equilibriums on the line of $x_1^* = a_{111}$ are listed in Table 5.2.

(iii) For $a_{222} < a_{121} < a_{122} < a_{223}$, the equilibriums of $x_1^* = a_{111}$ with $x_2^* = a_{121}$, a_{122}, a_{221}, a_{222}, a_{223} have the following properties, as shown in Fig. 5.3. Compared to the case of $a_{122} < a_{221}$, the saddle, source, or sink with hyperbolic and hyperbolic-secant flows on the series of equilibriums are changed. Only the locations of equilibriums are changed. The properties of the equilibriums on the line of $x_1^* = a_{111}$ are listed in Table 5.3.

(iv) For $a_{223} < a_{121}$, the equilibriums of $x_1^* = a_{111}$ with $x_2^* = a_{121}, a_{122}, a_{221}, a_{222}, a_{223}$ have the following properties, as shown in Fig. 5.4. Compared to the case of $a_{122} < a_{221}$, the saddle, source, or sink with hyperbolic and hyperbolic-secant flows exist on the series of equilibriums. Only the hyperbolic and hyperbolic-secant flows are switched. The properties of the equilibriums on the line of $x_1^* = a_{111}$ are listed in Table 5.4.

(v) For $a_{121} < a_{221}$ and $a_{122} \in (a_{221}, a_{222})$, the equilibriums of $x_1^* = a_{111}$ with $x_2^* = a_{121}$, a_{122}, a_{221}, a_{222}, a_{223} have the following properties, as shown in Fig. 5.5. Compared to the case of $a_{122} < a_{221}$, the two hyperbolic flows are separated by the saddle, and the two hyperbolic-secant flows are separated by the sink or source. source, and sink. The properties of the equilibriums on the line of $x_1^* = a_{111}$ are listed in Table 5.5.

(vi) For $a_{121} < a_{221}$ and $a_{122} \in (a_{222}, a_{223})$, the equilibriums of $x_1^* = a_{111}$ with $x_2^* = a_{121}$, a_{122}, a_{221}, a_{222}, a_{223} have the following properties, as shown in Fig. 5.6. The hyperbolic and hyperbolic-secant flows are separated by one of the saddle, source or sink. The properties of the equilibriums on the line of $x_1^* = a_{111}$ are listed in Table 5.6.

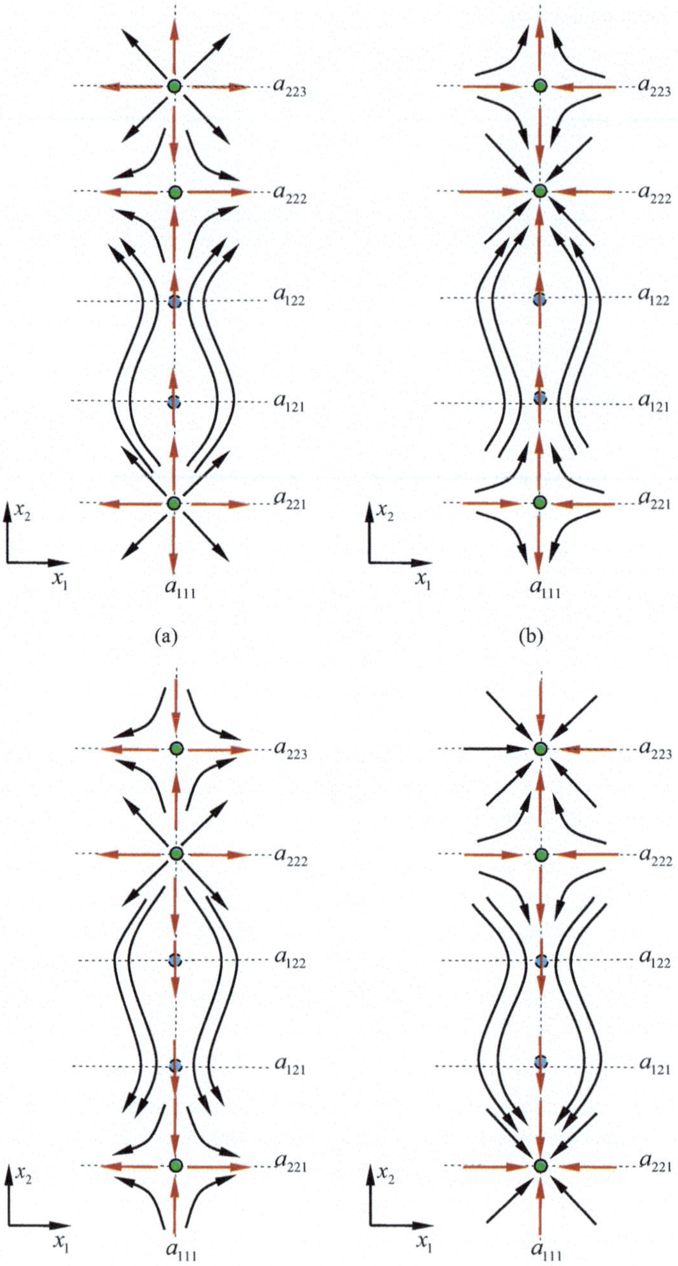

Fig. 5.2 Phase portraits ($a_{221} < a_{121} < a_{122} < a_{222}$) for two-dimensional systems on the x_1-direction with $x_1^* = a_{111}$ and on the x_2-direction with $x_2^* = a_{121}, a_{122}, a_{221}, a_{222}, a_{223}$. The equilibriums flows: (**a**) ($a_{110} > 0, a_{220} > 0$), (**b**) ($a_{110} < 0, a_{220} > 0$), (**c**) ($a_{110} > 0, a_{220} < 0$), (**d**) ($a_{110} < 0$, $a_{220} < 0$)

Table 5.2 Equilibriums on the line of $x_1^* = a_{111}$ for $a_{121}, a_{122} \in (a_{221}, a_{222})$

Fig.5.2	(a_{111}, x_2^*)	x_2^*				
	(a_{110}, a_{220})	a_{221}	a_{121}	a_{122}	a_{222}	a_{223}
(a)	(+,+)	(SO,SO)	(UP:DP,pF)	(DP:UP,pF)	(SO,SI)	(SO,SO)
(b)	(–,+)	(SI,SO)	(DP:UP,pF)	(UP:DP,pF)	(SI,SI)	(SI,SO)
(c)	(+,–)	(SO,SI)	(DP:UP,nF)	(UP:DP,nF)	(SO,SO)	(SO,SI)
(d)	(–,–)	(SI,SI)	(UP:DP,nF)	(DP:UP,nF)	(SI,SO)	(SI,SI)

(vii) For $a_{121} < a_{221}$ and $a_{223} < a_{122}$, the equilibriums of $x_1^* = a_{111}$ with $x_2^* = a_{121}$, a_{122}, a_{221}, a_{222}, a_{223} have the following properties, as shown in Fig. 5.7. The two hyperbolic or hyperbolic-secant flows are at the bottom and top of the equilibrium series. The properties of the equilibriums on the line of $x_1^* = a_{111}$ are listed in Table 5.7.

(viii) For $a_{121} \in (a_{221}, a_{222})$ and $a_{122} \in (a_{222}, a_{223})$, the equilibriums of $x_1^* = a_{111}$ with $x_2^* = a_{121}, a_{122}, a_{221}, a_{222}, a_{223}$ have the following properties, as shown in Fig. 5.8. Compared to the case of $a_{122} < a_{221}$, the source, sink, and source or sink, source and sink are separated by the two hyperbolic-secant flows. The three saddles are separated by the two hyperbolic flows. The properties of the equilibriums on the line of $x_1^* = a_{111}$ are listed in Table 5.8.

(ix) For $a_{121} \in (a_{221}, a_{222})$ and $a_{223} < a_{122}$, the equilibriums of $x_1^* = a_{111}$ with $x_2^* = a_{121}, a_{122}, a_{221}, a_{222}, a_{223}$ have the following properties, as shown in Fig. 5.9. The hyperbolic and hyperbolic-secant flows are separated by the sink, source and saddle. The properties of the equilibriums on the line of $x_1^* = a_{111}$ are listed in Table 5.9.

(x) For $a_{121} \in (a_{222}, a_{223})$ and $a_{223} < a_{122}$, the equilibriums of $x_1^* = a_{111}$ with $x_2^* = a_{121}$, a_{122}, a_{221}, a_{222}, a_{223} have the following properties, as shown in Fig. 5.10. The two hyperbolic flows are separated by the saddle, and the two hyperbolic-secant flows are separated by the source or sink. The properties of the equilibriums on the line of $x_1^* = a_{111}$ are listed in Table 5.10.

5.2 Inflection-Source and Sink Infinite-Equilibriums

For the above series of simple equilibriums, there are many switching bifurcations between two of the adjacent equilibrium series. The single infinite-equilibrium is discussed first, and then the double infinite-equilibriums are discussed for the switching bifurcations.

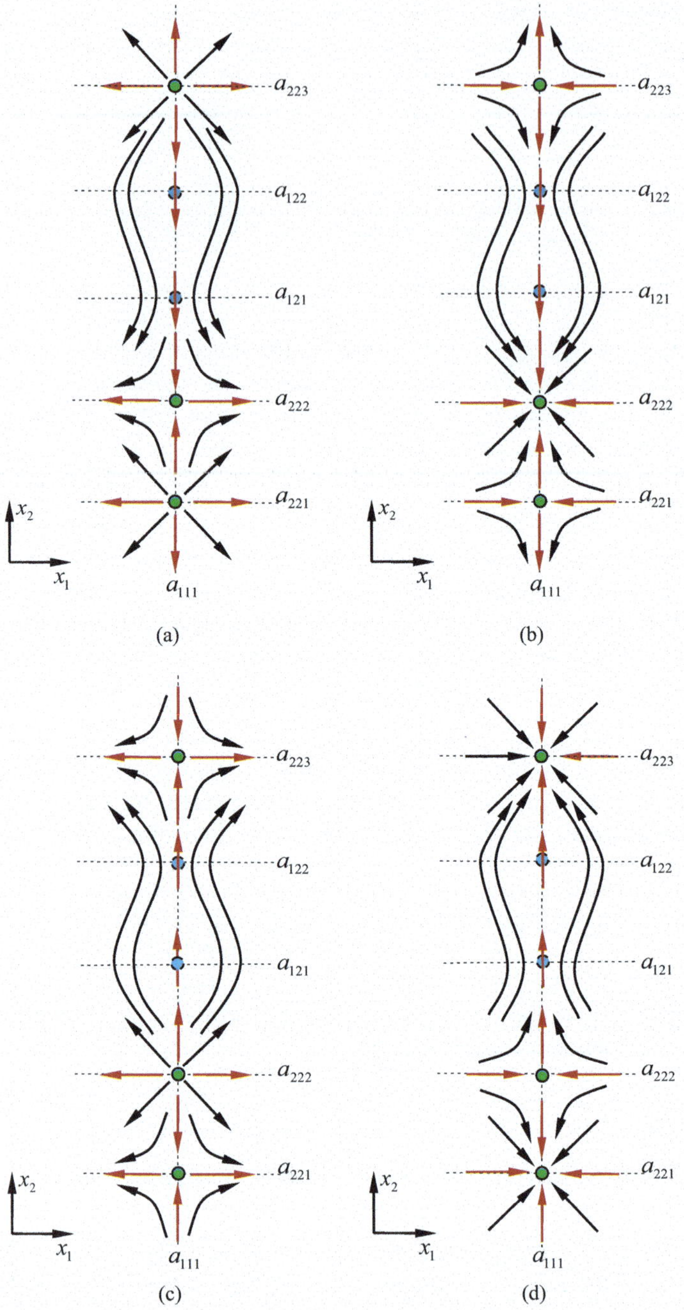

Fig. 5.3 Phase portraits ($a_{222} < a_{121} < a_{122} < a_{223}$) for two-dimensional systems on the x_1-direction with $x_1^* = a_{111}$ and on the x_2-direction with $x_2^* = a_{121}, a_{122}, a_{221}, a_{222}, a_{223}$. The equilibriums flows: **(a)** ($a_{110} > 0, a_{220} > 0$), **(b)** ($a_{110} < 0, a_{220} > 0$), **(c)** ($a_{110} > 0, a_{220} < 0$), **(d)** ($a_{110} < 0, a_{220} < 0$)

Table 5.3 Equilibriums on the line of $x_1^* = a_{111}$ for $a_{222} < a_{121} < a_{122} < a_{223}$

Fig.5.3	(a_{111}, x_2^*)	x_2^*				
	(a_{110}, a_{220})	a_{221}	a_{222}	a_{121}	a_{122}	a_{223}
(a)	(+,+)	(SO,SO)	(SO,SI)	(DP:UP,nF)	(UP:DP,nF)	(SO,SO)
(b)	(–,+)	(SI,SO)	(SI,SI)	(UP:DP,nF)	(DP:UP,nF)	(SI,SO)
(c)	(+,–)	(SO,SI)	(SO,SO)	(UP:DP,pF)	(DP:UP,pF)	(SO,SI)
(d)	(–,–)	(SI,SI)	(SI,SO)	(DP:UP,pF)	(UP:DP,pF)	(SI,SI)

5.2.1 *Single Inflection-Source and Sink Infinite-Equilibriums*

Consider a dynamical system as

$$\begin{aligned}\dot{x}_1 &= a_{110}(x_1 - a_{111})(x_2 - a_{121})(x_2 - a_{122}),\\ \dot{x}_2 &= a_{220}(x_2 - a_{221})(x_2 - a_{222})(x_2 - a_{223}),\end{aligned} \tag{5.3}$$

and the corresponding first integral manifold for $a_{12s_1} = a_{22l_1}$ ($l_1, l_2 \in \{1,2\}, l_1 \neq l_2$; $s_1, s_2, s_3 \in \{1,2,3\}, s_1 \neq s_2 \neq s_3$) is

$$\ln\frac{|x_1 - a_{111}|}{|x_{10} - a_{111}|} = \frac{a_{110}}{a_{220}}\sum\nolimits_{s_2=1, s_2 \neq s_1}^{3} \frac{(a_{22s_2} - a_{12l_2})}{(a_{22s_2} - a_{22s_3})} \ln\frac{|x_2 - a_{22s_2}|}{|x_{20} - a_{22s_2}|}. \tag{5.4}$$

Therefore, the 12 cases of equilibrium series with the inflection infinite-equilibriums will be presented. The infinite-equilibrium are for the switching bifurcations of two simple equilibriums.

(i) The phase portraits for the equilibrium series of $x_2^* = a_{121}, a_{122}, a_{221}, a_{222}, a_{223}$ with $x_1^* = a_{111}$ for $a_{221} = a_{121}$ and $a_{122} < a_{222}$ are presented in Fig. 5.11. The behaviors of equilibriums are listed in Table 5.11.

The inflection infinite-equilibrium of $x_2^* = a_{121} = a_{221}$ is summarized as follows.

$$\begin{aligned}(\bar{x}_1, a_{221}) &= \underbrace{(\text{II, SO})}_{\text{increasing-infelction source}} \quad \text{if } \bar{x}_1 \in (-\infty, a_{111}),\\ (\bar{x}_1, a_{221}) &= \underbrace{(\text{DI, SO})}_{\text{decreasing-inflection source}} \quad \text{if } \bar{x}_1 \in (a_{111}, \infty),\end{aligned} \tag{5.5}$$

for $a_{110} > 0, a_{220} > 0$;

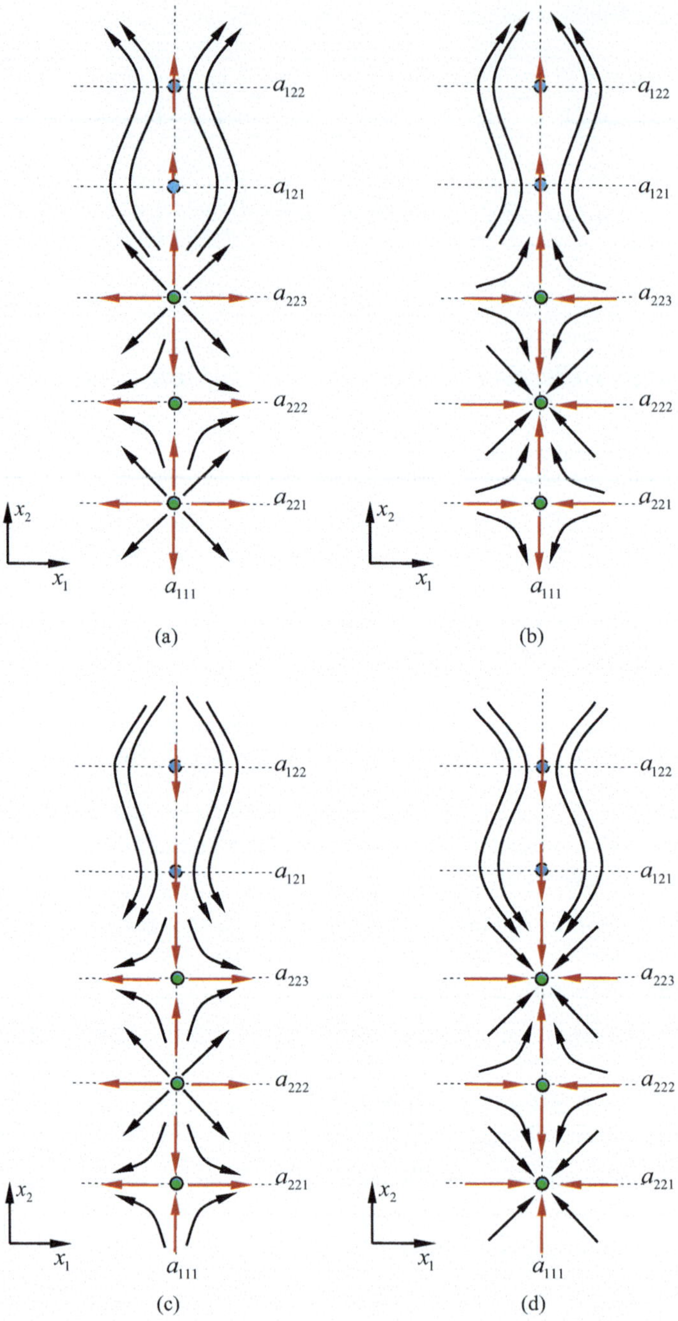

Fig. 5.4 Phase portraits ($a_{223} < a_{221}$) for two-dimensional systems on the x_1-direction with $x_1^* = a_{111}$ and on the x_2-direction with $x_2^* = a_{121}, a_{122}, a_{221}, a_{222}, a_{223}$. The equilibriums flows: (**a**) ($a_{110} > 0$, $a_{220} > 0$), (**b**) ($a_{110} < 0$, $a_{220} > 0$), (**c**) ($a_{110} > 0$, $a_{220} < 0$), (**d**) ($a_{110} < 0$, $a_{220} < 0$)

Table 5.4 Equilibriums on the line of $x_1^* = a_{111}$ for $a_{223} < a_{121}$

Fig.5.4	(a_{111}, x_2^*)	x_2^*				
	(a_{110}, a_{220})	a_{221}	a_{222}	a_{223}	a_{121}	a_{122}
(a)	(+,+)	(SO,SO)	(SO,SI)	(SO,SO)	(UP:DP,pF)	(DP:UP,pF)
(b)	(–,+)	(SI,SO)	(SI,SI)	(SI,SO)	(DP:UP,pF)	(UP:DP,pF)
(c)	(+,–)	(SO,SI)	(SO,SO)	(SO,SI)	(DP:UP,nF)	(UP:DP,nF)
(d)	(–,–)	(SI,SI)	(SI,SO)	(SI,SI)	(UP:DP,nF)	(DP:UP,nF)

$$
\begin{aligned}
(\overline{x}_1, a_{221}) &= \underbrace{(\text{DI, SO})}_{\text{decreasing-infelction source}} \quad \text{if } \overline{x}_1 \in (-\infty, a_{111}), \\
(\overline{x}_1, a_{221}) &= \underbrace{(\text{II, SO})}_{\text{increasing-inflection source}} \quad \text{if } \overline{x}_1 \in (a_{111}, \infty), \\
&\text{for } a_{110} < 0, a_{220} > 0;
\end{aligned}
\tag{5.6}
$$

$$
\begin{aligned}
(\overline{x}_1, a_{221}) &= \underbrace{(\text{DI, SI})}_{\text{decreasing-infelction sink}} \quad \text{if } \overline{x}_1 \in (-\infty, a_{111}), \\
(\overline{x}_1, a_{221}) &= \underbrace{(\text{II, SI})}_{\text{increasing-inflection sink}} \quad \text{if } \overline{x}_1 \in (a_{111}, \infty), \\
&\text{for } a_{110} > 0, a_{220} < 0;
\end{aligned}
\tag{5.7}
$$

$$
\begin{aligned}
(\overline{x}_1, a_{221}) &= \underbrace{(\text{II, SI})}_{\text{increasing-infelction sink}} \quad \text{if } \overline{x}_1 \in (-\infty, a_{111}), \\
(\overline{x}_1, a_{221}) &= \underbrace{(\text{DI, SI})}_{\text{decreasing-inflection sink}} \quad \text{if } \overline{x}_1 \in (a_{111}, \infty), \\
&\text{for } a_{110} < 0, a_{220} < 0.
\end{aligned}
\tag{5.8}
$$

The inflection-source and sink infinite-equilibriums of $x_2^* = a_{121} = a_{221}$ are the switching bifurcations of sink, source, and saddles with hyperbolic flows for $a_{221} \in (a_{121}, a_{122})$ and $a_{221} < a_{121}$ with $a_{122} < a_{222}$, as shown in Fig. 5.11. In addition, the saddle, sink, or source with hyperbolic or hyperbolic-secant flow exists. The infinite-equilibrium is on the bottom of the series of equilibriums.

(ii) The phase portraits for the equilibrium series of $x_2^* = a_{121}, a_{122}, a_{221}, a_{222}, a_{223}$ with $x_1^* = a_{111}$ for $a_{221} = a_{121}$ and $a_{122} \in (a_{222}, a_{223})$ are presented in Fig. 5.12. The behaviors of equilibriums are listed in Table 5.12. The inflection infinite-equilibrium of $x_2^* = a_{121} = a_{221}$ is the same as in Eqs. (5.5)–(5.8). The inflection source and sink infinite-equilibriums of $x_2^* = a_{121} = a_{221}$ are the switching bifurcations of sink, source, and saddles with hyperbolic flows for $a_{221} \in (a_{121}, a_{122})$ and $a_{221} < a_{121}$

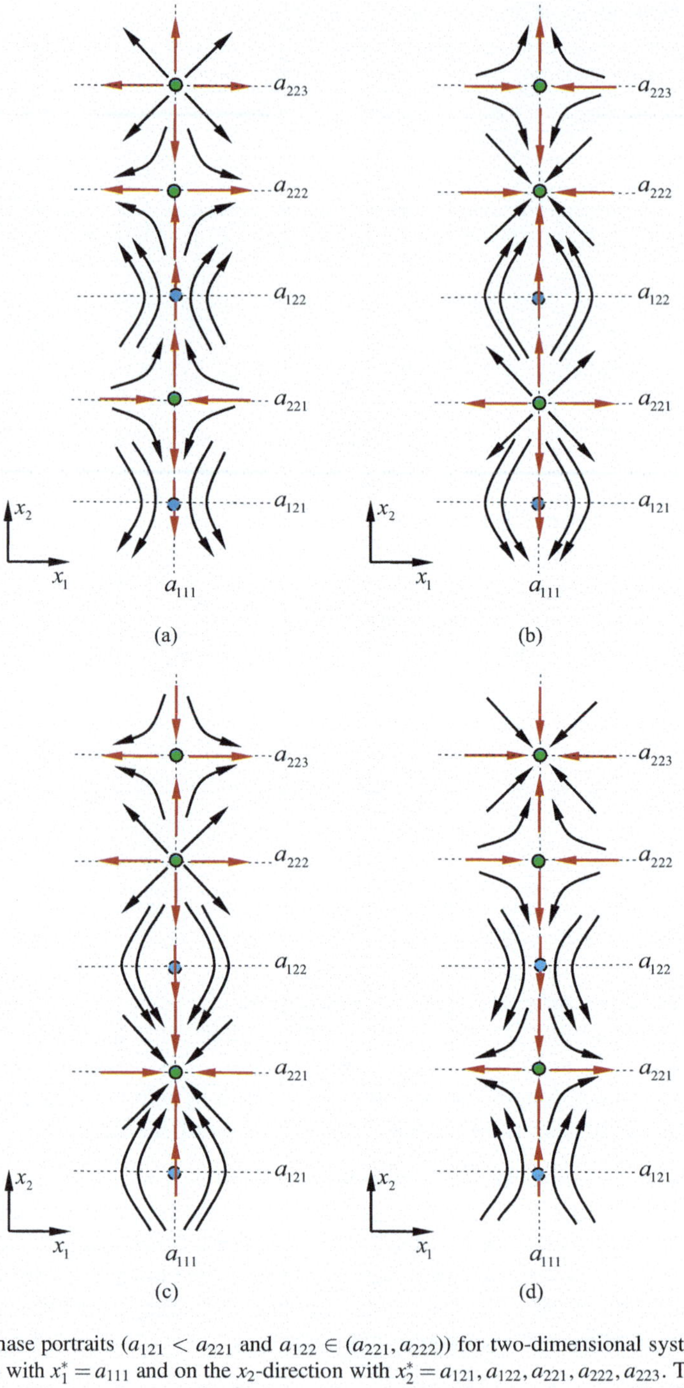

Fig. 5.5 Phase portraits ($a_{121} < a_{221}$ and $a_{122} \in (a_{221}, a_{222})$) for two-dimensional systems on the x_1-direction with $x_1^* = a_{111}$ and on the x_2-direction with $x_2^* = a_{121}, a_{122}, a_{221}, a_{222}, a_{223}$. The equilibriums series: **(a)** ($a_{110} > 0, a_{220} > 0$), **(b)** ($a_{110} < 0, a_{220} > 0$), **(c)** ($a_{110} > 0, a_{220} < 0$), **(d)** ($a_{110} < 0$, $a_{220} < 0$)

Table 5.5 Equilibriums on the line of $x_1^* = a_{111}$ for $a_{121} < a_{221}$ and $a_{122} \in (a_{221}, a_{222})$

Fig.5.5	(a_{111}, x_2^*)	x_2^*				
	(a_{110}, a_{220})	a_{121}	a_{221}	a_{122}	a_{222}	a_{223}
(a)	(+,+)	(DP:UP,nF)	(SI,SO)	(DP:UP,pF)	(SO,SI)	(SO,SO)
(b)	(–,+)	(UP:DP,nF)	(SO,SO)	(UP:DP,pF)	(SI,SI)	(SI,SO)
(c)	(+,–)	(UP:DP,pF)	(SI,SI)	(UP:DP,nF)	(SO,SO)	(SO,SI)
(d)	(–,–)	(DP:UP,pF)	(SO,SI)	(DP:UP,nF)	(SI,SO)	(SI,SI)

with $a_{122} \in (a_{222}, a_{223})$.The hyperbolic and hyperbolic-secant flows are changed because the locations of equilibriums are changed. The saddle and source become two saddles or sink and source, and the saddle and sink become two saddles or sink and source.

(iii) The phase portraits for the equilibrium series of $x_2^* = a_{121}, a_{122}, a_{221}, a_{222}, a_{223}$ with $x_1^* = a_{111}$ for $a_{221} = a_{121}$ and $a_{223} < a_{122}$ are presented in Fig. 5.13. The behaviors of equilibriums are listed in Table 5.13. The inflection infinite-equilibrium of $x_2^* = a_{121} = a_{221}$ is the same as in Eqs. (5.5)–(5.8). The inflection source and sink infinite-equilibriums of $x_2^* = a_{121} = a_{221}$ are the switching bifurcations of sink, source, and saddles with hyperbolic flows for $a_{221} \in (a_{121}, a_{122})$ and $a_{221} < a_{121}$ with $a_{223} < a_{122}$. The hyperbolic and hyperbolic-secant flows are changed because the locations of equilibriums are unchanged compared to $a_{122} < a_{222}$. The saddle and source become saddle and sink, and the saddle and sink become saddle and source.

(iv) The phase portraits for the equilibrium series of $x_2^* = a_{121}, a_{122}, a_{221}, a_{222}, a_{223}$ with $x_1^* = a_{111}$ for $a_{221} = a_{122}$ are presented in Fig. 5.14. The behaviors of equilibriums are listed in Table 5.14.

The inflection infinite-equilibrium of $x_2^* = a_{122} = a_{221}$ is summarized as follows.

$$
\begin{aligned}
&(\bar{x}_1, a_{221}) = \underbrace{(\text{DI, SO})}_{\text{decreasing-inflection source}} \quad \text{if } \bar{x}_1 \in (-\infty, a_{111}),\\
&(\bar{x}_1, a_{221}) = \underbrace{(\text{II, SO})}_{\text{increasing-infelction source}} \quad \text{if } \bar{x}_1 \in (a_{111}, \infty),\\
&\text{for } a_{110} > 0, a_{220} > 0;
\end{aligned}
\tag{5.9}
$$

$$
\begin{aligned}
&(\bar{x}_1, a_{221}) = \underbrace{(\text{II, SO})}_{\text{increasing-inflection source}} \quad \text{if } \bar{x}_1 \in (-\infty, a_{111}),\\
&(\bar{x}_1, a_{221}) = \underbrace{(\text{DI, SO})}_{\text{decreasing-infelction source}} \quad \text{if } \bar{x}_1 \in (a_{111}, \infty),\\
&\text{for } a_{110} < 0, a_{220} > 0;
\end{aligned}
\tag{5.10}
$$

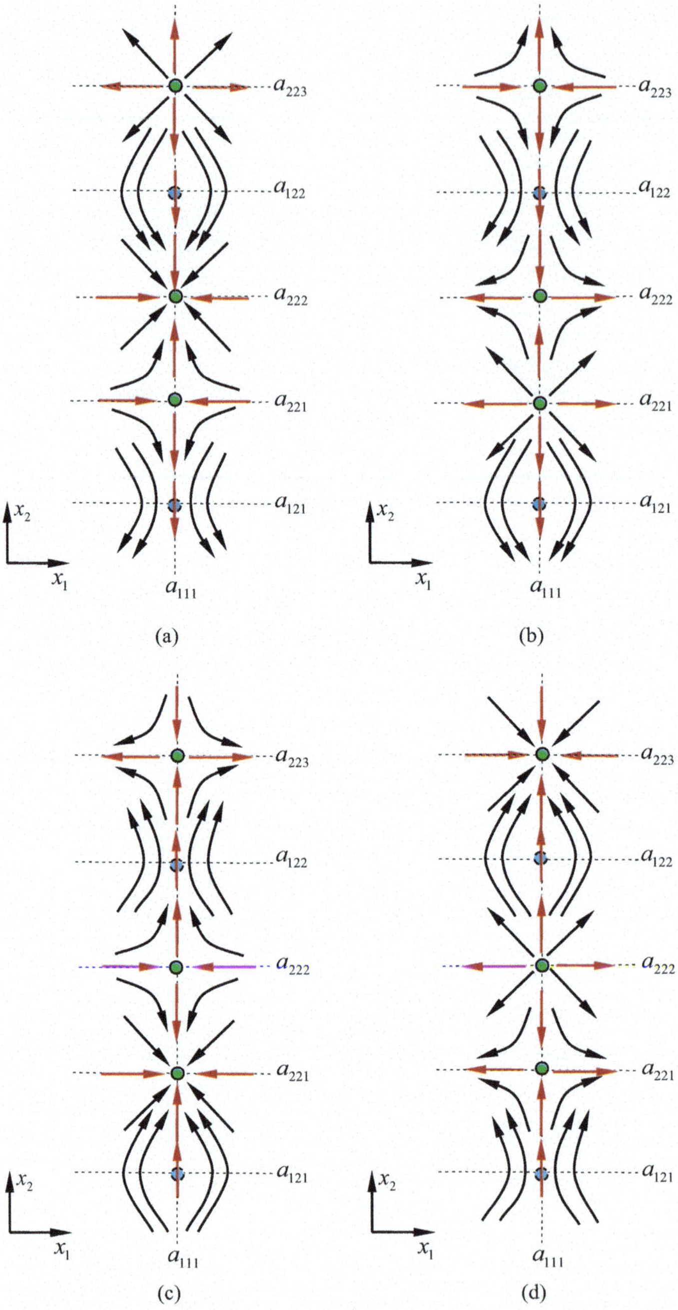

Fig. 5.6 Phase portraits ($a_{121} < a_{221}$ and $a_{122} \in (a_{222}, a_{223})$) for two-dimensional systems on the x_1-direction with $x_1^* = a_{111}$ and on the x_2-direction with $x_2^* = a_{121}, a_{122}, a_{221}, a_{222}, a_{223}$. The equilibriums flows: (**a**) ($a_{110} > 0$, $a_{220} > 0$), (**b**) ($a_{110} < 0$, $a_{220} > 0$), (**c**) ($a_{110} > 0$, $a_{220} < 0$), (**d**) ($a_{110} < 0$, $a_{220} < 0$)

Table 5.6 Equilibriums on the line of $x_1^* = a_{111}$ for $a_{121} < a_{221}$ and $a_{122} \in (a_{222}, a_{223})$

Fig.5.6	(a_{111}, x_2^*)	x_2^*				
	(a_{110}, a_{220})	a_{121}	a_{221}	a_{222}	a_{122}	a_{223}
(a)	(+,+)	(DP:UP,nF)	(SI,SO)	(SI,SI)	(UP:DP,nF)	(SO,SO)
(b)	(–,+)	(UP:DP,nF)	(SO,SO)	(SO,SI)	(DP:UP,nF)	(SI,SO)
(c)	(+,–)	(UP:DP,pF)	(SI,SI)	(SI,SO)	(DP:UP,pF)	(SO,SI)
(d)	(–,–)	(DP:UP,pF)	(SO,SI)	(SO,SO)	(UP:DP,pF)	(SI,SI)

$$
\begin{aligned}
&(\bar{x}_1, a_{221}) = \underbrace{(\text{II, SI})}_{\text{increasing-inflection sink}} \quad \text{if } \bar{x}_1 \in (-\infty, a_{111}),\\
&(\bar{x}_1, a_{221}) = \underbrace{(\text{DI, SI})}_{\text{decreasing-infelction sink}} \quad \text{if } \bar{x}_1 \in (a_{111}, \infty),\\
&\text{for } a_{110} > 0, a_{220} < 0;
\end{aligned}
\tag{5.11}
$$

$$
\begin{aligned}
&(\bar{x}_1, a_{221}) = \underbrace{(\text{DI, SI})}_{\text{decreasing-inflection sink}} \quad \text{if } \bar{x}_1 \in (-\infty, a_{111}),\\
&(\bar{x}_1, a_{221}) = \underbrace{(\text{II, SI})}_{\text{increasing-infelction sink}} \quad \text{if } \bar{x}_1 \in (a_{111}, \infty),\\
&\text{for } a_{110} < 0, a_{220} < 0.
\end{aligned}
\tag{5.12}
$$

The inflection-source and sink infinite-equilibriums of $x_2^* = a_{122} = a_{221}$ are the switching bifurcations of sink, source, and saddles with hyperbolic flows for $a_{221} \in (a_{121}, a_{122})$ and $a_{221} > a_{122}$, as shown in Fig. 5.14. In addition, the saddle, sink, or source with hyperbolic or hyperbolic-secant flow exists. The infinite-equilibrium is above the hyperbolic or hyperbolic-secant flow of $x_2^* = a_{121}$.

(v) The phase portraits for the equilibrium series of $x_2^* = a_{121}, a_{122}, a_{221}, a_{222}, a_{223}$ with$x_1^* = a_{111}$ for $a_{222} = a_{121}$ and $a_{122} < a_{223}$ are presented in Fig. 5.15. The behaviors of equilibriums are listed in Table 5.15.

The inflection infinite-equilibrium of $x_2^* = a_{222} = a_{121}$ is summarized as follows.

$$
\begin{aligned}
&(\bar{x}_1, a_{222}) = \underbrace{(\text{DI, SI})}_{\text{decreasing-inflection sink}} \quad \text{if } \bar{x}_1 \in (-\infty, a_{111}),\\
&(\bar{x}_1, a_{222}) = \underbrace{(\text{II, SI})}_{\text{increasing-infelction sink}} \quad \text{if } \bar{x}_1 \in (a_{111}, \infty),\\
&\text{for } a_{110} > 0, a_{220} > 0;
\end{aligned}
\tag{5.13}
$$

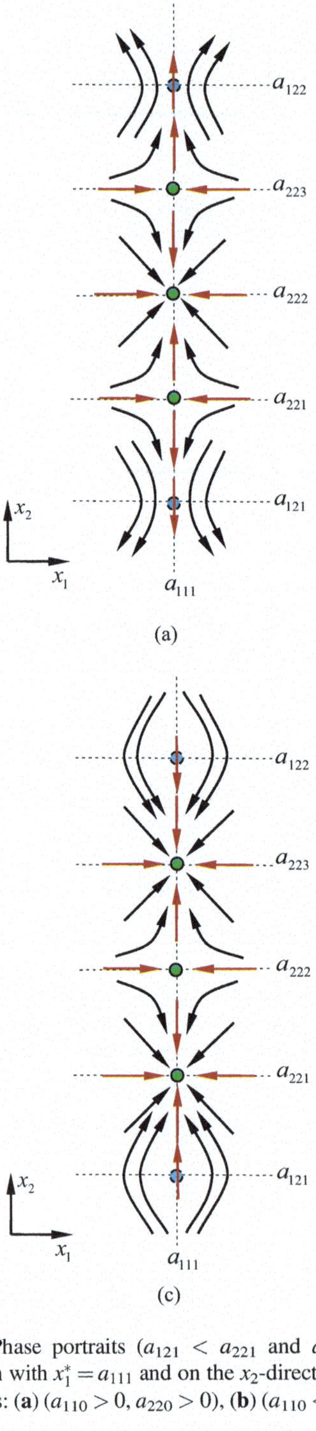

Fig. 5.7 Phase portraits ($a_{121} < a_{221}$ and $a_{223} < a_{122}$) for two-dimensional systems on the x_1-direction with $x_1^* = a_{111}$ and on the x_2-direction with $x_2^* = a_{121}, a_{122}, a_{221}, a_{222}, a_{223}$. The equilibriums flows: (**a**) ($a_{110} > 0, a_{220} > 0$), (**b**) ($a_{110} < 0, a_{220} > 0$), (**c**) ($a_{110} > 0, a_{220} < 0$), (**d**) ($a_{110} < 0, a_{220} < 0$)

Table 5.7 Equilibriums on the line of $x_1^* = a_{111}$ for $a_{121} < a_{221}$ and $a_{223} < a_{122}$

Fig.5.7	(a_{111}, x_2^*)	x_2^*				
	(a_{110}, a_{220})	a_{121}	a_{221}	a_{222}	a_{223}	a_{122}
(a)	(+,+)	(DP:UP,nF)	(SI,SO)	(SI,SI)	(SI,SO)	(DP:UP,pF)
(b)	(–,+)	(UP:DP,nF)	(SO,SO)	(SO,SI)	(SO,SO)	(UP:DP,pF)
(c)	(+,–)	(UP:DP,pF)	(SI,SI)	(SI,SO)	(SI,SI)	(UP:DP,nF)
(d)	(–,–)	(DP:UP,pF)	(SO,SI)	(SO,SO)	(SO,SI)	(DP:UP,nF)

$$
\begin{aligned}
&(\bar{x}_1, a_{222}) = \underbrace{(\text{II, SI})}_{\text{increasing-inflection sink}} \quad \text{if } \bar{x}_1 \in (-\infty, a_{111}),\\
&(\bar{x}_1, a_{222}) = \underbrace{(\text{DI, SI})}_{\text{decreasing-infelction sink}} \quad \text{if } \bar{x}_1 \in (a_{111}, \infty),\\
&\text{for } a_{110} < 0, a_{220} > 0;
\end{aligned} \tag{5.14}
$$

$$
\begin{aligned}
&(\bar{x}_1, a_{222}) = \underbrace{(\text{II, SO})}_{\text{increasing-inflection source}} \quad \text{if } \bar{x}_1 \in (-\infty, a_{111}),\\
&(\bar{x}_1, a_{222}) = \underbrace{(\text{DI, SO})}_{\text{decreasing-infelction source}} \quad \text{if } \bar{x}_1 \in (a_{111}, \infty),\\
&\text{for } a_{110} > 0, a_{220} < 0;
\end{aligned} \tag{5.15}
$$

$$
\begin{aligned}
&(\bar{x}_1, a_{222}) = \underbrace{(\text{DI, SO})}_{\text{decreasing-inflection source}} \quad \text{if } \bar{x}_1 \in (-\infty, a_{111}),\\
&(\bar{x}_1, a_{222}) = \underbrace{(\text{II, SO})}_{\text{increasing-infelction source}} \quad \text{if } \bar{x}_1 \in (a_{111}, \infty),\\
&\text{for } a_{110} < 0, a_{220} < 0.
\end{aligned} \tag{5.16}
$$

The inflection source and sink infinite-equilibriums of $x_2^* = a_{222} = a_{121}$ are the switching bifurcations of sink, source, and saddles with hyperbolic flows for $a_{121} \in (a_{221}, a_{222})$ and $a_{121} > a_{222}$ with $a_{122} < a_{223}$, as shown in Fig. 5.15. The two saddles, or two sinks, or two sources with hyperbolic or hyperbolic-secant flow exist. The infinite-equilibrium is above the saddle, sink, or source of $x_2^* = a_{221}$.

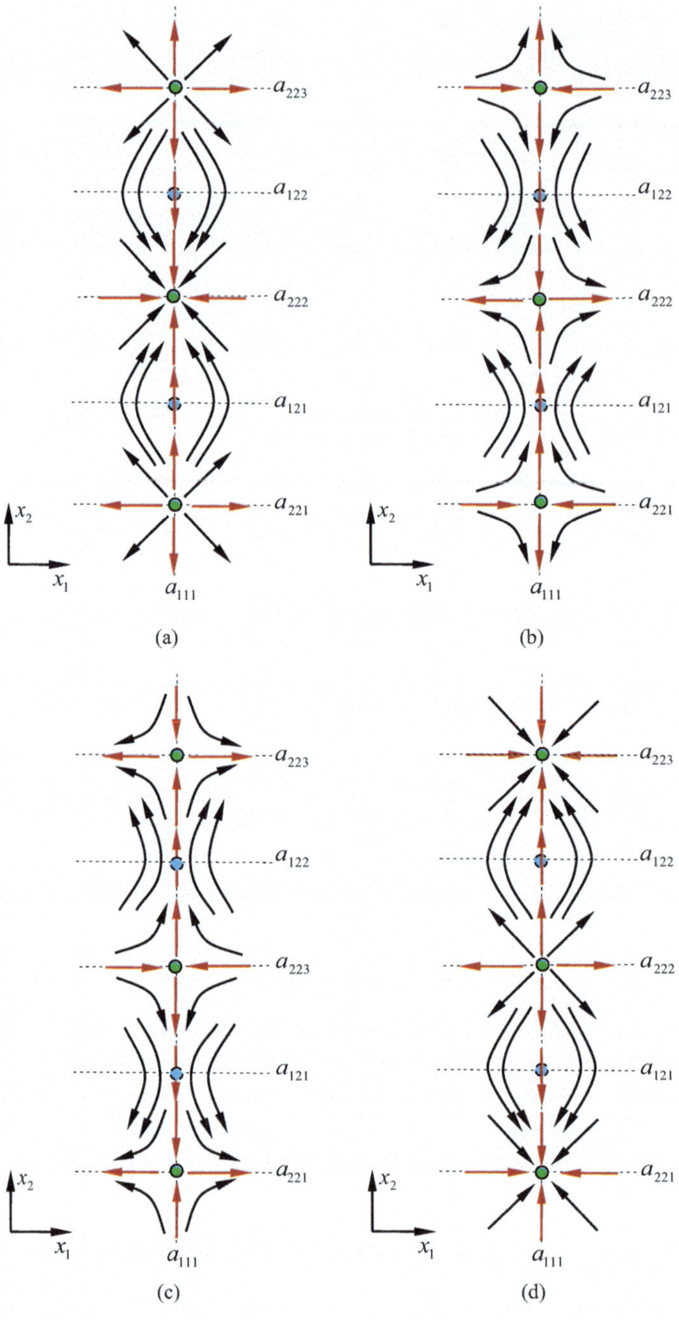

Fig. 5.8 Phase portraits ($a_{121} \in (a_{221}, a_{222})$, $a_{122} \in (a_{222}, a_{223})$) for two-dimensional systems on the x_1-direction with $x_1^* = a_{111}$ and on the x_2-direction with $x_2^* = a_{121}, a_{122}, a_{221}, a_{222}, a_{223}$. The equilibriums series: (**a**) ($a_{110} > 0, a_{220} > 0$), (**b**) ($a_{110} < 0, a_{220} > 0$), (**c**) ($a_{110} > 0, a_{220} < 0$), (**d**) ($a_{110} < 0$, $a_{220} < 0$)

Table 5.8 Equilibriums on the line of $x_1^* = a_{111}$ for $a_{121} \in (a_{221}, a_{222})$ and $a_{122} \in (a_{222}, a_{223})$

Fig.5.8	(a_{111}, x_2^*)	x_2^*				
	(a_{110}, a_{220})	a_{121}	a_{221}	a_{222}	a_{122}	a_{223}
(a)	(+,+)	(SO,SO)	(DP:UP,pF)	(SI,SI)	(DP:UP,nF)	(SO,SO)
(b)	(−,+)	(SI,SO)	(UP:DP,pF)	(SO,SI)	(UP:DP,nF)	(SI,SO)
(c)	(+,−)	(SO,SI)	(UP:DP,nF)	(SI,SO)	(UP:DP,pF)	(SO,SI)
(d)	(−,−)	(SI,SI)	(DP:UP,nF)	(SO,SO)	(DP:UP,pF)	(SI,SI)

(vi) The phase portraits for the equilibrium series of $x_2^* = a_{121}, a_{122}, a_{221}, a_{222}, a_{223}$ with $x_1^* = a_{111}$ for $a_{222} = a_{121}$ and $a_{223} < a_{122}$ are presented in Fig. 5.16. The behaviors of equilibriums are listed in Table 5.16. The inflection infinite-equilibrium of $x_2^* = a_{222} = a_{121}$ is the same as in Eqs. (5.13)–(5.16). The inflection source and sink infinite-equilibriums of $x_2^* = a_{222} = a_{121}$ are the switching bifurcations of sink, source, and saddles with hyperbolic flows for $a_{121} \in (a_{221}, a_{222})$ and $a_{121} > a_{222}$ with $a_{223} < a_{122}$, as shown in Fig. 5.16. The two saddles, two sinks, or two sources with hyperbolic or hyperbolic-secant flow exist. The infinite-equilibrium is above the saddle, sink, or source of $x_2^* = a_{221}$.

(vii) The phase portraits for the equilibrium series of $x_2^* = a_{121}, a_{122}, a_{221}, a_{222}, a_{223}$ with $x_1^* = a_{111}$ for $a_{222} = a_{122}$ and $a_{121} < a_{221}$ are presented in Fig. 5.17. The behaviors of equilibriums are listed in Table 5.17.

The inflection infinite-equilibrium of $x_2^* = a_{222} = a_{122}$ is summarized as follows.

$$
\begin{aligned}
&(\bar{x}_1, a_{222}) = \underbrace{(\text{II, SI})}_{\text{increasing-infelction sink}} && \text{if } \bar{x}_1 \in (-\infty, a_{111}),\\
&(\bar{x}_1, a_{222}) = \underbrace{(\text{DI, SI})}_{\text{decreasing-inflection sink}} && \text{if } \bar{x}_1 \in (a_{111}, \infty),\\
&\text{for } a_{110} > 0, a_{220} > 0;
\end{aligned}
\tag{5.17}
$$

$$
\begin{aligned}
&(\bar{x}_1, a_{222}) = \underbrace{(\text{DI, SI})}_{\text{decreasing-infelction sink}} && \text{if } \bar{x}_1 \in (-\infty, a_{111}),\\
&(\bar{x}_1, a_{222}) = \underbrace{(\text{II, SI})}_{\text{increasing-inflection sink}} && \text{if } \bar{x}_1 \in (a_{111}, \infty),\\
&\text{for } a_{110} < 0, a_{220} > 0.
\end{aligned}
\tag{5.18}
$$

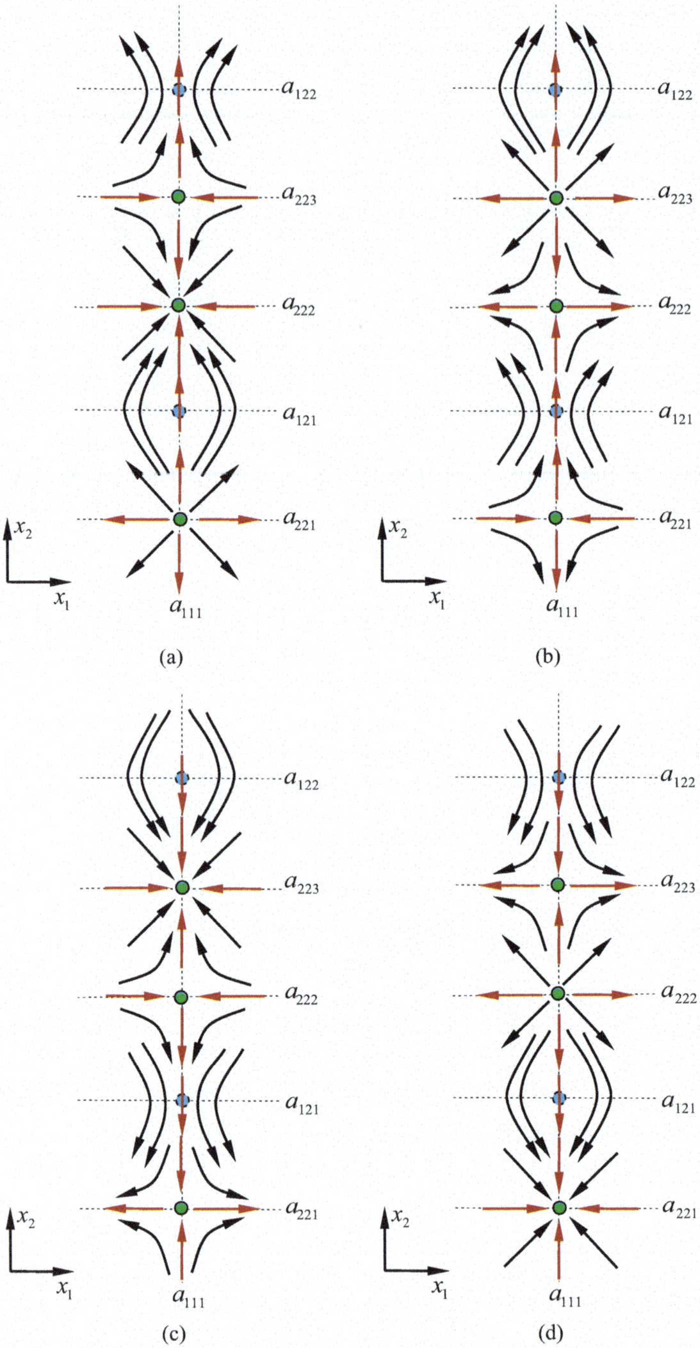

Fig. 5.9 Phase portraits ($a_{121} \in (a_{221}, a_{222})$, $a_{223} < a_{122}$) for two-dimensional systems on the x_1-direction with $x_1^* = a_{111}$ and on the x_2-direction with $x_2^* = a_{121}, a_{122}, a_{221}, a_{222}, a_{223}$. The equilibriums flows: (**a**) ($a_{110} > 0$, $a_{220} > 0$), (**b**) ($a_{110} < 0$, $a_{220} > 0$), (**c**) ($a_{110} > 0$, $a_{220} < 0$), (**d**) ($a_{110} < 0$, $a_{220} < 0$)

Table 5.9 Equilibriums on the line of $x_1^* = a_{111}$ for $a_{121} \in (a_{221}, a_{222})$ and $a_{223} > a_{122}$

Fig.5.9	(a_{111}, x_2^*)	x_2^*				
	(a_{110}, a_{220})	a_{221}	a_{121}	a_{222}	a_{223}	a_{122}
(a)	(+,+)	(SO,SO)	(UP:DP,pF)	(SI,SI)	(SI,SO)	(DP:UP,pF)
(b)	(–,+)	(SI,SO)	(DP:UP,pF)	(SO,SI)	(SO,SO)	(UP:DP,pF)
(c)	(+,–)	(SO,SI)	(DP:UP,nF)	(SI,SO)	(SI,SI)	(UP:DP,nF)
(d)	(–,–)	(SI,SI)	(UP:DP,nF)	(SO,SO)	(SO,SI)	(DP:UP,nF)

$$
\begin{aligned}
&(\bar{x}_1, a_{222}) = \underbrace{(\text{DI, SO})}_{\text{decreasing-infelction source}} \quad \text{if } \bar{x}_1 \in (-\infty, a_{111}),\\
&(\bar{x}_1, a_{222}) = \underbrace{(\text{II, SO})}_{\text{increasing-inflection source}} \quad \text{if } \bar{x}_1 \in (a_{111}, \infty),\\
&\text{for } a_{110} > 0, a_{220} < 0;
\end{aligned} \tag{5.19}
$$

$$
\begin{aligned}
&(\bar{x}_1, a_{222}) = \underbrace{(\text{II, SO})}_{\text{increasing-infelction source}} \quad \text{if } \bar{x}_1 \in (-\infty, a_{111}),\\
&(\bar{x}_1, a_{222}) = \underbrace{(\text{DI, SO})}_{\text{decreasing-inflection source}} \quad \text{if } \bar{x}_1 \in (a_{111}, \infty),\\
&\text{for } a_{110} < 0, a_{220} < 0.
\end{aligned} \tag{5.20}
$$

The inflection source and sink infinite-equilibriums of $x_2^* = a_{222} = a_{121}$ are the switching bifurcations of sink, source and saddles with hyperbolic flows for $a_{122} \in (a_{221}, a_{222})$ and $a_{122} \in (a_{222}, a_{223})$ with $a_{121} < a_{221}$, as shown in Fig. 5.17. The saddle and sink or source with hyperbolic or hyperbolic-secant flow exist. The infinite-equilibrium is below the saddle, sink, or source of $x_2^* = a_{223}$.

(viii) The phase portraits for the equilibrium series of $x_2^* = a_{121}, a_{122}, a_{221}, a_{222}, a_{223}$ with $x_1^* = a_{111}$ for $a_{222} = a_{122}$ and $a_{221} < a_{121}$ are presented in Fig. 5.18. The behaviors of equilibriums are listed in Table 5.18. The inflection infinite-equilibrium of $x_2^* = a_{222} = a_{122}$ is the same as in Eqs. (5.17)–(5.20). The inflection source and sink infinite-equilibriums of $x_2^* = a_{222} = a_{121}$ are the switching bifurcations of sink, source, and saddles with hyperbolic flows for $a_{122} \in (a_{221}, a_{222})$ and $a_{122} > a_{222}$ with $a_{221} < a_{121}$, as shown in Fig. 5.18. The two saddles, two sinks, or two sources with hyperbolic or hyperbolic-secant flow exist.

(ix) The phase portraits for the equilibrium series of $x_2^* = a_{121}, a_{122}, a_{221}, a_{222}, a_{223}$ with $x_1^* = a_{111}$ for $a_{223} = a_{121}$ are presented in Fig. 5.19. The behaviors of equilibriums are listed in Table 5.19.

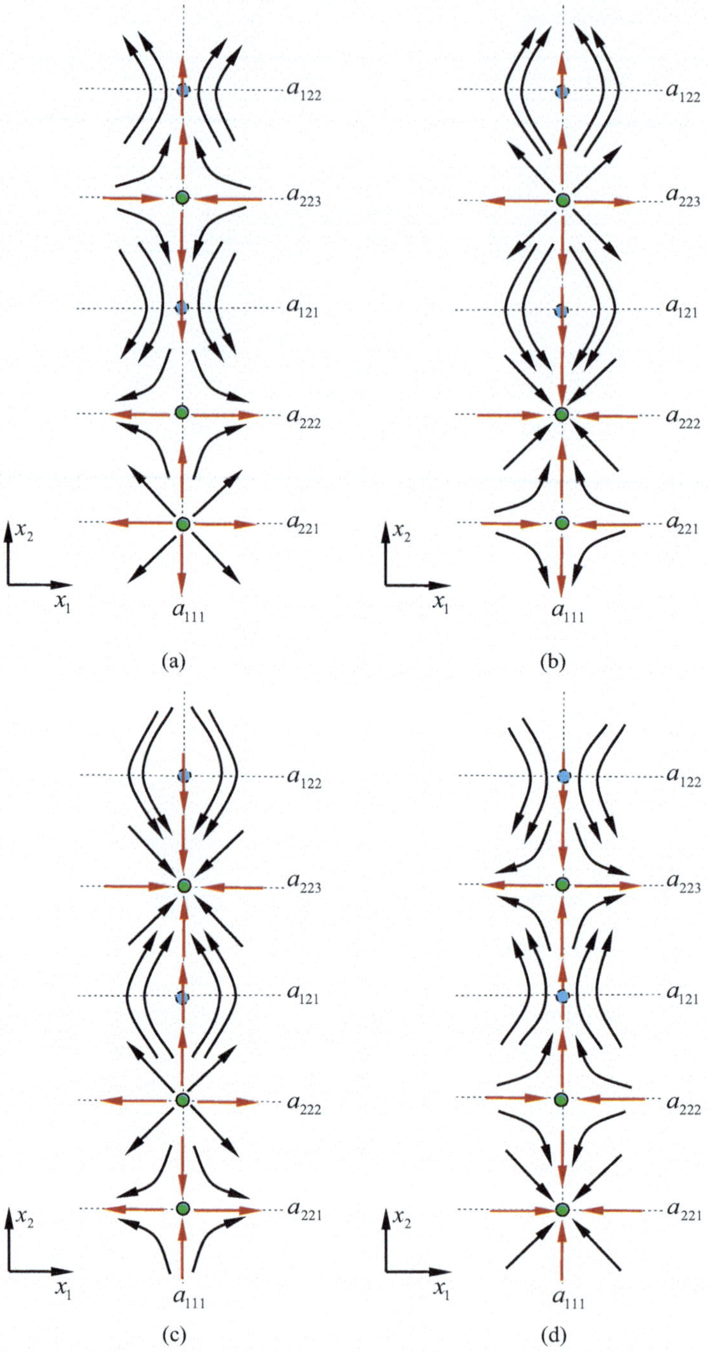

Fig. 5.10 Phase portraits ($a_{121} \in (a_{222}, a_{221})$, $a_{223} < a_{122}$) for two-dimensional systems on the x_1-direction with $x_1^* = a_{111}$ and on the x_2-direction with $x_2^* = a_{121}, a_{122}, a_{221}, a_{222}, a_{223}$. The equilibriums flows: (**a**) ($a_{110} > 0, a_{220} > 0$), (**b**) ($a_{110} < 0, a_{220} > 0$), (**c**) ($a_{110} > 0, a_{220} < 0$), (**d**) ($a_{110} < 0$, $a_{220} < 0$)

Table 5.10 Equilibriums on the line of $x_1^* = a_{111}$ for $a_{121} \in (a_{221}, a_{222})$ and $a_{223} > a_{122}$

Fig.5.10	(a_{111}, x_2^*)	x_2^*				
	(a_{110}, a_{220})	a_{221}	a_{222}	a_{121}	a_{223}	a_{122}
(a)	(+,+)	(SO,SO)	(SO,SI)	(DP:UP,nF)	(SI,SO)	(DP:UP,pF)
(b)	(–,+)	(SI,SO)	(SI,SI)	(UP:DP,nF)	(SO,SO)	(UP:DP,pF)
(c)	(+,–)	(SO,SI)	(SO,SO)	(UP:DP,pF)	(SI,SI)	(UP:DP,nF)
(d)	(–,–)	(SI,SI)	(SI,SO)	(DP:UP,pF)	(SO,SI)	(DP:UP,nF)

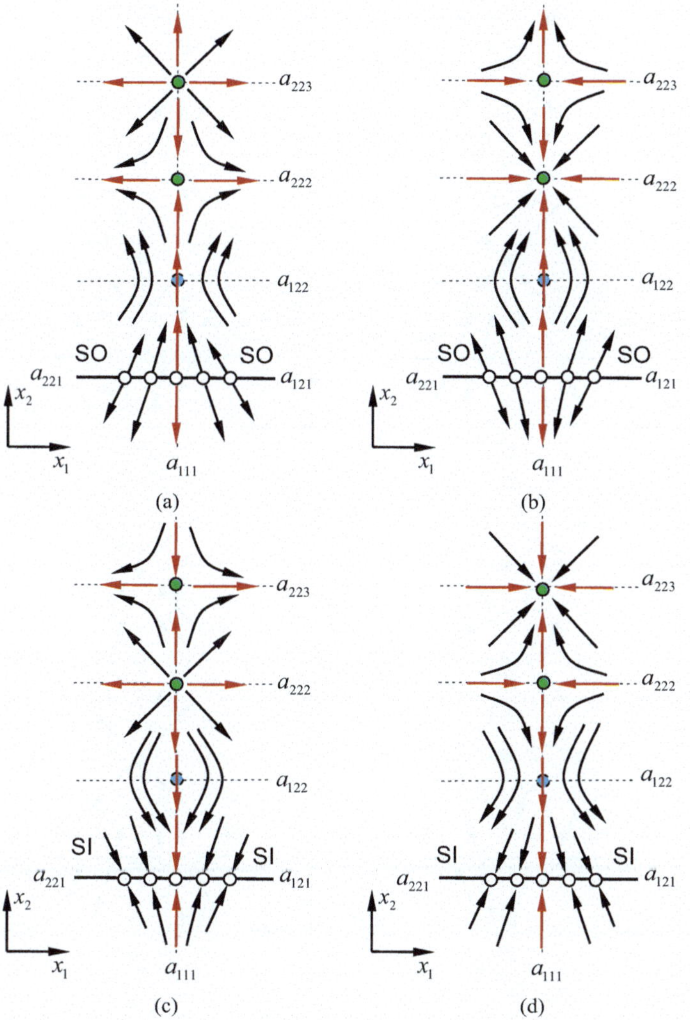

Fig. 5.11 Phase portraits ($a_{121} = a_{221}$, $a_{122} < a_{222}$) for two-dimensional systems on the x_1-direction with $x_1^* = a_{111}$ and on the x_2-direction with $x_2^* = a_{121}, a_{122}, a_{221}, a_{222}, a_{223}$. **(a)** ($a_{110} > 0$, $a_{220} > 0$), **(b)** ($a_{110} < 0$, $a_{220} > 0$), **(c)** ($a_{110} > 0$, $a_{220} < 0$), **(d)** ($a_{110} < 0$, $a_{220} < 0$)

Table 5.11 Equilibriums on the line of $x_1^* = a_{111}$ for $a_{121} = a_{221}$ and $a_{122} < a_{222}$

Fig. 5.11	(a_{111}, x_2^*)	x_2^*			
	(a_{110}, a_{220})	$a_{221} = a_{121}$	a_{122}	a_{222}	a_{223}
(a)	(+,+)	(II:DI,SO)	(DP:UP,pF)	(SO,SI)	(SO,SO)
(b)	(−,+)	(DI:II,SO)	(UP:DP,pF)	(SI,SI)	(SI,SO)
(c)	(+,−)	(DI:II,SI)	(UP:DP,nF)	(SO,SO)	(SO,SI)
(d)	(−,−)	(II:DI,SI)	(DP:UP,nF)	(SI,SO)	(SI,SI)

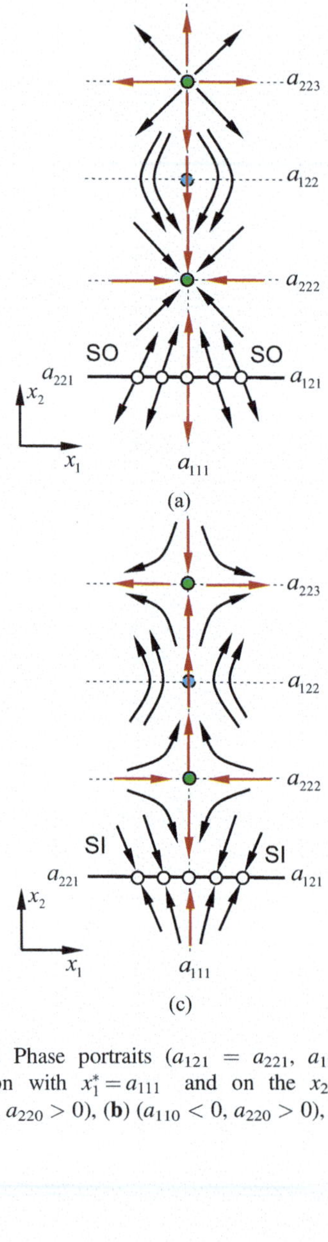

Fig. 5.12 Phase portraits ($a_{121} = a_{221}$, $a_{122} < a_{223}$) for two-dimensional systems on the x_1-direction with $x_1^* = a_{111}$ and on the x_2-direction with $x_2^* = a_{121}, a_{122}, a_{221}, a_{222}, a_{223}$. **(a)** ($a_{110} > 0$, $a_{220} > 0$), **(b)** ($a_{110} < 0$, $a_{220} > 0$), **(c)** ($a_{110} > 0$, $a_{220} < 0$), **(d)** ($a_{110} < 0$, $a_{220} < 0$)

Table 5.12 Equilibriums on the line of $x_1^* = a_{111}$ for $a_{121} = a_{221}$ and $a_{122} \in (a_{222}, a_{223})$

	(a_{111}, x_2^*)	x_2^*			
Fig. 5.12	(a_{110}, a_{220})	$a_{221} = a_{121}$	a_{222}	a_{122}	a_{223}
(a)	(+,+)	(II:DI,SO)	(SI,SI)	(UP:DP,nF)	(SO,SO)
(b)	(−,+)	(DI:II,SO)	(SO,SI)	(DP:UP,nF)	(SI,SO)
(c)	(+,−)	(DI:II,SI)	(SI,SO)	(DP:UP,pF)	(SO,SI)
(d)	(−,−)	(II:DI,SI)	(SO,SO)	(UP:DP,pF)	(SI,SI)

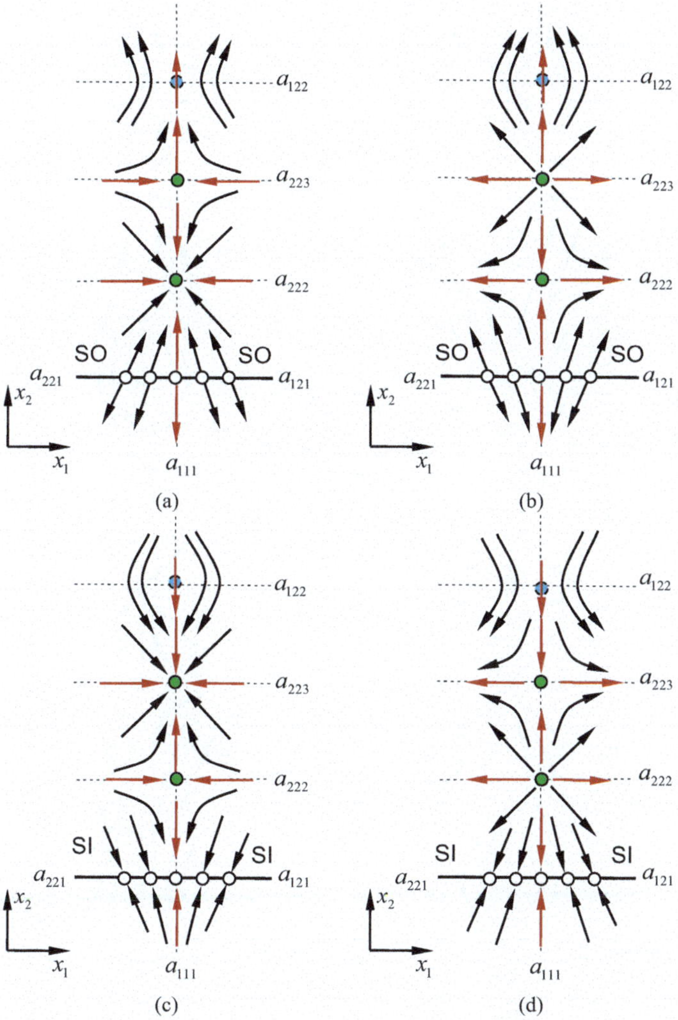

Fig. 5.13 Phase portraits ($a_{121} = a_{221}$, $a_{223} = a_{122}$) for two-dimensional systems on the x_1-direction with $x_1^* = a_{111}$ and on the x_2-direction with $x_2^* = a_{121}, a_{122}, a_{221}, a_{222}, a_{223}$. (**a**) ($a_{110} > 0$, $a_{220} > 0$), (**b**) ($a_{110} < 0$, $a_{220} > 0$), (**c**) ($a_{110} > 0$, $a_{220} < 0$), (**d**) ($a_{110} < 0$, $a_{220} < 0$)

Table 5.13 Equilibriums on the line of $x_1^* = a_{111}$ for $a_{121} = a_{221}$ and $a_{223} < a_{122}$

	(a_{111}, x_2^*)	x_2^*			
Fig. 5.13	(a_{110}, a_{220})	$a_{221} = a_{121}$	a_{222}	a_{223}	a_{122}
(a)	(+,+)	(II:DI,SO)	(SI,SI)	(SI,SO)	(DP:UP,pF)
(b)	(−,+)	(DI:II,SO)	(SO,SI)	(SO,SO)	(UP:DP,pF)
(c)	(+,−)	(DI:II,SI)	(SI,SO)	(SI,SI)	(UP:DP,nF)
(d)	(−,−)	(II:DI,SI)	(SO,SO)	(SO,SI)	(DP:UP,nF)

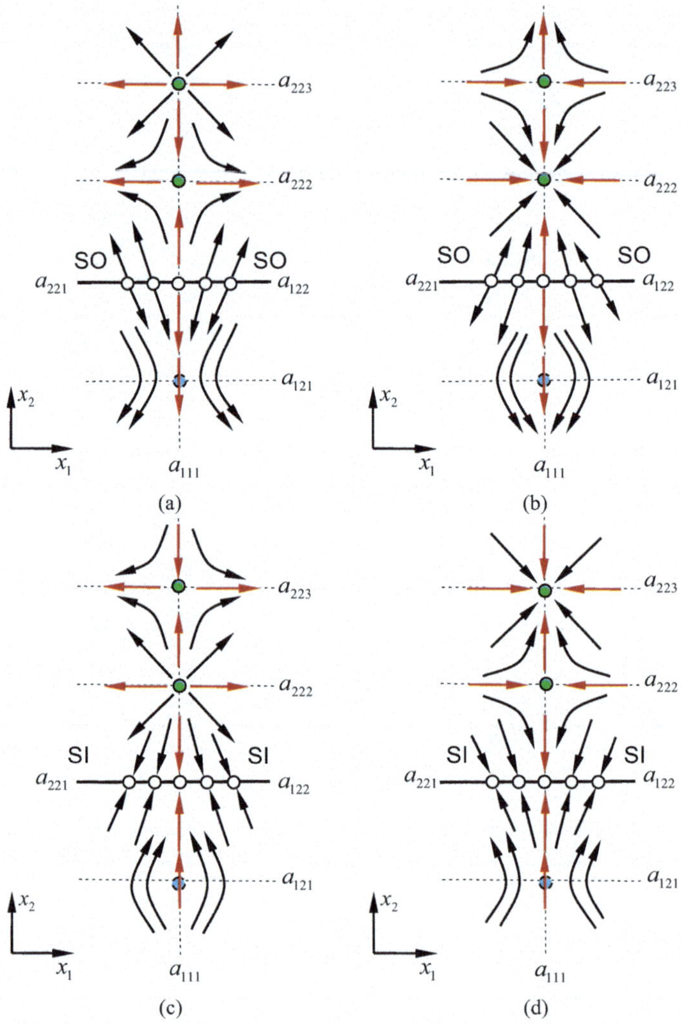

Fig. 5.14 Phase portraits ($a_{122} = a_{221}$) for two-dimensional systems on the x_1-direction with $x_1^* = a_{111}$ and on the x_2-direction with $x_2^* = a_{121}, a_{122}, a_{221}, a_{222}, a_{223}$. (**a**) ($a_{110} > 0$, $a_{220} > 0$), (**b**) ($a_{110} < 0$, $a_{220} > 0$), (**c**) ($a_{110} > 0$, $a_{220} < 0$), (**d**) ($a_{110} < 0$, $a_{220} < 0$)

Table 5.14 Equilibriums on the line of $x_1^* = a_{111}$ for $a_{122} = a_{221}$

Fig. 5.14	(a_{111}, x_2^*)	x_2^*			
	(a_{110}, a_{220})	a_{121}	$a_{122} = a_{221}$	a_{222}	a_{223}
(a)	(+,+)	(DP:UP,nF)	(DI:II,SO)	(SO,SI)	(SO,SO)
(b)	(−,+)	(UP:DP,nF)	(II:DI,SO)	(SI,SI)	(SI,SO)
(c)	(+,−)	(UP:DP,pF)	(II:DI,SI)	(SO,SO)	(SO,SI)
(d)	(−,−)	(DP:UP,pF)	(DI:II,SI)	(SI,SO)	(SI,SI)

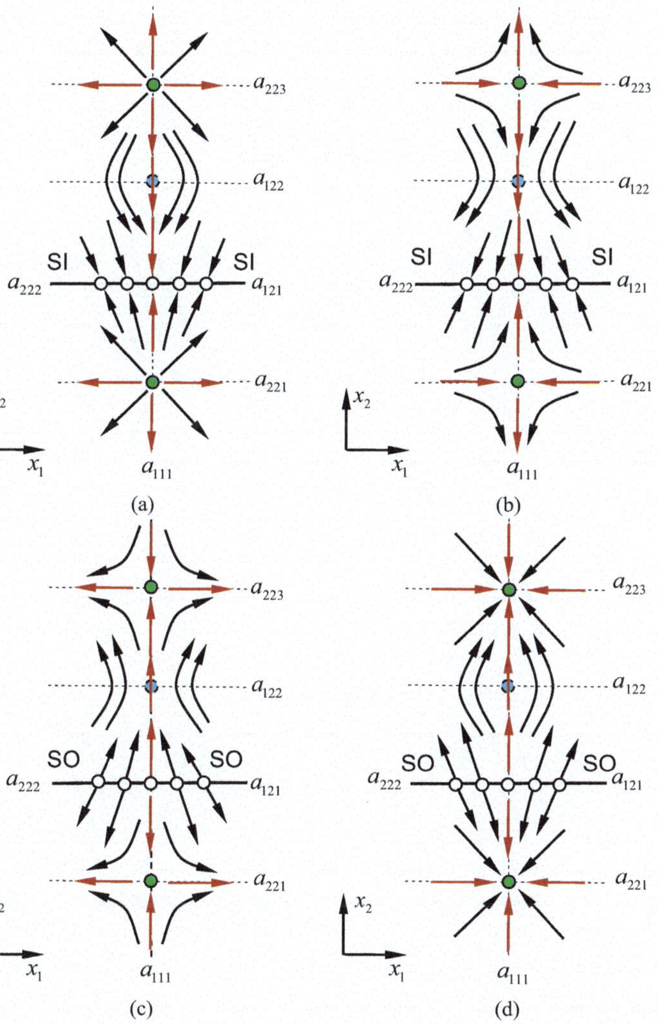

Fig. 5.15 Phase portraits ($a_{121} = a_{222}$, $a_{122} < a_{223}$) for two-dimensional systems on the x_1-direction with $x_1^* = a_{111}$ and on the x_2-direction with $x_2^* = a_{121}, a_{122}, a_{221}, a_{222}, a_{223}$. (**a**) ($a_{110} > 0$, $a_{220} > 0$), (**b**) ($a_{110} < 0$, $a_{220} > 0$), (**c**) ($a_{110} > 0$, $a_{220} < 0$), (**d**) ($a_{110} < 0$, $a_{220} < 0$)

Table 5.15 Equilibriums on the line of $x_1^* = a_{111}$ for $a_{222} = a_{121}$ and $a_{122} < a_{223}$

	(a_{111}, x_2^*)	x_2^*			
Fig. 5.15	(a_{110}, a_{220})	a_{221}	$a_{222} = a_{121}$	a_{122}	a_{223}
(a)	(+,+)	(SO,SO)	(DI:II,SI)	(UP:DP,nF)	(SO,SO)
(b)	(−,+)	(SI,SO)	(II:DI,SI)	(DP:UP,nF)	(SI,SO)
(c)	(+,−)	(SO,SI)	(II:DI,SO)	(DP:UP,pF)	(SO,SI)
(d)	(−,−)	(SI,SI)	(DI:II,SO)	(UP:DP,pF)	(SI,SI)

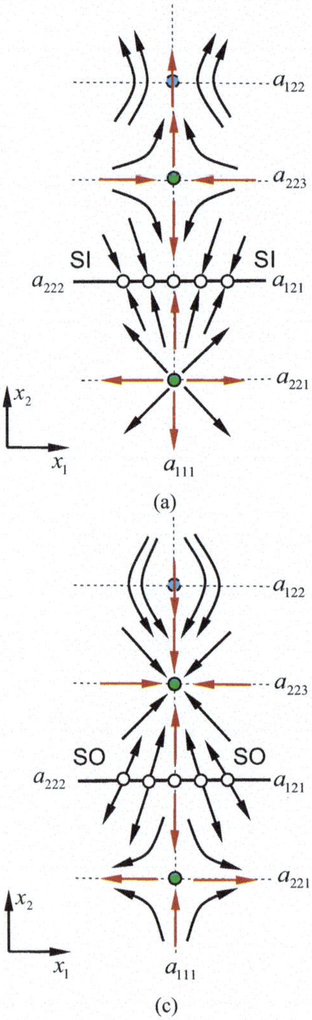

Fig. 5.16 Phase portraits ($a_{121} = a_{222}$, $a_{223} < a_{122}$) for two-dimensional systems on the x_1-direction with $x_1^* = a_{111}$ and on the x_2-direction with $x_2^* = a_{121}, a_{122}, a_{221}, a_{222}$. **(a)** ($a_{110} > 0$, $a_{220} > 0$), **(b)** ($a_{110} < 0$, $a_{220} > 0$), **(c)** ($a_{110} > 0$, $a_{220} < 0$), **(d)** ($a_{110} < 0$, $a_{220} < 0$)

Table 5.16 Equilibriums on the line of $x_1^* = a_{111}$ for $a_{222} = a_{121}$ and $a_{223} < a_{122}$

	(a_{111}, x_2^*)	x_2^*			
Fig. 5.16	(a_{110}, a_{220})	a_{221}	$a_{222} = a_{121}$	a_{223}	a_{122}
(a)	(+,+)	(SO,SO)	(DI:II,SI)	(SI,SO)	(DP:UP,pF)
(b)	(−,+)	(SI,SO)	(II:DI,SI)	(SO,SO)	(UP:DP,pF)
(c)	(+,−)	(SO,SI)	(II:DI,SO)	(SI,SI)	(UP:DP,nF)
(d)	(−,−)	(SI,SI)	(DI:II,SO)	(SO,SI)	(DP:UP,nF)

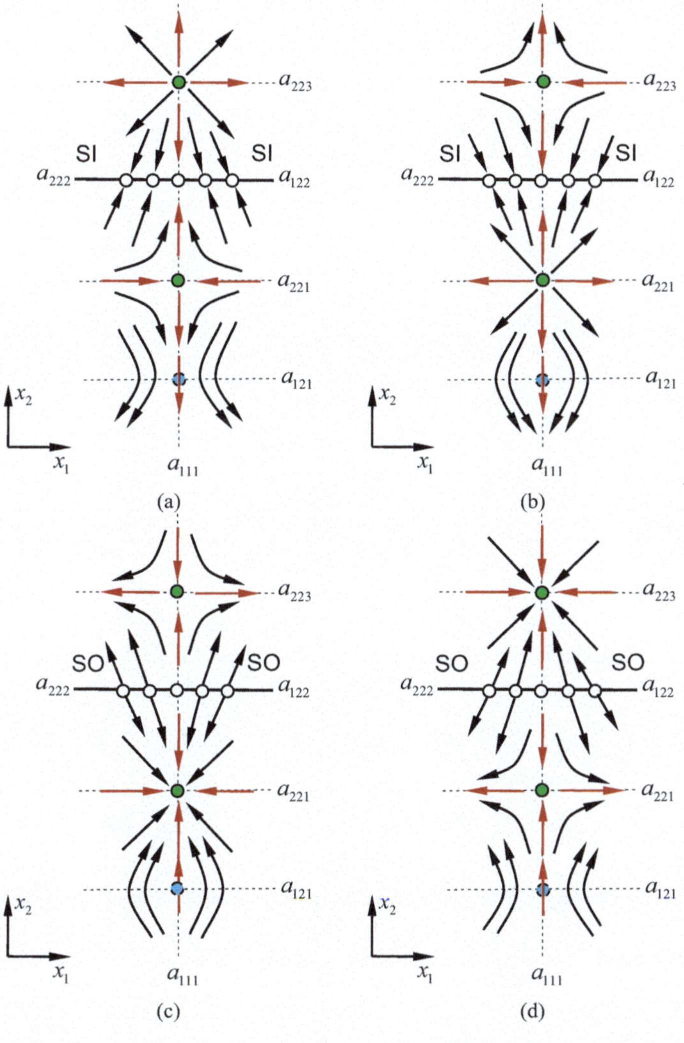

Fig. 5.17 Phase portraits ($a_{122} = a_{222}$, $a_{121} < a_{221}$) for two-dimensional systems on the x_1-direction with $x_1^* = a_{111}$ and on the x_2-direction with $x_2^* = a_{121}, a_{122}, a_{221}, a_{222}$. (**a**) ($a_{110} > 0$, $a_{220} > 0$), (**b**) ($a_{110} < 0$, $a_{220} > 0$), (**c**) ($a_{110} > 0$, $a_{220} < 0$), (**d**) ($a_{110} < 0$, $a_{220} < 0$)

Table 5.17 Equilibriums on the line of $x_1^* = a_{111}$ for $a_{222} = a_{122}$ and $a_{121} < a_{221}$

	(a_{111}, x_2^*)	x_2^*			
Fig. 5.17	(a_{110}, a_{220})	a_{121}	a_{221}	$a_{222} = a_{122}$	a_{223}
(a)	(+,+)	(DP:UP,nF)	(SI,SO)	(II:DI,SI)	(SO,SO)
(b)	(−,+)	(UP:DP,nF)	(SO,SO)	(DI:II,SI)	(SI,SO)
(c)	(+,−)	(UP:DP,pF)	(SI,SI)	(DI:II,SO)	(SO,SI)
(d)	(−,−)	(DP:UP,pF)	(SO,SI)	(II:DI,SO)	(SI,SI)

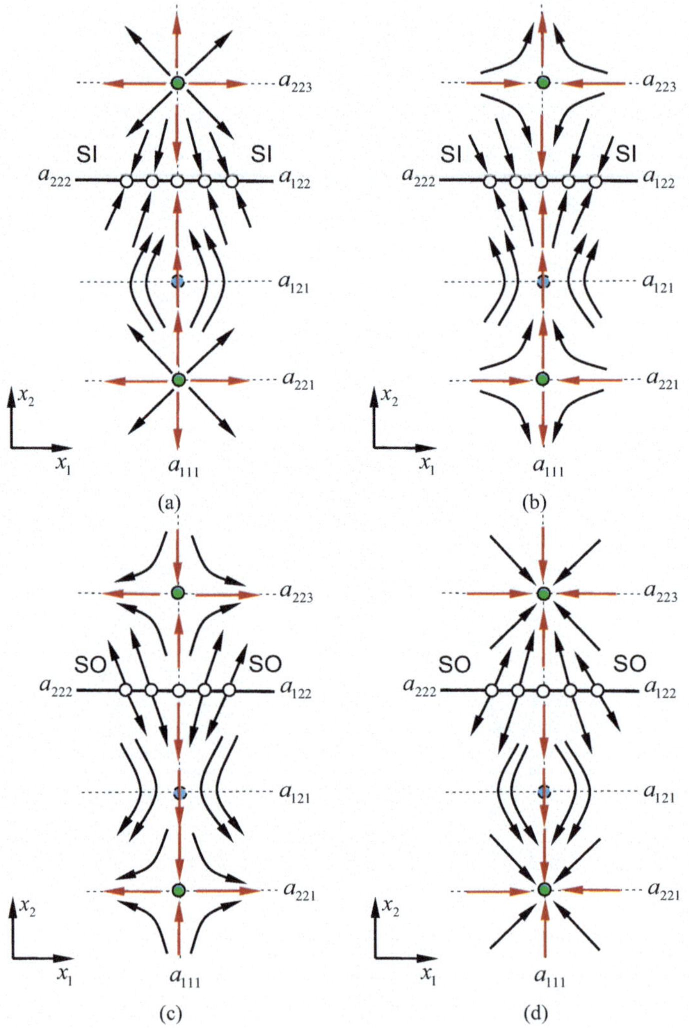

Fig. 5.18 Phase portraits ($a_{122} = a_{222}$, $a_{221} < a_{121}$) for two-dimensional systems on the x_1-direction with $x_1^* = a_{111}$ and on the x_2-direction with $x_2^* = a_{121}, a_{122}, a_{221}, a_{222}$. **(a)** ($a_{110} > 0$, $a_{220} > 0$), **(b)** ($a_{110} < 0$, $a_{220} > 0$), **(c)** ($a_{110} > 0$, $a_{220} < 0$), **(d)** ($a_{110} < 0$, $a_{220} < 0$)

Table 5.18 Equilibriums on the line of $x_1^* = a_{111}$ for $a_{222} = a_{122}$ and $a_{221} < a_{121}$

Fig. 5.18	(a_{111}, x_2^*)	x_2^*			
	(a_{110}, a_{220})	a_{221}	a_{121}	$a_{222} = a_{121}$	a_{223}
(a)	(+,+)	(SO,SO)	(UP:DP,pF)	(II:DI,SI)	(SO,SO)
(b)	(−,+)	(SI,SO)	(DP:UP,pF)	(DI:II,SI)	(SI,SO)
(c)	(+,−)	(SO,SI)	(DP:UP,nF)	(DI:II,SO)	(SO,SI)
(d)	(−,−)	(SI,SI)	(UP:DP,nF)	(II:DI,SO)	(SI,SI)

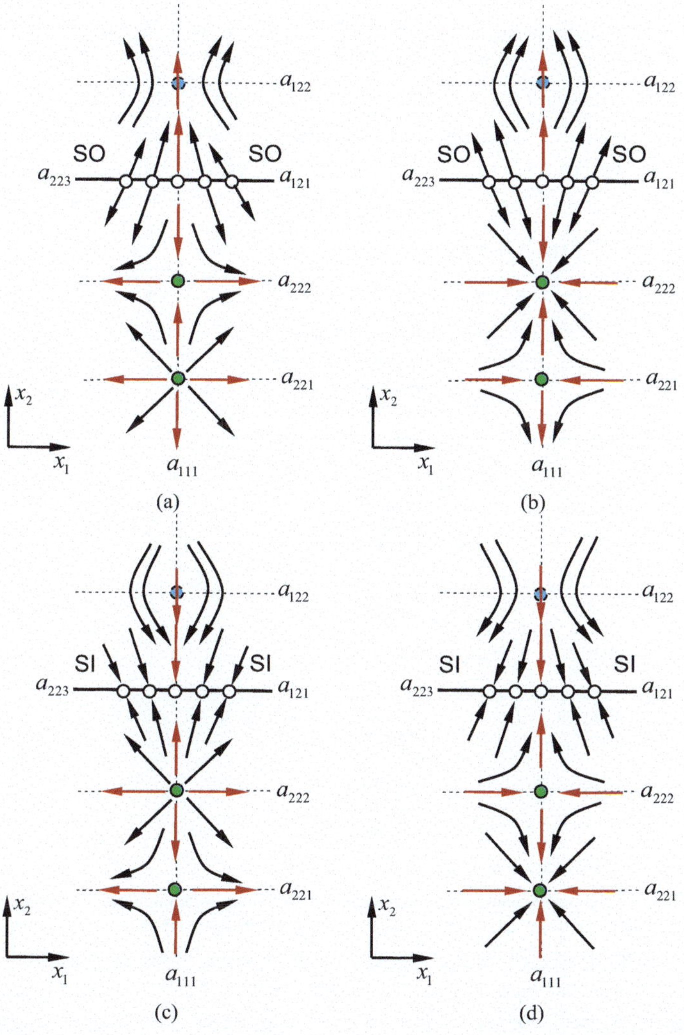

Fig. 5.19 Phase portraits ($a_{223} = a_{121}$) for two-dimensional systems on the x_1-direction with $x_1^* = a_{111}$ and on the x_2-direction with $x_2^* = a_{121}, a_{122}, a_{221}, a_{222}$. (**a**) ($a_{110} > 0$, $a_{220} > 0$), (**b**) ($a_{110} < 0$, $a_{220} > 0$), (**c**) ($a_{110} > 0$, $a_{220} < 0$), (**d**) ($a_{110} < 0$, $a_{220} < 0$)

Table 5.19 Equilibriums on the line of $x_1^* = a_{111}$ for $a_{223} = a_{121}$

	(a_{111}, x_2^*)	x_2^*			
Fig. 5.19	(a_{110}, a_{220})	a_{221}	a_{222}	$a_{223} = a_{121}$	a_{121}
(a)	(+,+)	(SO,SO)	(SO,SI)	(II:DI,SO)	(DP:UP,pF)
(b)	(−,+)	(SI,SO)	(SI,SI)	(DI:II,SO)	(UP:DP,pF)
(c)	(+,−)	(SO,SI)	(SO,SO)	(DI:II,SI)	(UP:DP,nF)
(d)	(−,−)	(SI,SI)	(SI,SO)	(II:DI,SI)	(DP:UP,nF)

The inflection infinite-equilibrium of $x_2^* = a_{223} = a_{121}$ is summarized as follows.

$$\begin{aligned} (\bar{x}_1, a_{223}) &= \underbrace{(\text{II, SO})}_{\text{increasing-inflection source}} \quad \text{if } \bar{x}_1 \in (-\infty, a_{111}), \\ (\bar{x}_1, a_{223}) &= \underbrace{(\text{DI, SO})}_{\text{decreasing-infelction source}} \quad \text{if } \bar{x}_1 \in (a_{111}, \infty), \\ &\text{for } a_{110} > 0, a_{220} > 0; \end{aligned} \tag{5.21}$$

$$\begin{aligned} (\bar{x}_1, a_{223}) &= \underbrace{(\text{DI, SO})}_{\text{decreasing-inflection source}} \quad \text{if } \bar{x}_1 \in (-\infty, a_{111}), \\ (\bar{x}_1, a_{223}) &= \underbrace{(\text{II, SO})}_{\text{increasing-infelction source}} \quad \text{if } \bar{x}_1 \in (a_{111}, \infty), \\ &\text{for } a_{110} < 0, a_{220} > 0; \end{aligned} \tag{5.22}$$

$$\begin{aligned} (\bar{x}_1, a_{223}) &= \underbrace{(\text{DI, SI})}_{\text{decreasing-inflection sink}} \quad \text{if } \bar{x}_1 \in (-\infty, a_{111}), \\ (\bar{x}_1, a_{223}) &= \underbrace{(\text{II, SI})}_{\text{increasing-infelction sink}} \quad \text{if } \bar{x}_1 \in (a_{111}, \infty), \\ &\text{for } a_{110} > 0, a_{220} < 0; \end{aligned} \tag{5.23}$$

$$\begin{aligned} (\bar{x}_1, a_{223}) &= \underbrace{(\text{II, SI})}_{\text{increasing-inflection sink}} \quad \text{if } \bar{x}_1 \in (-\infty, a_{111}), \\ (\bar{x}_1, a_{223}) &= \underbrace{(\text{DI, SI})}_{\text{decreasing-infelction sink}} \quad \text{if } \bar{x}_1 \in (a_{111}, \infty), \\ &\text{for } a_{110} < 0, a_{220} < 0. \end{aligned} \tag{5.24}$$

The inflection source and sink infinite-equilibriums of $x_2^* = a_{223} = a_{121}$ are the switching bifurcations of sink, source, and saddles with hyperbolic flows for $a_{121} \in (a_{222}, a_{223})$ and $a_{223} < a_{121}$, as shown in Fig. 5.19. The saddle and sink or source with hyperbolic or hyperbolic-secant flow exist. The infinite-equilibrium is below the hyperbolic or hyperbolic-secant flow of $x_2^* = a_{122}$.

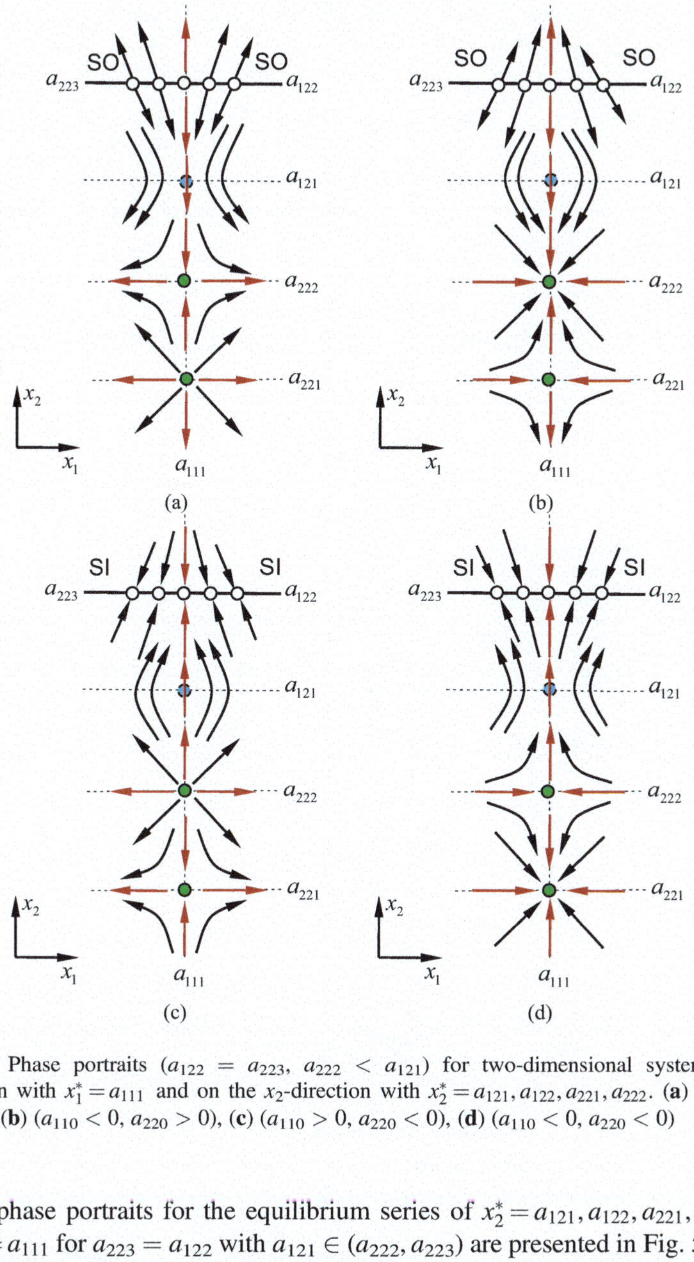

Fig. 5.20 Phase portraits ($a_{122} = a_{223}$, $a_{222} < a_{121}$) for two-dimensional systems on the x_1-direction with $x_1^* = a_{111}$ and on the x_2-direction with $x_2^* = a_{121}, a_{122}, a_{221}, a_{222}$. (**a**) ($a_{110} > 0$, $a_{220} > 0$), (**b**) ($a_{110} < 0$, $a_{220} > 0$), (**c**) ($a_{110} > 0$, $a_{220} < 0$), (**d**) ($a_{110} < 0$, $a_{220} < 0$)

(x) The phase portraits for the equilibrium series of $x_2^* = a_{121}, a_{122}, a_{221}, a_{222}, a_{223}$ with $x_1^* = a_{111}$ for $a_{223} = a_{122}$ with $a_{121} \in (a_{222}, a_{223})$ are presented in Fig. 5.20. The behaviors of equilibriums are listed in Table 5.20.

Table 5.20 Equilibriums on the line of $x_1^* = a_{111}$ for $a_{223} = a_{122}$ and $a_{121} \in (a_{222}, a_{223})$

	(a_{111}, x_2^*)	x_2^*			
Fig. 5.20	(a_{110}, a_{220})	a_{221}	a_{222}	a_{121}	$a_{223} = a_{122}$
(a)	(+,+)	(SO,SO)	(SO,SI)	(DP:UP,nF)	(DI:II,SO)
(b)	(−,+)	(SI,SO)	(SI,SI)	(UP:DP,nF)	(II:DI,SO)
(c)	(+,−)	(SO,SI)	(SO,SO)	(UP:DP,pF)	(II:DI,SI)
(d)	(−,−)	(SI,SI)	(SI,SO)	(DP:UP,pF)	(DI:II,SI)

The inflection infinite-equilibrium of $x_2^* = a_{223} = a_{122}$ is summarized as follows.

$$\begin{aligned}
&(\bar{x}_1, a_{223}) = \underbrace{(\text{DI, SO})}_{\text{decreasing-inflection source}} \quad \text{if } \bar{x}_1 \in (-\infty, a_{111}),\\
&(\bar{x}_1, a_{223}) = \underbrace{(\text{II, SO})}_{\text{increasing-infelction source}} \quad \text{if } \bar{x}_1 \subset (a_{111}, \infty),\\
&\text{for } a_{110} > 0, a_{220} > 0;
\end{aligned} \tag{5.25}$$

$$\begin{aligned}
&(\bar{x}_1, a_{223}) = \underbrace{(\text{II, SO})}_{\text{increasing-inflection source}} \quad \text{if } \bar{x}_1 \in (-\infty, a_{111}),\\
&(\bar{x}_1, a_{223}) = \underbrace{(\text{DI, SO})}_{\text{decreasing-infelction source}} \quad \text{if } \bar{x}_1 \in (a_{111}, \infty),\\
&\text{for } a_{110} < 0, a_{220} > 0;
\end{aligned} \tag{5.26}$$

$$\begin{aligned}
&(\bar{x}_1, a_{223}) = \underbrace{(\text{II, SI})}_{\text{increasing-inflection sink}} \quad \text{if } \bar{x}_1 \in (-\infty, a_{111}),\\
&(\bar{x}_1, a_{223}) = \underbrace{(\text{DI, SI})}_{\text{decreasing-infelction sink}} \quad \text{if } \bar{x}_1 \in (a_{111}, \infty),\\
&\text{for } a_{110} > 0, a_{220} < 0;
\end{aligned} \tag{5.27}$$

$$\begin{aligned}
&(\bar{x}_1, a_{223}) = \underbrace{(\text{DI, SI})}_{\text{decreasing-inflection sink}} \quad \text{if } \bar{x}_1 \in (-\infty, a_{111}),\\
&(\bar{x}_1, a_{223}) = \underbrace{(\text{II, SI})}_{\text{increasing-infelction sink}} \quad \text{if } \bar{x}_1 \in (a_{111}, \infty),\\
&\text{for } a_{110} < 0, a_{220} < 0.
\end{aligned} \tag{5.28}$$

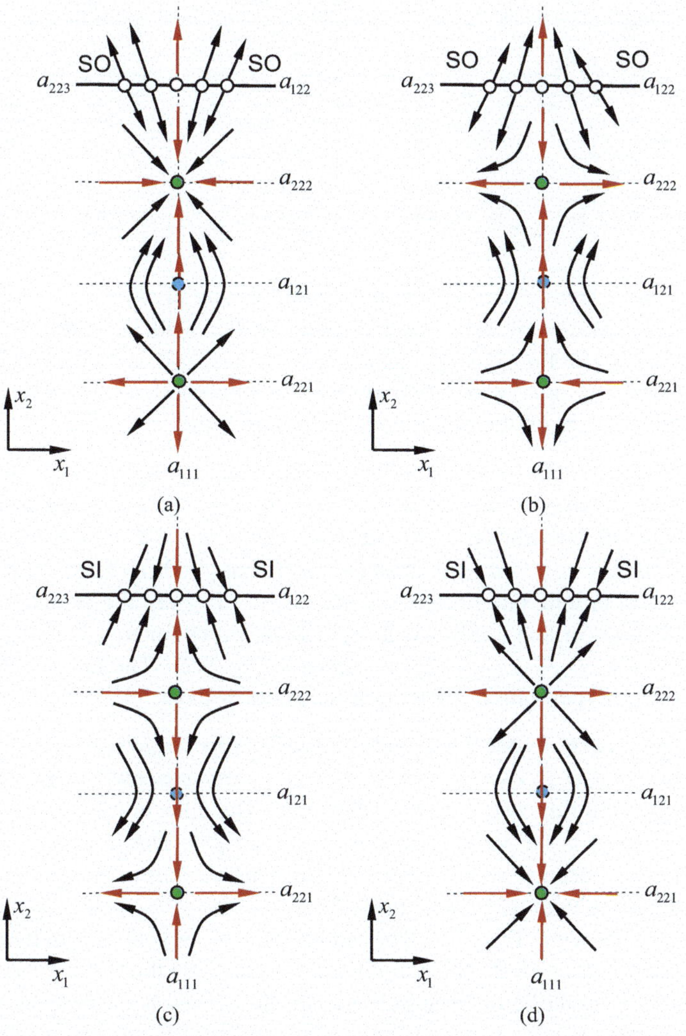

Fig. 5.21 Phase portraits ($a_{122} = a_{223}$, $a_{121} \in (a_{221}, a_{222})$) for two-dimensional systems on the x_1-direction with $x_1^* = a_{111}$ and on the x_2-direction with $x_2^* = a_{121}, a_{122}, a_{221}, a_{222}$. (**a**) ($a_{110} > 0$, $a_{220} > 0$), (**b**) ($a_{110} < 0$, $a_{220} > 0$), (**c**) ($a_{110} > 0$, $a_{220} < 0$), (**d**) ($a_{110} < 0$, $a_{220} < 0$)

Table 5.21 Equilibriums on the line of $x_1^* = a_{111}$ for $a_{223} = a_{122}$ and $a_{121} \in (a_{221}, a_{222})$

	(a_{111}, x_2^*)	x_2^*			
Fig. 5.21	(a_{110}, a_{220})	a_{221}	a_{121}	a_{222}	$a_{223} = a_{122}$
(a)	(+,+)	(SO,SO)	(UP:DP,pF)	(SI,SI)	(DI:II,SO)
(b)	(−,+)	(SI,SO)	(DP:UP,pF)	(SO,SI)	(II:DI,SO)
(c)	(+,−)	(SO,SI)	(DP:UP,nF)	(SI,SO)	(II:DI,SI)
(d)	(−,−)	(SI,SI)	(UP:DP,nF)	(SO,SO)	(DI:II,SI)

The inflection source and sink infinite-equilibriums of $x_2^* = a_{223} = a_{122}$ are the switching bifurcations of sink, source, and saddles with hyperbolic flows for $a_{122} \in (a_{222}, a_{223})$ and $a_{122} > a_{223}$ with $a_{121} \in (a_{222}, a_{223})$, as shown in Fig. 5.20. The saddle and sink or source with hyperbolic or hyperbolic-secant flow exist. The infinite-equilibrium is on the top of the equilibrium series.

(xi) The phase portraits for the equilibrium series of $x_2^* = a_{121}, a_{122}, a_{221}, a_{222}, a_{223}$ with $x_1^* = a_{111}$ for $a_{223} = a_{122}$ with $a_{121} \in (a_{221}, a_{222})$ are presented in Fig. 5.21. The behaviors of equilibriums are listed in Table 5.21. The inflection infinite-equilibrium of $x_2^* = a_{223} = a_{122}$ is the same as in Eqs. (5.25)–(5.28). The inflection source and sink infinite-equilibriums of $x_2^* = a_{223} = a_{122}$ are the switching bifurcations of sink, source, and saddles with hyperbolic flows for $a_{122} \in (a_{222}, a_{223})$ and $a_{122} > a_{223}$ with $a_{121} \in (a_{221}, a_{222})$, as shown in Fig. 5.21. The two saddles or sink and source with hyperbolic or hyperbolic-secant flow exist.

(xii) The phase portraits for the equilibrium series of $x_2^* = a_{121}, a_{122}, a_{221}, a_{222}, a_{223}$ with $x_1^* = a_{111}$ for $a_{223} = a_{122}$ with $a_{121} < a_{221}$ are presented in Fig. 5.22. The behaviors of equilibriums are listed in Table 5.22. The inflection infinite-equilibrium of $x_2^* = a_{223} = a_{122}$ is the same as in Eqs. (5.21)–(5.24). The inflection-source and sink infinite-equilibriums of $x_2^* = a_{223} = a_{122}$ are the switching bifurcations of sink, source, and saddles with hyperbolic flows for $a_{122} \in (a_{222}, a_{223})$ and $a_{122} > a_{223}$ with $a_{121} < a_{221}$, as shown in Fig. 5.21. The saddle and sink or source with hyperbolic or hyperbolic-secant flow exist. The infinite-equilibrium is on the top of the equilibrium series, and the hyperbolic or hyperbolic-secant flows are on the bottom of equilibrium series.

5.2.2 Paralleled Double-Switching Bifurcations

Consider a dynamical system as

$$\begin{aligned} \dot{x}_1 &= a_{110}(x_1 - a_{111})(x_2 - a_{121})(x_2 - a_{122}), \\ \dot{x}_2 &= a_{220}(x_2 - a_{221})(x_2 - a_{222})(x_2 - a_{223}), \end{aligned} \tag{5.29}$$

and the corresponding first integral manifold for $a_{12s_1} = a_{22l_1}$ and $a_{12s_2} = a_{22l_2}$ (l_1, $l_2 \in \{1, 2\}$, $l_1 \neq l_2$; s_1, s_2, $s_3 \in \{1, 2, 3\}$, $s_1 \neq s_2 \neq s_3$) is

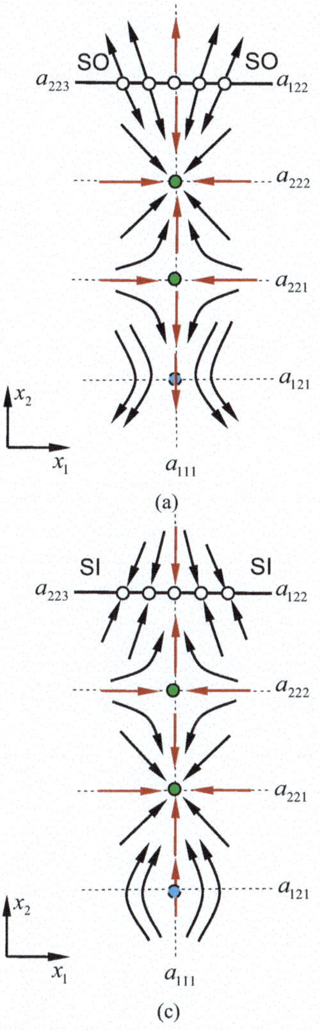

Fig. 5.22 Phase portraits ($a_{122} = a_{223}$, $a_{121} < a_{221}$) for two-dimensional systems on the x_1-direction with $x_1^* = a_{111}$ and on the x_2-direction with $x_2^* = a_{121}, a_{122}, a_{221}, a_{222}$. (**a**) ($a_{110} > 0$, $a_{220} > 0$), (**b**) ($a_{110} < 0$, $a_{220} > 0$), (**c**) ($a_{110} > 0$, $a_{220} < 0$), (**d**) ($a_{110} < 0$, $a_{220} < 0$)

Table 5.22 Equilibriums on the line of $x_1^* = a_{111}$ for $a_{223} = a_{122}$ and $a_{121} < a_{221}$

Fig. 5.22	(a_{111}, x_2^*)	x_2^*			
	(a_{110}, a_{220})	a_{121}	a_{221}	a_{222}	$a_{223} = a_{122}$
(a)	(+,+)	(DP:UP,nF)	(SI,SO)	(SI,SI)	(DI:II,SO)
(b)	(−,+)	(UP:DP,nF)	(SO,SO)	(SO,SI)	(II:DI,SO)
(c)	(+,−)	(UP:DP,pF)	(SI,SI)	(SI,SO)	(II:DI,SI)
(d)	(−,−)	(DP:UP,pF)	(SO,SI)	(SO,SO)	(DI:II,SI)

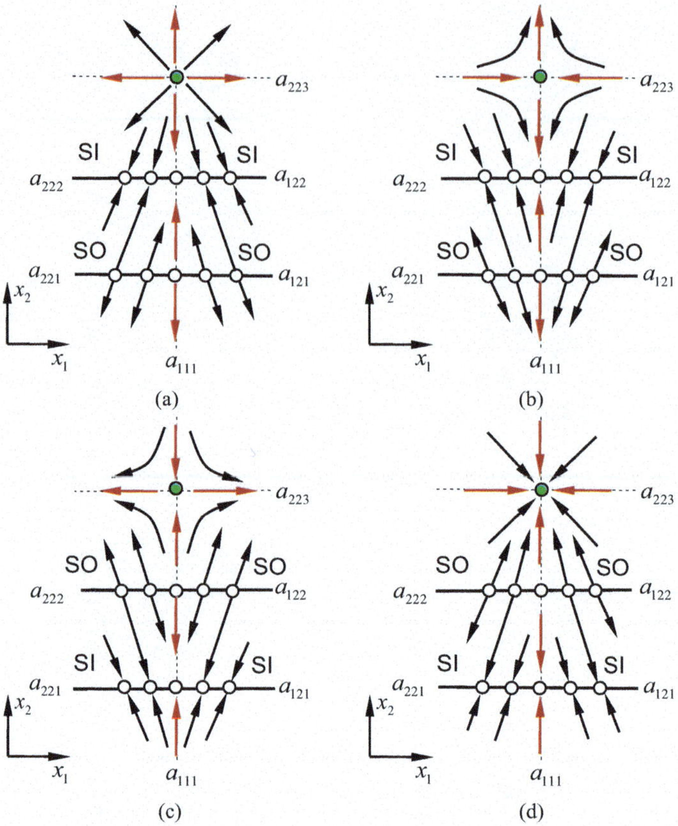

Fig. 5.23 Phase portraits ($a_{121} = a_{221}$, $a_{122} = a_{222}$) for two-dimensional systems on the x_1-direction with $x_1^* = a_{111}$ and on the x_2-direction with $x_2^* = a_{121}, a_{122}, a_{221}, a_{222}$. (**a**) ($a_{110} > 0$, $a_{220} > 0$), (**b**) ($a_{110} < 0$, $a_{220} > 0$), (**c**) ($a_{110} > 0$, $a_{220} < 0$), (**d**) ($a_{110} < 0$, $a_{220} < 0$)

Table 5.23 Equilibriums on the line of $x_1^* = a_{111}$ for $a_{121} = a_{221}$ and $a_{122} = a_{222}$

Fig. 5.23	(a_{111}, x_2^*) (a_{110}, a_{220})	x_2^* $a_{221} = a_{121}$	 $a_{222} = a_{122}$	 a_{223}
(a)	(+,+)	(II:DI,SO)	(II:DI,SI)	(SO,SO)
(b)	(−,+)	(DI:II,SO)	(DI:II,SI)	(SI,SO)
(c)	(+,−)	(DI:II,SI)	(DI:II,SO)	(SO,SI)
(d)	(−,−)	(II:DI,SI)	(II:DI,SO)	(SI,SI)

$$\ln \frac{|x_1 - a_{111}|}{|x_{10} - a_{111}|} = \frac{a_{110}}{a_{220}} \ln \frac{|x_2 - a_{22s_3}|}{|x_{20} - a_{22s_3}|}. \tag{5.30}$$

Table 5.24 Double infinite-equilibriums of $a_{121} = a_{221}$ and $a_{122} = a_{222}$

Fig. 5.23	(a_{110}, a_{220})	$\bar{x}_1$ for $(\bar{x}_1, a_{221})$		$\bar{x}_1$ for $(\bar{x}_1, a_{222})$	
		$(-\infty, a_{111})$	(a_{111}, ∞)	$(-\infty, a_{111})$	(a_{111}, ∞)
(a)	(+,+)	(II,SO)	(DI,SO)	(II,SI)	(DI,SI)
(b)	(−,+)	(DI,SO)	(II,SO)	(DI,SI)	(II,SI)
(c)	(+,−)	(DI,SI)	(II,SI)	(DI,SO)	(II,SO)
(d)	(−,−)	(II,SI)	(DI,SI)	(II,SO)	(DI,SO)

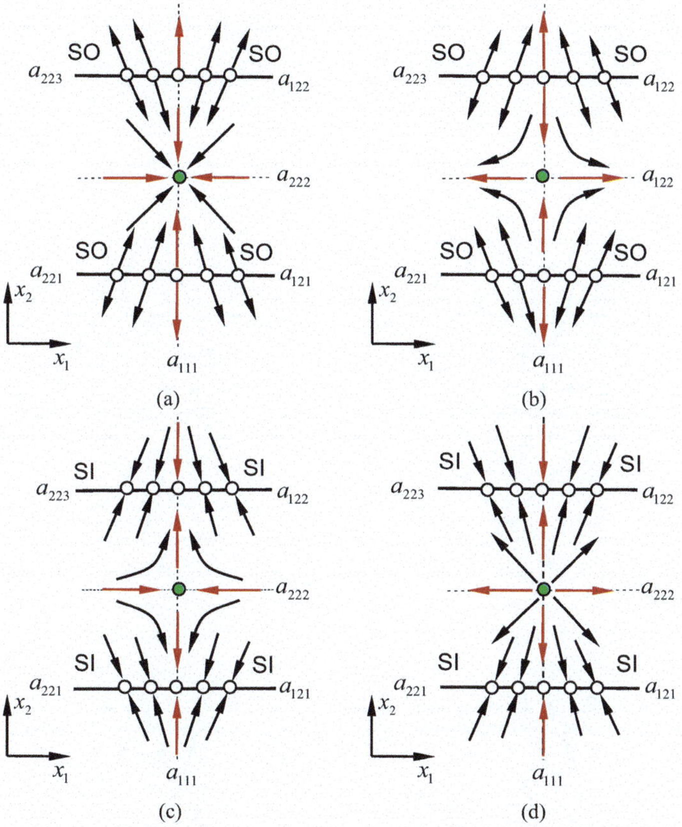

Fig. 5.24 Phase portraits ($a_{121} = a_{221}$, $a_{122} = a_{223}$) for two-dimensional systems on the x_1-direction with $x_1^* = a_{111}$ and on the x_2-direction with $x_2^* = a_{121}, a_{122}, a_{221}, a_{222}$. (**a**) ($a_{110} > 0$, $a_{220} > 0$), (**b**) ($a_{110} < 0$, $a_{220} > 0$), (**c**) ($a_{110} > 0$, $a_{220} < 0$), (**d**) ($a_{110} < 0$, $a_{220} < 0$)

Table 5.25 Equilibriums on the line of $x_1^* = a_{111}$ for $a_{121} = a_{221}$ and $a_{122} = a_{223}$

	(a_{111}, x_2^*)	x_2^*		
Fig. 5.24	(a_{110}, a_{220})	$a_{221} = a_{121}$	a_{222}	$a_{223} = a_{122}$
(a)	(+,+)	(II:DI,SO)	(SI,SI)	(DI:II,SO)
(b)	(−,+)	(DI:II,SO)	(SO,SI)	(II:DI,SO)
(c)	(+,−)	(DI:II,SI)	(SI,SO)	(II:DI,SI)
(d)	(−,−)	(II:DI,SI)	(SO,SO)	(DI:II,SI)

Table 5.26 Double infinite-equilibriums of $a_{121} = a_{221}$ and $a_{122} = a_{223}$

		$\bar{x}_1$ for $(\bar{x}_1, a_{221})$		$\bar{x}_1$ for $(\bar{x}_1, a_{222})$	
Fig. 5.24	(a_{110}, a_{220})	$(-\infty, a_{111})$	(a_{111}, ∞)	$(-\infty, a_{111})$	(a_{111}, ∞)
(a)	(+,+)	(II,SO)	(DI,SO)	(DI,SO)	(II,SO)
(b)	(−,+)	(DI,SO)	(II,SO)	(II,SO)	(DI,SO)
(c)	(+,−)	(DI,SI)	(II,SI)	(II,SI)	(DI,SI)
(d)	(−,−)	(II,SI)	(DI,SI)	(DI,SI)	(II,SI)

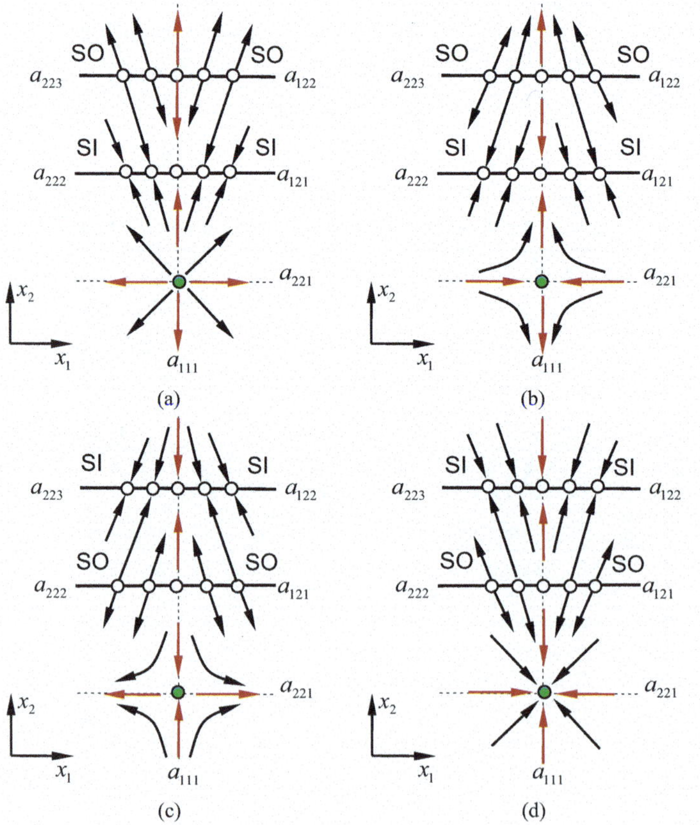

Fig. 5.25 Phase portraits ($a_{121} = a_{222}$, $a_{122} = a_{223}$) for two-dimensional systems on the x_1-direction with $x_1^* = a_{111}$ and on the x_2-direction with $x_2^* = a_{121}, a_{122}, a_{221}, a_{222}$. **(a)** ($a_{110} > 0$, $a_{220} > 0$), **(b)** ($a_{110} < 0$, $a_{220} > 0$), **(c)** ($a_{110} > 0$, $a_{220} < 0$), **(d)** ($a_{110} < 0$, $a_{220} < 0$)